Acoustical Oceanography

Ocean Engineering: A Wiley Series

EDITOR:
MICHAEL E. McCORMICK, Ph. D.
U.S. Naval Academy
ASSOCIATE EDITOR:
RAMESWAR BHATTACHARYYA, D. Ingr.
U.S. Naval Academy

Michael E. McCormick	Ocean Engineering Wave Mechanics
John B. Woodward	Marine Gas Turbines
H. O. Berteaux	Buoy Engineering
Clarence S. Clay and Herman Medwin	Acoustical Oceanography: Principles and Applications

ACOUSTICAL OCEANOGRAPHY:

PRINCIPLES AND APPLICATIONS

CLARENCE S. CLAY
Professor of Geophysics
University of Wisconsin,
Madison, Wisconsin

HERMAN MEDWIN
Professor of Physics
Naval Postgraduate School
Monterey, California

A WILEY-INTERSCIENCE PUBLICATION

JOHN WILEY & SONS,   New York • London • Sydney • Toronto

Library of Congress Cataloging in Publication Data

Clay, Clarence Samuel, 1923–
  Acoustical oceanography.

  (Ocean engineering, a Wiley series)
  "A Wiley-Interscience publication."
  Includes bibliographies and index.
  1. Underwater acoustics. 2. Sea-water—Acoustic
properties. 3. Oceanography. I. Medwin, Herman,
1920–   joint author. II. Title.

QC242.2.C55   551.4'601   77-1133
ISBN 0-471-16041-5

Printed in the United States of America

10 9 8 7 6 5 4 3

To the memory of

CARL ECKART

Scientist, Teacher, Giant among Men

SERIES PREFACE

Ocean engineering is both old and new. It is old in that man has concerned himself with specific problems in the ocean for thousands of years. Ship building, prevention of beach erosion, and construction of offshore structures are just a few of the specialties that have been developed by engineers over the ages. Until recently, however, these efforts tended to be restricted to specific areas. Within the past decade an attempt has been made to coordinate the activities of all technologists in ocean work, calling the entire field "ocean engineering." Here we have its newness.

Ocean Engineering: A Wiley Series has been created to introduce engineers and scientists to the various areas of ocean engineering. Books in this series are so written as to enable engineers and scientists easily to learn the fundamental principles and techniques of a specialty other than their own. The books can also serve as textbooks in advanced undergraduate and introductory graduate courses. The topics to be covered in this series include ocean engineering wave mechanics, marine corrosion, coastal engineering, dynamics of marine vehicles, offshore structures, and geotechnical or seafloor engineering. We think that this series fills a great need in the literature of ocean technology.

MICHAEL E. McCORMICK, EDITOR
RAMESWAR BHATTACHARYYA, ASSOCIATE EDITOR

November 1972

vii

PREFACE

We have chosen the title to emphasize the usefulness of underwater sound in solving oceanographic problems: "principles" because we give an introductory development of the theory and techniques; "applications" because we apply the theory to the location and identification of bodies and blobs in the sea, to features on the sea floor, and to the physical characterization of the layers beneath the bottom.

A year of college physics is sufficient background to begin the book. The first year of calculus and a willingness to learn is sufficient for most of the derivations. The development moves from an intuitive use of Huygens' principle at the beginning, through wave and ray methods, to theoretical and practical acoustical oceanography at the end.

Since acoustical oceanography requires a mixture of many skills, we have organized the book to give the reader many choices. The first five chapters present sound in the ocean, sonar systems, and signal processing. Then, the reader can choose specialized areas such as nonlinear acoustics, objects and creatures in the sea, the sea floor, waveguides for sound, and the scattering of sound by rough surfaces. The applications are the parametric source, fish population estimates, the earth beneath the sea, and sonographs of the sea floor. For especially curious readers, appendices give details and extend the material.

Sound is the only effective way to communicate or to sense and identify at long range under the surface of the ocean. We invite marine biologists, geophysicists and geologists, oceanographers, and ocean engineers to join us in using sound to probe the sea and its boundaries.

CLARENCE S. CLAY
HERMAN MEDWIN

Madison, Wisconsin
Monterey, California
November 1976

ACKNOWLEDGMENTS

The following people helped us in too many ways to list individually:

Arthur B. Baggeroer	Richard K. Johnson
W. E. Batzler	G. L. Maynard
Jane Clay	Ruth McCormick
Paul Dombrowski	Eileen Medwin
C. R. Dunlap	David Mintzer
W. A. Friedl	Werner Neubauer
Brenda Griffing	G. V. Pickwell
E. C. Haderlie	Isadore Rudnick
E. L. Hamilton	J. E. Sanders
Kay Hanson	Paul Wolff
J. B. Hersey	

Students at the University of Wisconsin and Naval Postgraduate School

C. S. C.
H. M.

CONTENTS

Appendixes

A1. Formulas From Mathematics 369

A2. Wave Propagation 376

A3. Diffraction Effects and Losses to the Medium 408

A4. Fourier Transformations and Applications 424

Acoustical Oceanography

AN ACOUSTICAL VIEW OF THE SEA

1.1. HOW IT ALL BEGAN

It is a rare person whose pulse is not stirred by the dramatic sight of the restless surface of the sea. The chaotic sea surface is a limitless source of inspiration to poet, painter, and musician, alike.

But what lies beneath this churning surface? How can we probe the depths of the sea? The physical and chemical oceanographers know that we use Nansen bottles to retrieve samples for shipboard or laboratory studies of the sea chemistry. Instruments that are sensitive to temperatures, static pressure, electrical conductivity, material particle velocity, sound speed, and light penetration can be lowered from a ship, floated from a buoy, suspended from an ocean tower, or allowed to drift with the undersea currents and reveal their information by acoustic command. The marine biologist can use plankton nets of different sieve sizes to sort out the vegetable and small animal populations, and fish trawls or lines to obtain a rough census of the larger inhabitants.

If only we could *see* what is beneath the ocean surface. But visible light is scattered and extinguished by the very richness of the dense sea life. In some turbid locations, even with intense illumination, it is impossible to see your hand at the end of your outstretched arm. Without searchlights

1

humans are "blind as a bat" at ocean depths greater than about 100 m even in the cleanest water.

But bats navigate, communicate, and sense their prey and their surroundings by acoustic echo ranging. The bat has developed a high-pitched source, a matched receiver, and a sophisticated signal processor that puts to shame the tapping white cane of the blind human. And these magnificent aptitudes of the bat appear to be part of the armament of several species of ocean life as well, notably the porpoise, which also uses short bursts of high frequency sound to communicate and to navigate.

In 1912 the collision of the steamship *Titanic* with an iceberg and the subsequent loss of hundreds of lives triggered man's use of sound for sensing in the sea. Within a month, L. R. Richardson, who had already filed for a British patent for echo ranging with airborne sound, filed another patent application (10 May 1912), this time for "detecting the presence of large objects under water by means of the echo of compressional waves having a wavelength in water of 30 cm or less ... directed in a beam ... by a projector ..." (Hunt, 1954). R. A. Fessenden, who had been working on the same problem, filed for a U.S. patent on 29 January 1913 and succeeded in detecting an iceberg at a distance of 2 miles on 27 April 1914. The basic idea in all of this was that knowledge of the speed of sound in water, and the time of travel of the sound, permits the calculation of the distance to the reflector.

Meanwhile, in Europe, the urgent need for a submarine detector in World War I stimulated the imagination of a young Russian engineer, Constantin Chilowsky, who worked with the French physicist P. Langevin to develop an underwater source that transmitted sound across the Seine in Paris during the winter of 1915–1916. The British Allies, hearing of the French activities and successes, assigned Dr. Robert W. Boyle to organize a group to attack the same problem, and by summer 1916 the British had duplicated the French success. A major step forward was made when Langevin used as his sound source a sandwich of a plate of the piezoelectric material, quartz, between two plates of steel. The high intensities of this sound generator enabled transmission to a range of 8 km and produced the first detection of an echo from a submarine. The practical use of sound in the sea had been demonstrated by 1918, but the extensive use of sonar (*so*und *na*vigation and *r*anging) for submarine detection had to await World War II.

Although military use of sound in the ocean was quiescent during the 25 year period between the great wars, commercial exploitation was quite active. Echo ranging to the sea floor was an obvious way to determine the depth of water, and the depth sounder was soon born. If echoes could be obtained from submarines, they might also be received from schools of fish. A British fishing boat captain, R. Balls, was using a sound "echometer" in the early 1930s, and the Norwegian O. Sund described "Echo Sounding in Fisheries Research" in 1935 (Nature **135**, 956).

With the development of electronics, all these acoustic sensors became more effective, and with the growing knowledge of acoustic signal processing and display, greater ranges could be achieved and smaller targets could be identified. Now, as these words are being written, acoustics has finally provided eyes to sense the sea, its inhabitants, its plant life and its garbage, and the surface and bottom that contain them. High frequency sonars can count marine life only millimeters in dimension, and low frequency sounds can be identified at ranges of thousands of miles.

To use sound as a tool in the sea, we need to learn about the speed with which it travels, its rate of attenuation, how it bends around obstacles and into shadows, and the way it is reflected, scattered, and transmitted at boundaries. All these acoustical phenomena depend on physical characteristics of the ocean volume, its surface, and its bottom— subjects that are the traditional realm of the physical oceanographer and geophysicist. Understanding of sound propagation at sea also requires some of the knowledge amassed by the marine biologist. This chapter surveys the acoustically significant characteristics of the sea. We proceed to consider the sea and its boundaries from the point of view of a sound wave.

1.2. THE STRATIFICATION OF THE SEA

The most important acoustical parameter of the ocean is the speed of sound. The earliest measurement of sound speed in water was probably the work of Colladon and Sturm (1827) performed in Nov. 1826, Lake Geneva, Switzerland. Figure 1.2.1 is a diagram of their experiment. The average speed at sea is approximately 1.5 km/s (ca 5000 ft/s). However because the speed of sound increases with increasing water temperature, salinity, and pressure, it changes significantly with season, time of day, depth, geographical position, and proximity to rivers and melting ice. A simplified form of the dependence is given in (1.2.1).

$$c = 1449.2 + 4.6T - 0.055T^2 + 0.00029T^3$$
$$+ (1.34 - 0.010T)(S - 35) + 0.016z \quad \textbf{(1.2.1)}$$

where c = speed (m/s)
   T = temperature (°C)
   S = salinity, (parts per thousand: ‰)
   z = depth (m)

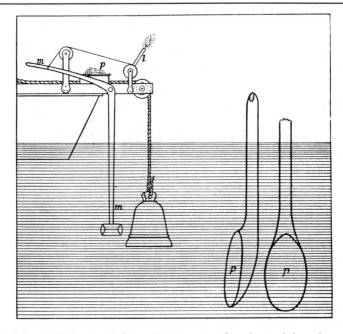

Figure 1.2.1. Colladon and Sturm's apparatus for determining the speed of sound in water. Two boats were stationed 13,487 m apart in Lake Geneva, Switzerland. A bell suspended from one of the boats was struck under water by means of a lever m, which at the same moment caused the candle l to ignite the powder p and set off a visible flash of light. An observer in the second boat with a listening tube measured the time that elapsed between the flash of the light and the sound of the bell. The mean of a number of such observations gave a value of 9.4 s. The temperature of the water was 8°C. The experimental speed of sound in freshwater at 8°C was therefore 13,487/9.4 or 1435 m/s. The results were reported in detail in 1827 and 1828. The empirical value given by (1.2.1) is 1438 m/s. The near agreement to the fourth significant figure is fortuitous; only two significant figures are justified by the time measurement. (Millikan et al., 1965.)

Direct, *in situ* data on the speed of sound are obtained by use of a sound "velocimeter," which essentially measures the time required for a very high frequency beam (of the order of 1 MHz) to traverse a short path (of the order of tens of centimeters). The sound velocimeter is easily used and is accurate to 0.1 m/s.

Chapter 3 describes how spatial variations of sound speed cause wave bending. Our need here is to give a sampling of the variation of these parameters in the world's oceans.

Consider Fig. 1.2.2, which shows the temperature profiles as a function of depth at a location in the Pacific, and Fig. 1.2.3, which gives salinity profiles at the same place. The most prominent features of these two charts are the almost parallel, horizontal isotherms (constant temperature lines) and isohalines (constant salinity lines). Since the lines of constant depth are horizontal in stable water, all three constituents of the sound speed are approximately horizontally stratified. Therefore the sound speed itself is approximately horizontally stratified. This fact of ocean physics has a major effect on sound propagation. It is illustrated in Fig.

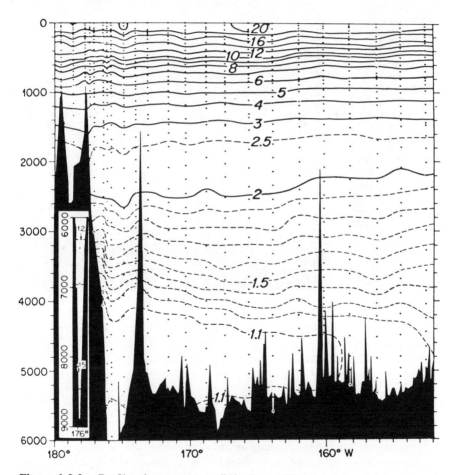

Figure 1.2.2. Profile of temperature (°C) along 28°S near western boundary of deep South Pacific, depths in meters, July 3–18, 1967. Tonga–Kermadec Trench continued in inset. (Warren, 1970).

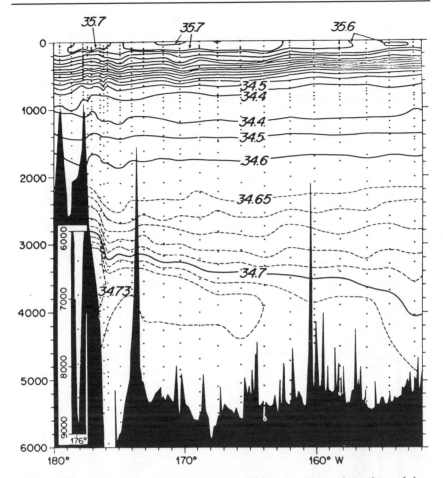

Figure 1.2.3. Profile of salinity (‰) along 28°S near western boundary of deep South Pacific, depths in meters. July 3–18, 1967. Tonga–Kermadec Trench continued in inset. (Warren, 1970.)

1.2.4. As Chapter 3 shows, the stratified ocean may produce convergent or divergent beams and shadow regions, depending on the variation of speed with depth.

Since the dependence of sound speed on depth is of critical importance in the spreading of sound, it is valuable to notice how the temperature profiles can be changed by the input of energy from the sun (insolation) and by the nocturnal radiation of heat from the ocean. Figure 1.2.5 represents the near-surface effect, with the near-surface temperatures

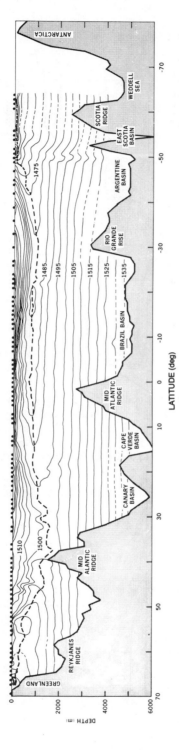

Figure 1.2.4. North–south section of sound channel structure in the North and South Atlantic along the 30.50°W meridian. Sound speeds are in meters per second and the approximate sound channel axis depth is indicated by a heavy dashed line. The axis is along sound speed minima. Data positions and interpolated isovelocity contours are represented by dots and broken lines; respectively. (Northrop and Colborn, 1974.)

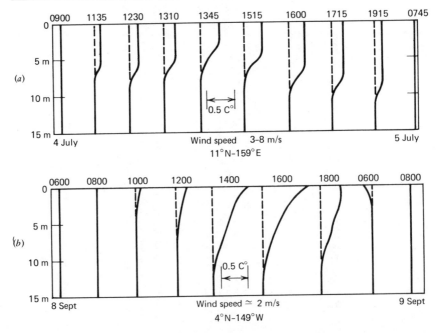

Figure 1.2.5. Two temperature gradient profiles. (*a*) Mixed layer persists during the day, becoming warmer because of midday heating and reaching greater depths as a result of mixing due to high wind speeds. (*b*) Wind speed low, so that the mixing is not as effective. At 0600 of the second day, the time of minimum winds, there is a temperature inversion due to nighttime cooling. (Eckart, 1968.)

varying with the time of day, under the control of energy input and output. The amount of the energy and its penetration downward depend on local conditions and season of the year. When cooling takes place, the cooler, more dense, surface waters descend and are replaced by the lower, slightly warmer, waters. The resultant mixing can create a near-surface layer of virtually isothermal water called the "mixed layer." Well-mixed waters have virtually the same sound speed at all points (i.e., are homogeneous). However, deviations in the gradient of the sound speed can profoundly affect the transmission, as we show in Section 3.2.

Below the mixed layer lies the "thermocline," in which the temperature rapidly decreases with increasing depth. Continual mixing forces, particularly during storms, gradually increase the thickness of the mixed layer at the cost of the eroding thermocline. The manner in which sound speeds may depend on season and latitude and longitude is illustrated in Fig. 1.2.6.

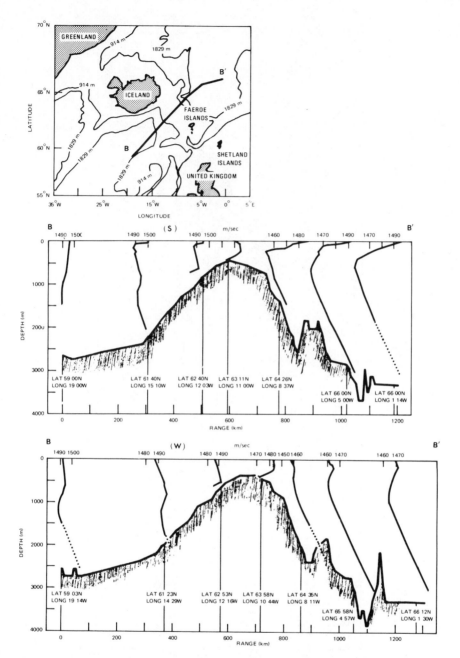

Figure 1.2.6. Selected sound speed profiles across the Greenland–Iceland–Faeroe gap for summer (*S*) (*Middle*) and winter (*W*) (*Bottom*). Location of profile *B–B'* is given in map at top. Shaded area marks ocean bottom profile. (Northrop and Colborn, 1974.)

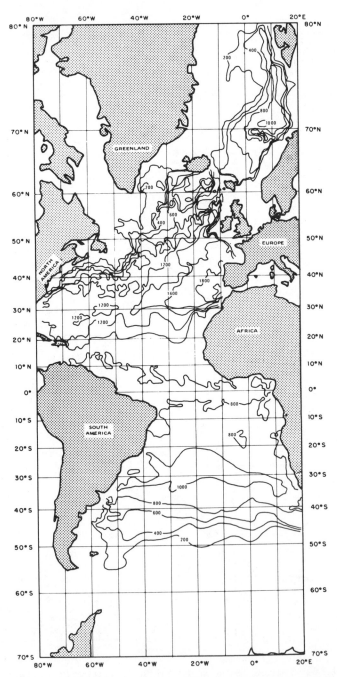

Figure 1.2.7. Deep channel sound axis depth contours for the Atlantic Ocean. Channel depths, at contour intervals of 200 m, range from near the surface in the polar regions to 1800 m in the North Atlantic (Northrop and Colborn, 1974.)

Below the thermocline there is a region of constant temperature. Since the sound speed increases with increasing depth, the top of the isothermal region (bottom of the thermocline) is a minimum in the sound speed profile. This level is the axis of the deep sound channel, within which sound energy is more or less confined by refraction (Figs. 1.2.4 and 1.2.7).

To clarify our use of the words "speed" (a scalar) and "velocity" (a vector quantity), let us use the acoustician's term "sound speed" when the propagation of sound is at the same rate in all directions (isotropic), as in the ocean volume. However when we discuss propagation through sediments and solids at the sea floor, which are generally anisotropic situations, we adopt the language of the geophysicist and speak of the "sound velocity" through these materials. The two terms are sometimes used interchangeably in the literature. Finally we note that the instrument used to measure sound speed at sea is called a sound velocimeter, perhaps because the term "speedometer" has automotive connotations.

1.3.   THE RESTLESS SEA

As it proceeds away from its source, a sound at sea encounters changing values of temperature, salinity, and density, as well as free and entrained objects and bubbles. These acoustically significant quantities are constant neither in space nor in time. There are near-surface variations, due to the underwater effect of surface waves and turbulence, which show perceptible changes during a time scale of the order of seconds. These are superimposed on changes due to internal waves with periodicities of the order of minutes to hours, tidal (12 hour), diel* (24 hour), and seasonal changes.

We know that near the ocean surface, depending on the time of day and the meteorological history at the location, there is often a layer of well-mixed, "isothermal" water. The name reflects the property of common bathythermographs (BTs) which shows such a region as having a uniform temperature, independent of depth. In fact, research probes with finer temperature resolution and shorter time constants have revealed that the crudely isothermal region is characterized by drifting patches of water of slightly different sound speeds. "Microstructure" at a fixed point is shown by data such as in Fig. 1.3.1. Temperature fluctuations near the surface are a function of time of day and depth of measurement. The

---

* Bioacousticians have begun to use the word "diel" for 24 hour cyclic effects, and to reserve the words "diurnal" and "nocturnal" for daytime and nighttime effects, respectively.

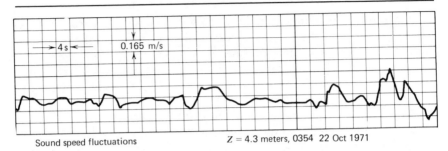

Sound speed fluctuations Z = 4.3 meters, 0354   22 Oct 1971

Figure 1.3.1. Sound speed fluctuations at a fixed point in well-mixed water. (Seymour, 1972.)

short-time temperature fluctuations are generally greatest near the surface and greatest during the late afternoon, particularly during sunny periods.

Within and below the thermocline there are large regions of the ocean that possess for extended periods of time an ordered structure of alternate layers of turbulent and relatively stable water. About one-half of the ocean's regions below the mixed upper layer may show this very extensive active layered character. The active regions contain identifiable "layers," warmer than the liquid above or below them, ranging from thickness sometimes as little as 10 cm or less and extending for hundreds of meters, up to layers of thickness 10 to 15 m extending for tens of kilometers. The layers are separated by thin "sheets" of relatively steeper temperature gradients of order 0.01°C/cm. Temperature inversions of the same steepness with compensating salinity variations are found in these active regions. And scattered throughout the active volume are intermittent and patchy regions, again, larger in extent than in thickness. It appears that turbulence microstructure in these layers is always accompanied by temperature microstructure. On the other hand, investigators have found temperature microstructure virtually free of turbulence microstructure and that may represent evidence of a former turbulent microstructure. Figure 1.3.2 is an example of layered microstructure with its characteristic steplike character.

Frontal regions of moving water masses have also been found at sea. These structures parallel the meteorologist's well-developed picture of air mass behavior. For example, one might find a front with an angle of advance of 1°, in which quiescent warm water is being uplifted by a mass of highly turbulent, colder lower water.

Internal waves, generated within the ocean mass, are analogous to ocean surface waves. They are volume gravity waves having their maximum vertical displacement amplitude at a plane where the density is

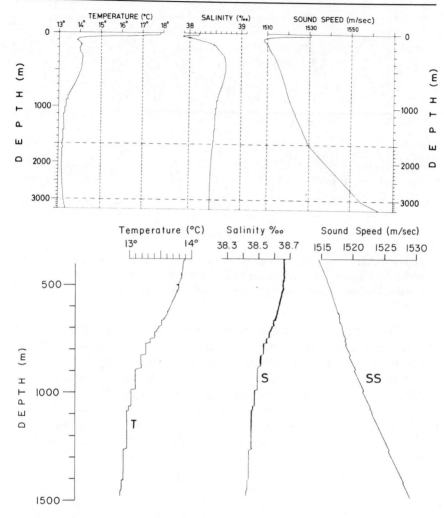

Figure 1.3.2. (*Top*) Gross profiles of temperature, salinity, and sound speed. (*Bottom*) Magnified profiles for depths 400 to 1500 m showing layered micros-tructure. (Mellberg and Johannessen, 1973.)

changing most rapidly with depth or between two water masses of different densities. But the motion of an internal wave is detectable far above and below this interface. All the characteristics of the water mass change at a point through which internal waves pass. For example, it is not unusual to find temperature fluctuations of a few degrees with periodicities of the order of minutes or hours. In the two examples of Fig.

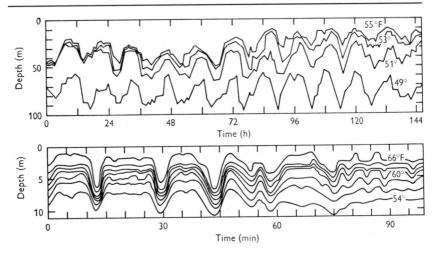

Figure 1.3.3. Isotherms in internal wave motions. (*Top*) dominant component with tidal frequency and amplitude about 15 m. (*Bottom*) Oscillations in a shallow thermocline with a considerably higher dominant frequency. (Phillips, 1966.)

1.3.3 one features tidal periodicities of predominantly 12 hours, the other occurs in shallow water and has prominent cycles of about 15 minutes duration. Generally a spectrum of frequencies is present.

Although material velocity variations usually have a lesser acoustic effect than temperature variations, the motion of the medium is significant in certain circumstances and may even overwhelm the temperature influence. For a simple long-crested (swell) wave in deep water, the vertical and horizontal components of the material velocity and the displacement are almost totally orbital near the surface (Fig. 1.3.4). The depth dependence of the displacement amplitude a is given by

$$a = a_o \exp\left(\frac{-2\pi z}{\Lambda}\right) \tag{1.3.1}$$

where Λ = surface wavelength
z = depth
a_o = orbital displacement amplitude at the surface
$\exp(\ )$ = exponential notation, appendix A1.3.1

Underwater motion that is defined by (1.3.1) may be called the "coherent" component of the oscillation. Superimposed on the coherent is the "incoherent" turbulent motion often, and for simplicity, assumed to be independent of direction (isotropic). The region in which three-dimensionally isotropic, or horizontally isotropic, turbulence exists is a

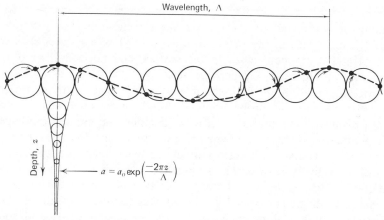

Figure 1.3.4. Cross section of ocean wave traveling from left to right. The material displacement and particle velocity decrease exponentially with increasing distance from the surface [see (1.3.1)].

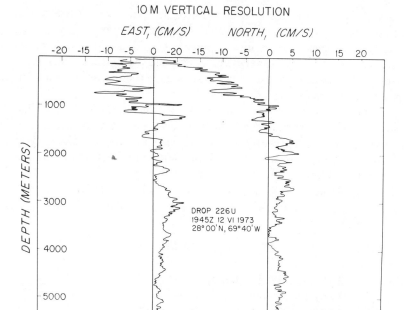

Figure 1.3.5. Profile of horizontal velocity of water motions versus depth. (Sanford, 1974.)

15

function of depth and of sea surface wave conditions. For example, a study (Dunn, 1965) in the Menai Straits during low sea states showed a large orbital component close to the surface and horizontally isotropic turbulence at depths of 3 m and greater.

Horizontal motion of the sea, if the speed varies with depth, can markedly influence the propagation of sound. In air when a wind blows from a distant sound source toward the listener, the greater wind speed aloft bends the sound rays downward and greatly increases the loudness of the sound. The same effect occurs at sea when there is a variation of ocean flow velocity with depth, a so-called shear flow, as in Fig. 1.3.5. The graph indicates flows of the order of 10 cm/s, varying with depth, in the Sargasso Sea. The speed of sound was almost constant at these particular depths. Under such conditions the sound paths and sound intensity are very sensitive to the speed and direction of the shear flow.

1.4. THE SEA SURFACE

Stressed by the turbulent winds above it, the ocean surface can be tweaked to tiny patches of cat's-paws or whipped into a frenzy of raging, mountainous seas. The smallest ripples are evidence that the shape and movement of the disturbance are controlled by surface tension at the water–air interface. These "capillary waves" have short wavelengths (generally less than ca 2 cm).

"Gravity" waves, the larger and more obvious ocean waves, oscillate in response to the restoring force of their own weight as they rise and fall relative to the average water level. The wavelengths of the gravity waves are greater than those of the capillaries and may reach dimensions up to thousands of meters in intense storms in the open sea.

The combination of surface tension and gravity forces produces capillary–gravity waves, whose frequency–wavelength relation in "deep water" is given by

$$\Omega^2 = \kappa g + \frac{\alpha \kappa^3}{\rho} \qquad (1.4.1)$$

where $\Omega = 2\pi f =$ angular frequency
$\kappa = 2\pi/\Lambda =$ wave number
$f =$ frequency (cycles/s: Hz)
$\Lambda =$ wavelength (m)
$g =$ acceleration of gravity $= 9.8$ m/s^2
$\alpha =$ surface tension ($\approx 7.4 \times 10^{-2}$ N/m)
$\rho =$ water density (kg/m^3)

Physical oceanographers generally define a measurement as being in deep water, thus unaffected by the bottom, when the water depth D is

$$D > \frac{\Lambda}{2} \tag{1.4.2}$$

At that depth (1.3.1) shows that orbital velocity (and displacements) are down to 4% of the surface value.

For small wavelengths, large κ, the second term of (1.4.1) dominates and the capillary wave relation is obtained

$$\Omega^2 = \frac{\alpha \kappa^3}{\rho} \tag{1.4.3}$$

On the other hand, for large wavelengths (1.4.1) gives us the relation for deep-water gravity waves

$$\Omega^2 = \kappa g \tag{1.4.4}$$

To get a better feel for surface wave propagation, we can use the well-known relations for phase velocity c

$$c = f\Lambda \quad \text{or} \quad \Omega = \kappa c \tag{1.4.5}$$

where the velocity is measured along the direction of propagation. Then, for capillary waves, the phase velocity is found to increase with *decreasing* wavelength

$$c^2 = \frac{\alpha \kappa}{\rho} \tag{1.4.6}$$

For gravity waves the speed increases with *increasing* wavelength according to

$$c^2 = \frac{g}{\kappa} \tag{1.4.7}$$

The minimum wave velocity is 23.1 cm/s, at wavelength 1.73 cm, which serves as the border between capillary and gravity waves. The phase velocity for capillary–gravity waves is shown for a wide range of wavelengths in Fig. 1.4.1.

The description of the ocean surface has been vastly simplified when we talk of only one wave frequency, as previously. The true sea surface condition is more accurately described as the superposition of wave trains from the capillaries up through the gravity waves. These wave trains have different amplitudes and may be travelling in different directions, and it is

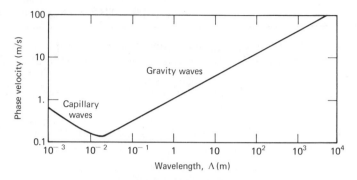

Figure 1.4.1. Phase velocity of capillary and gravity surface waves as a function of wavelength. Deep water is assumed, depth $> \Lambda/2$.

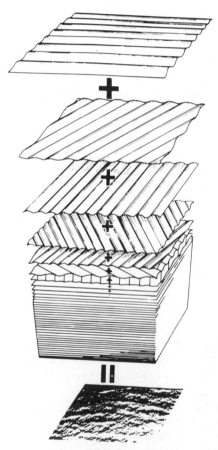

Figure 1.4.2. A sea may be represented by the sum of many simple wave trains of different wave lengths moving in different directions. (Pierson, 1955.)

their sum that should be used to describe the surface. Such a superposition of the elementary waves is presented in Fig. 1.4.2. The full description of ocean wave frequencies and their amplitudes is called the "frequency spectrum" of the surface height.

Even the superposition of simple corrugations as in Fig. 1.4.2 is not quite enough. The corrugations of the real sea may not be parallel or "long-crested." Furthermore, the addition of the components cannot be done in a simple linear manner. That is, $1+1$ is not exactly 2 in the nonlinear world of the ocean surface. The frequency spectrum of the ocean surface displacement is nevertheless a useful description.

After a wave spectrum has been generated by a storm system at sea, the various components move out from the storm area, diverging and attenuating as they go. Since the higher frequencies attenuate more rapidly than the lower ones, the wave spectrum is continually filtered as it propagates. Toward the end of its life, at great distances from its origin, the wave system has only its lowest frequencies, longest wavelengths, and is commonly called "swell."

It is possible to bounce underwater sound from an ocean surface and, by the change of the sound frequency and amplitude, to characterize the statistics and the spectrum of the surface. This sensing can be done at close range or remotely.

1.5. THE OCEAN BOTTOM: SEA FLOOR SPREADING
AND THE FORMATION OF THE SEA FLOOR

At the end of World War II, marine geophysicists and geologists began the comprehensive mapping of the sea floor and the measuring of the structure of ocean basins. We use results of their efforts to give a simplified description of the sea floor and the processes that form it. This simplified structure is the basis of our acoustical examination of the properties of the materials and character of the sea floor. Because so many individuals have contributed, we omit specific references.

Geologists, geophysicists, and geochemists explain the formation of the sea floor by combining sedimentation theory and the processes known as "sea floor spreading," "continental drift," and "global tectonics." Magma from the Earth's interior rises to the surface at midocean ridges, solidifies, and creates new sea floor (Fig. 1.5.1). The sea floor moves away from the ridge like a conveyor belt and catches the sediments that fall on it. On the east coast of the North American continent, the sea floor joins the continent and both move away from the mid-Atlantic ridge. The sea floor descends into the Earth's interior at the trenches. On a geological time

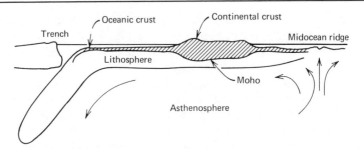

Figure 1.5.1. Sketch of lithosphere and convection cells in asthenosphere. The Mohorovicic discontinuity at the base of the crust is labeled "Moho."

scale the process is rapid, and the sea floor spreading rate may be as high as 10 cm/year. The oldest parts of the sea floor are less than 200 million years old, a fraction of the ages of the oceans and continents.

These ideas, which originated with the work of Alfred Wegener at the beginning of the century, were controversial until enough geophysical data were accumulated to permit investigators to establish that the continents *are moving* and the sea floors *are young.*

On the basis of heat flow measurements and earthquake seismology, we now believe that the energy for sea floor spreading comes from the heat of radioactive decay in the interior of the earth. The heat energy drives convection cells in the asthenosphere (Fig. 1.5.1). This layer has no rigidity on the time scale of continental drift. The drag of the convection currents causes the outer layer, the lithosphere, to move. The continents and ocean basins are on the six major lithospheric plates. The lithospheric plates slide past one another and where they collide one plate overrides the other to form a trench. The overridden plate descends into the asthenosphere and is the location of deep focus earthquakes.

The midocean ridges, where the magma is ascending and the new sea floor originates, are the sources of shallow focus earthquakes and volcanic activity. On the basis of marine seismic measurements (see across), we believe that the solidifying magma forms the basement rock, oceanic layer, and upper mantle (Fig. 1.5.2). The lithosphere consists of these layers and is about 70 to 100 km thick. The volcanism at the midocean ridge brings basalt to the sea floor seawater interface and an ooze of iron-manganese oxides forms. While these processes are occurring, sediments from other sources blanket the sea floor.

Very fine clay particles, originally from erosion on land, slowly fall to the bottom. Marine plants and animals form skeltons or shells of silica (SiO_2) or calcium carbonate ($CaCO_3$). The remains of these also fall to

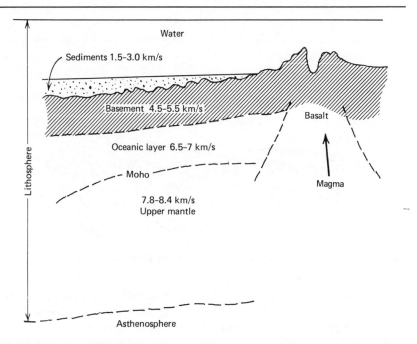

Figure 1.5.2. Structure and sound velocities of the sea floor. Adapted from J. I. Ewing. (1969).

the bottom. Calcium carbonate is soluble, and the amount of $CaCO_3$ that reaches the bottom depends on the supersaturation of $CaCO_3$ in seawater. The waters of the Atlantic Ocean are supersaturated with calcite, the less soluble form of $CaCO_3$, above 4500 m; below this depth, which is called the saturation horizon or compensation depth, calcite dissolves. The saturation horizon in the Pacific ranges from 400 to 3500 m.

The falling particles bury the iron-manganese ooze, and the sedimentary blanket thickens as the sea floor moves away from the ridge, (Fig. 1.5.3). As long as the sea floor is above the saturation horizon, a $CaCO_3$-rich ooze falls on the area. When the sea floor moves beneath the saturation horizon, the clay and silica components become predominant. Under areas having low biological productivity the sediment is mainly red (brown) deep sea clay. The silica component is large under areas having high biological productivity. These major layers are intermixed with thin layers of volcanic origin.

On the continental side of ocean basins, erosional products also move down into the ocean basins. Here the velocity of the water in the rivers is important because high velocities are needed to carry large particles. As

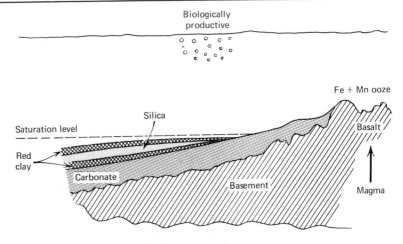

Figure 1.5.3.   Sediment structure.

the velocity of the water decreases at river mouths, the sediments sink to the bottom—the coarser particles first, fine particles later. The sediments form the beaches and river deltas along the sea coast.

The sediments build up along the continental slope until they become unstable. Sometimes a unit breaks off and slides down the slope (slump); at other times the sediments lose their rigidity, become a fluid, and flow down the slope (Fig. 1.5.4). Flows of the second type are called *turbidity*

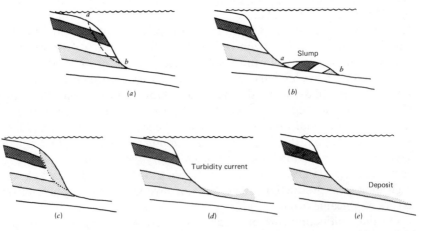

Figure 1.5.4.   Sedimentation processes. (a) and (b) a slump, (c), (d) and (e) Turbidity current and deposit.

currents and can move coarse sediments such as sand hundreds of kilometers into ocean basins. The extent of flow depends on the structure of the basin.

Because the reflection and transmission of a sound wave at a sedimentary plane depends on the density and speed of sound of the material, accoustical probing at all frequencies becomes a means for identifying the superficial sedimentary layer. Greater penetration is provided by using lower frequencies and even buried layers can be located and similarly identified.

When the ocean bottom is not a smooth plane but a rough surface, the scatter of sound furnishes the clue for characterization of the bottom over vast areas of roughness. Again, change of sound frequency permits an estimate of different scales of roughness. The evaluation can range from the most gross peaks and valleys that can be described by low frequency sounds at large ranges to the smallest sand ripples and nodules that can be identified by high frequencies at shorter ranges.

1.6. LIFE IN THE SEA

Until this point our dissection of the ocean, in search of parameters important to sound propagation, has been dominated by the physical characteristics of the sea. There are important biological realms to consider as well. Figure 1.6.1 is a schematic representation of the biomass of the ocean from the acoustical point of view of plants or animals. The presentation is in the form of a pyramid because the smaller bodies at the base are generally much more numerous than the larger ones at the apex. From the acoustical point of view, plankton and nekton represent "pieces" of the ocean at which the density and compressibility differ from those of the surrounding liquid. Therefore, to the acoustician, distributions of marine life represent distributions of inhomogeneities that may be of the type we want to study (e.g., fish) or they may be smaller bodies that by their scatter of sound, create a vast "fog" or "smoke screen" that prevents us from identifying our target. As we see later, whether a body backscatters sound effectively depends on how its density and compressibility differ from those of the surrounding ocean and how the size of the body compares to the acoustic wavelength. If the sound wavelength being used is very large compared to, say, a copepod of dimension 2 mm there will be minimal backscatter and it will be virtually impossible to "see" a single copepod. By using the relation $\lambda = c/f$, where λ is the sound wavelength and f is the sound frequency, we can calculate roughly the minimum frequency that is effective in causing backscatter from a single

Optimum Detection Frequency(3) for Bodies with Resonant Bubbles	Marine Plants or Animals	Equivalent Diameter (1)	Minimum Effective Detection Frequency for Non-Resonant Bodies (2)
10 to 3 Hz	Largest Nekton: whales and sharks	2 to 6 m	250 to 75 Hz
100 to 10 Hz	Larger Nekton and Largest Plankton: rat-tails, deep sea cods, tuna, scyphostone	0.2 to 2 m	2500 to 250 Hz
1000 to 100 Hz	Smallest Nekton and Larger Plankton: myctophids (6) stomiatoids, hatchet fishes	2 to 20 cm	25 to 2.5 kHz
10 to 1 kHz	Megaplankton: euphausiid (6), amphipod, chaetognath, some fish larvae	2 to 20 mm	250 to 25 kHz
100 to 10 kHz	Macroplankton: copepods	0.2 to 2 mm	2500 to 250 kHz
1000 to 100 kHz	Microphytoplankton: dinoflagellates and diatoms (4) Microzooplankton: radiolarians, foraminiferan and ciliates	20 to 200μ	25 to 2.5 MHz
10 to 1 MHz	Nanoplankton: flagellates, coccolithophores and diatoms	2 to 20μ	250 to 25 MHz
> 10 MHz	Ultrananoplankton:bacterioplankton	< 2μ	> 250 MHz

Figure 1.6.1. The biomass pyramid. Notes: (1) The equivalent diameter for detectability of a single nonresonant body is the diameter of the equivalent spherical volume of the animal. The equivalent biomass is approximately the density of water times the volume of the animal. (2) The minimum effective detection frequency for nonresonant bodies is the frequency below which sound diffraction makes the body appear smaller than it is. Substantially reduced backscatter (Rayleigh scatter) occurs when the sound wavelength in water is greater than the equivalent circumference. For this column we use the criterion $f > c/(\pi$ Diameter) for Rayleigh scatter. (3) The resonant bubble detection frequency has been calculated for sea level; for depth z meters multiply by $(1 + 0.1z)^{1/2}$. It is assumed that the marine animal or plant carries a spherical free air bubble of volume 5% of the total biovolume. Only a fraction of the marine animals possess a swim bladder that is filled with gas and will resonate to sound. In some species of myctophids the swim bladder is gas filled, in others it is lipid filled. This condition depends on the age of the animal, within a given species. The resonance frequency of lipid filled cavities is very much greater than that of gasfilled cavities. (4) Maximum photosynthetic activity (and bubble production) occurs in the smaller phytoplankton species with equivalent diameters 5 to 50 μ. (5) Plankton categories adapted from Parsons and Takahashi (1973). (6) The capability of swimming in horizontal water currents, which distinguishes nekton from plankton, is not clearly possessed by some animals (e.g., euphausiid and myctophid.)

inhabitant of the biomass pyramid. The minimum effective frequency scale to the right of the pyramid was calculated by using the equivalent circumference of the marine life equal to the wavelength as a crude criterion for detectability. A large concentration of small animals may, of course, be detectable at much lower frequencies than those in Fig. 1.6.1. The technique for making such calculations is given later.

An important exception to the backscatter rule that requires search wavelengths small compared to the scatterer is the very great scatter from a resonant bubble in water. There are two such cases: the free bubble, and the bubble used for buoyancy in many species of marine animals. Free gas bubbles, which occur in relatively high numbers within the top 10 m of the sea surface, play an effective role in scavenging nondissolved materials from the upper ocean. Because sound is preferentially absorbed and scattered by resonant bubbles, acoustical probes are the definitive way to conduct a census of these ocean microbubbles. The resonance frequency of a free gas bubble is given by

$$f = \frac{1}{2\pi a}\left(\frac{3\gamma P_A}{\rho}\right)^{1/2} \qquad (1.6.1)$$

where a = bubble radius
  γ = ratio of specific heats for bubble gas ($= 1.4$ for air)
  P_A = ambient pressure at bubble depth
  ρ = density of water ($\simeq 1035$ kg/m^3 for seawater)
This formula was used to calculate the bubble detection frequency in Fig. 1.6.1.

Fish that carry gas bubbles for flotation are particularly easy to detect if the search frequency is close to the bubble resonance frequency. The vast number of bubble-carrying species of marine animals can be summarized best by paraphrasing N. B. Marshall (1970). A gas-filled swim bladder is present in the adult stage of species comprising about a third of the mesopelagic fauna (ca 150 to 1000 m depth) notably in myctophids (ca 180 or 200 species), stomiatoids (ca 30 or 250 species), gonostomatids (excluding *cyclothone*, spp.), hatchetfish, most trichiuroids (ca 40 species), and some melamphaids. The swim bladder, a purely hydrostatic organ in these animals, may occupy up to 5% of the body volume, presumably enough to eliminate their weight in water and make them neutrally buoyant. The swim bladder is either absent or regressed in the adults of all (ca 150) species of bathypelagic fishes, (1000 to 4000 m). Except for certain squaloid sharks, chimaeroids, aleocephalids, and ateleopids, fishes of the benthopelagic fauna (living near the bottom) have a well-developed, gas-filled swim bladder. Of some 750 species, the main groups are rattails (Macrouridae, ca 300 species), deep-sea cods (Moridae, ca 70 species), and brotulids (ca 250 species).

PROBLEMS

1.2.1. Your ship is in an estuary carrying fresh water into the sea. Using the suspended silt as a clue, you find that the upper 5 m is nearly fresh water of temperature 20°C. Below is ocean water of $S = 20$ ppt, temperature 15°C. What is the sound speed profile to 20 m depth?

1.2.2. Using the data of Figs. 1.2.2 and 1.2.3 calculate the sound speed profile to the bottom at longitude 169°W, 28°S in July 1967.

1.3.1. Calculate the displacement amplitude of a neutrally buoyant float at depths 10, 20, 50, 100 m for swell of wavelength 150 m. Assume that the crest to trough amplitude at the surface is 2 m.

1.4.1. Calculate the frequency (Hz) and the angular frequency Ω of the wave of Problem 1.3.1. Is the surface tension effect negligible? What is the phase velocity?

1.5.1. Assume a uniform sedimentation rate, over a large area over a long time, of 1 mm/1000 years. Assume that the sea floor is spreading 3 cm/year. Draw a graph of the sediment thickness (over the basement) from ridge axis to a distance of 1000 km.

1.5.2. Suppose sediments on a continental slope are shaken by an earthquake, become fluidlike, and start to flow down the slope because the density is about 10% higher than that of the water. Assuming that the flat abyssal plane onto which it flows is 1 km deeper, estimate the maximum velocity of the turbidity flow when it reaches the abyssal plane. Assume negligible friction. The velocities are surprisingly large. Ewing and Heezen's discussion of the Grand Bank earthquake can be found in introductory books on oceanography and marine geology. *Hint.* Consider the potential and kinetic energies of the process.

SUGGESTED READING

W. S. Broecker, *Chemical Oceanography*, Harcourt Brace Jovanovich, New York, 1974.

B. C. Heezen, *The Floors of the Oceans*, Vol. I, *The North Atlantic*. Geological Society of America Special Paper No. 65, New York, 1959.

F. V. Hunt, *Electroacoustics. The Analysis of Transduction and Its Historical Background*, Harvard University Press and John Wiley, New York, 1954. Chapter 1 is a fascinating account of the discovery and development of sound sources and receivers from the work of Benjamin Franklin in the eighteenth century to the hi fi of the twentieth.

J. A. Jacobs, R. D. Russell, and J. Tuzo Wilson, *Physics and Geology*, 2nd ed., McGraw-Hill, New York, 1974.

M. J. Keen, *An Introduction to Marine Geology*, Pergamon Press, New York, 1968.

O. M. Phillips, *The Dynamics of the Upper Ocean*, Cambridge University Press, New York, 1966. This is an advanced monograph for fluid dynamicists and physical oceanographers.

H. Takeuchi, S. Uyeda, and H. Kanamori, *Debate About the Earth*, Freeman & Cooper, San Francisco, 1967.

W. S. Von Arx, *An Introduction to Physical Oceanography*, Addison-Wesley, Reading, Mass. 1962.

J. Tuzo Wilson, ed., *Continents Adrift*. In *Readings from* Scientific American, W. H. Freeman, San Francisco, 1970.

TWO

SOUND WAVES AND HUYGENS' PRINCIPLE

The core materials for acoustical studies of the ocean are the processes by which acoustic waves propagate through the water, interact at the boundaries, and scatter at objects. Three levels of understanding are needed: the first level is a qualitative description of the processes. Initially we use Huygens' principle to develop qualitative descriptions of the process and to establish a basis for the ray tracing methods. The second level involves quantitative descriptions of the processes. Quantitative methods are based on solutions of the wave equation, and either the results or implications of the results appear throughout acoustical oceanography. The third level is intermixed between the first two levels because it involves the adaptations of simplified quantitative models to the real ocean.

2.1. IMPULSIVE DISTURBANCES

A sound wave is a density disturbance that travels through the medium. The following experiment demonstrates the phenomenon. A source in the medium is caused to make a tiny impulsive expansion. A sequence of snapshots is taken, showing the instantaneous density of the medium (Fig. 2.1.1). The disturbance moves outward from the source and travels at

29

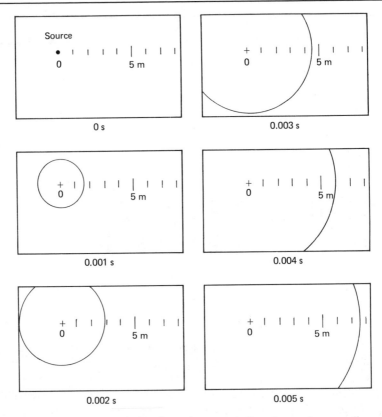

Figure 2.1.1. Sequence of sketches of a sound disturbance in a medium. The impulse wave front is a slight increase of the density of the medium.

constant speed. Later in the chapter we use the "laws of physics" to explain simply how the disturbance travels. For the present, we assume the experimental evidence—namely, after the medium is disturbed by a source, disturbances travel away from the source. The distance is proportional to the travel time.

2.2. HUYGENS' PRINCIPLE

The conceptual base of almost all the ways of describing traveling disturbances in a complicated medium was given by Christian Huygens, Dutch physicist-astronomer (1629–1695). His hypothesis is: each point on an advancing wave front can be considered as a source of secondary

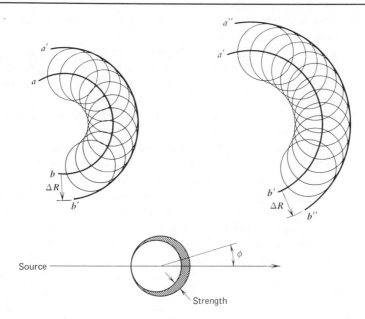

Figure 2.2.1. (*Top*) Huygens' principle for radiation from a point source. (*Bottom*) Effect of the obliquity factor in altering the strength of the radiation from the wavelets.

waves, which move forward as spherical wavelets in an isotropic medium. The outer surface that envelops all these wavelets constitutes the new wave front (Fig. 2.2.1).

As stated, the principle tacitly assumes that the wavelets have maximum intensity in the direction toward which the wave front is moving and zero in the backward direction. Stokes (1849) derived an obliquity factor for the wavelet density amplitude (Fig. 2.2.1).

$$\text{amplitude} \sim \left(\cos\frac{\phi}{2}\right)^2 \sim \frac{1+\cos\phi}{2} \tag{2.2.1}$$

The effect of the obliquity factor is to reduce the intensity of all waves propagating in directions other than normal to the wave front, and in particular to eliminate the backward wave. This, of course, is exactly what is observed.

In the short time Δt, the disturbance from each of the secondary sources on the wave front ab travels a distance ΔR. The outgoing portions of the wavelets coalesce to form the new wave front $a'b'$ and which in turn becomes a secondary source for the wave front $a''b''$. The strength of the wavelet is maximum in the direction away from the source and zero in the backward direction.

The student usually encounters Huygens' principle in physical optics, which emphasizes the diffraction and interference aspects of continuous waves (CW), such as the light from a laser. The sources used in geophysics and underwater sound measurements are often impulsive, like an explosion, or can be approximated as being impulsive. The application of Huygens' principle to an impulsive wave front is particularly simple and physically direct. The wave front expands radially until it encounters either a medium having a different sound speed or an object.

The development can be demonstrated graphically. Instead of separate sketches of sequential positions of the wave front as in Fig. 2.1.1 or 2.2.1, the sketches for different times have been combined in Fig. 2.2.2a. We ignore the boundary temporarily. Each successive position is indicated by 1, 2, 3, ... The time intervals Δt are uniform, and the wave front travels a constant distance ΔR during each interval. In the ray direction, normal to the wave front for the isotropic medium, the distance of advance of the wave front is given by

$$\Delta R = c\,\Delta t \tag{2.2.2}$$

where c is the sound speed. For clarity, the wavelet constructions are omitted.

2.2.1. Reflection at a Plane Interface

Figure 2.2.2b, the Huygens' construction of the interaction of the wave at the plane boundary, suggests that the wave front of the reflection is expanding from a source beneath the reflecting interface. The apparent source after reflection, called the image source, is labeled $\mathscr{R}$I in Fig. 2.2.2c. The $\mathscr{R}$ stands for the reflection coefficient, the fraction of the incident density wave that is reflected. The $\mathscr{R}$ reminds us that the wave front is a reflected wave, even though for a perfectly reflecting boundary $\mathscr{R} = 1$. A way to treat the image and the real source is illustrated in Fig. 2.2.2c. As in Fig. 2.2.2a, the real wave front is assumed to travel away from the real source and into "image space." The image wave of the proper strength is simultaneously initiated at the image point and travels into real space to become the reflected wave.

2.2.2. Refraction at a Plane Interface: Snell's Law

For the next example we assume that the impulse wave front is a plane wave over our region of observation. It could have come from a very

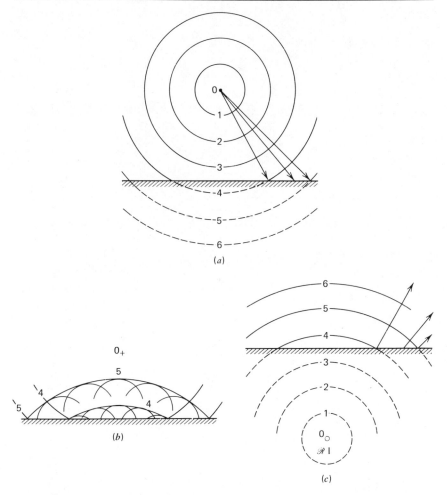

Figure 2.2.2. Successive positions of an impulse wave front for source and image of strength $\mathfrak{R}I$. The dashed lines are in "image space." For simplicity, the reflection is assumed to be perfect, $\mathfrak{R} = 1$. The rays are perpendicular to the wave fronts.

distant source, which would mean that the curvature of the wave front is negligible in our viewing range. The wave is incident on the plane boundary of two media having sound speeds c_1 and c_2, as in Fig. 2.2.3. Either c_1 or c_2 may be greater, but the figure is drawn for $c_2 > c_1$; the reader can sketch a figure for $c_1 > c_2$. Successive positions of the wave fronts are shown as they move from medium 1 and into medium 2. For

34

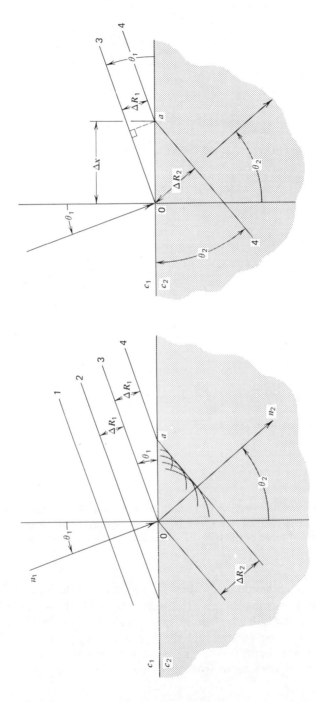

Figure 2.2.3. Refraction of a plane wave. The impulse wave front is shown at times $\Delta t, 2\Delta t, \ldots, n_1$ and n_2 are the normals to the wave fronts; $c_2 > c_1$ as in the case of sound going from air to water or from water to most sediments.

the illustration, the wavelets for wave front position 4 were drawn as originating at the interface of c_1 and c_2. The reflected wave front is omitted for simplicity. In the time Δt, the wave front has moved a distance ΔR_1 in medium 1 and ΔR_2 in medium 2. In the same time, the contact of the wave front at the interface moved from 0 to a, a distance Δx along the x axis. The angles are measured between the rays and the normal to the interface or between the wave front and the interface. The propagation distances in the two media are

$$\Delta R_1 = \Delta x \sin \theta_1 \quad \text{and} \quad \Delta R_2 = \Delta x \sin \theta_2$$

The speeds are

$$c_1 = \frac{\Delta R_1}{\Delta t} \quad \text{and} \quad c_2 = \frac{\Delta R_2}{\Delta t}$$

Therefore,

$$\frac{\sin \theta_1}{c_1} = \frac{\sin \theta_2}{c_2} \tag{2.2.3}$$

This is well-known as Snell's law. We use Snell's law again in Section 2.9 and throughout Chapter 3.

2.2.3. Diffraction

The reader might think that the next obvious choices for examples of Huygens' principle would be the scattering of sound by objects such as points or lines. The drawings showing the wavelet radiating away from the point after the incident wave front has passed the point would be easy to make. Unfortunately ideal points and lines have scattered waves of zero amplitude. More importantly these examples would give a misleading picture of the nature of diffraction.

Let us assume that the source is above a thin, semi-infinite plane reflector that permits part of the wave to be transmitted. The situation is illustrated in Fig. 2.2.4, where the plane extends from the edge at B, infinitely to the left, and behind and in front of the page. The components behave as follows: (1) The wave front spreads away from the source and interacts with the reflector. (2) The interactions become sources of Huygens' wavelets at the reflector. (3) The envelopes of the wavelets at the reflector become the transmitted and the reflected waves. (4) Beyond the edge of the reflector, the outgoing wave continues. (5) The envelope of the Huygens wavelets originating at the edge (spherical surfaces D_5 and then D_6) form a wave front. The limiting wave front that appears to spread from the edge is referred to as the "diffracted" or "boundary

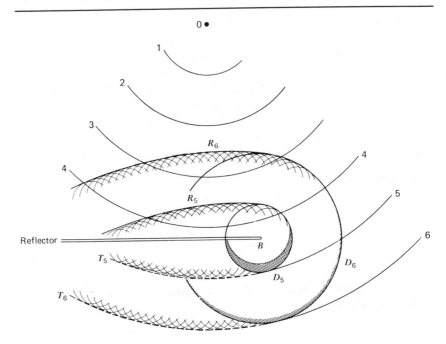

Figure 2.2.4. Huygens' wavelet construction for a semi-infinite partially reflecting plane. The positions of the impulse wave front are given at equal increments of time. The diffracted wave fronts (D_5, D_6) are shaded. The transmitted waves (T_5, T_6) are indicated by dashed lines, and reflected waves (R_5, R_6). are shown solid; both are shown as envelopes of wavelets.

wave." This wave is observed as a separate arrival. On recalling the obliquity factor (2.2.1), we realize that the wavelets are most intense in the reflected and transmitted directions and weakest to the sides. The diffracted wave is strongest in the direction of propagation, but because of its late arrival and unique direction of propagation, it is more readily detected in all other directions. The boundary wave exists because there is a reflecting plane to the left of the edge B and none to the right. Wave forms and amplitudes of boundary waves are discussed in Chapter 10.

2.3. CONTINUOUS WAVE DISTURBANCES: SINUSOIDAL SIGNALS

2.3.1. Variations in Space or Time

A sinusoidally excited source expands and contracts repeatedly. The resulting condensations (density increases) and rarefactions (decreases) in

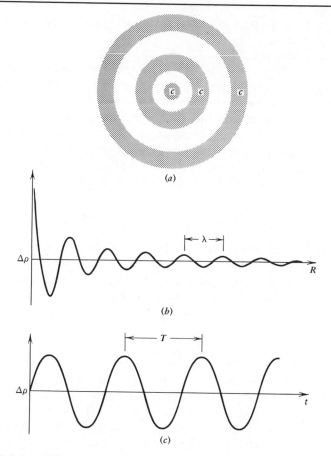

Figure 2.3.1. CW source. (*a*) Spatial disturbances of the density are alternately condensations *C* and rarefraction. The distance between the adjacent maxima is the wavelength λ. (*b*) The dependence of density change Δρ on *R*. The amplitude decreases rapidly near the source and very slowly at great distance. (*c*) Temporal dependence (*t*) of the CW source. The time between adjacent maxima is the period *T*.

the medium move away from the source at the sound speed *c*, as would the disturbance from an impulsive source. An instantaneous picture of the density fluctuations would resemble the sketch in Fig. 2.3.1*a*. This disturbance is called a continuous wave (CW), and it comes from a CW* source.

---

*The IEEE definition of (CW) is: "Waves, the successive oscillations of which are identical under steady-state conditions." We often use the definition in the more restricted sense in which CW is a sinusoidal or cosinusoidal wave.

The distance between adjacent condensations along the direction of travel is the wavelength λ.

The disturbances sketched in Fig. 2.3.1a radiate outward from a source which is small compared to λ. As a condensation moves outward, its energy is spread over larger and larger spheres. Correspondingly its amplitude decreases. Later we show that the amplitude decreases as $1/R$, where R is the distance from the source. Frequently we make measurements in a small region at large distances from the source. Within that region, the wave fronts appear to be plane and have negligible change of amplitude. The distance between crests continues to be λ. The spatial dependence of the density fluctuation is, for example

$$\Delta\rho = a \sin \frac{2\pi R}{\lambda} \quad \text{or} \quad \Delta\rho = b \cos \frac{2\pi R}{\lambda} \qquad (2.3.1)$$

where a and b are amplitudes of the density fluctuations. This is the "plane wave approximation."

The time between adjacent crests passing any fixed point in the medium is the period T (Fig. 2.3.1c). The temporal dependence of the density fluctuation is

$$\Delta\rho = a \sin 2\pi ft \quad \text{or} \quad \Delta\rho = b \cos 2\pi ft \qquad (2.3.2)$$

where f is the frequency of the oscillation. The frequency is measured in cycles per seconds or hertz (Hz). The sine and cosine functions are periodic; that is, they repeat themselves for every increment of 2π [e.g., $\sin(\theta + 2n\pi) = \sin\theta$ and $\cos(\theta + 2n\pi) = \cos\theta$]. The application of the periodicity of the functions to (2.3.2) gives the period T of the function. The functions repeat for two adjacent times t_1 and $t_2 = t_1 + T$ such that $2\pi ft_2 - 2\pi ft_1 = 2\pi$. Therefore

$$T = \frac{1}{f} \ [\text{s}] \qquad \textbf{(2.3.3)}$$

Functions that repeat periodically are also called harmonic functions. The simplest harmonic functions are sine and cosine functions. Recalling Fig. 2.3.1, in the time T the disturbance has moved the distance λ. The speed at which it travels, the sound speed c, is

$$c = \frac{\lambda}{T} = f\lambda \qquad \textbf{(2.3.4)}$$

With the units of λ being meters, c will be in meters per second.

The 2π multiplying ft puts that product in units of radians. It is customary to absorb the 2π into the frequency by defining ($\equiv$) the angular frequency as follows:

$$\omega = 2\pi f \text{ rad/s} \qquad \text{or} \qquad \omega = \frac{2\pi}{T} \qquad (2.3.5)$$

The sinusoidal function that describes the spatial dependence of changes of density is $\sin(2\pi R/\lambda)$. On comparing this with $\sin(2\pi t/T)$, we see that $2\pi/\lambda$ which is called the wave number k, is analogous to $2\pi/T$, the angular frequency. The relationships of the wave number to f and c are derived from $c = f\lambda$. The wave number is given by

$$k \equiv \frac{2\pi}{\lambda} \ [m^{-1}]$$

$$k = \frac{2\pi f}{c} \qquad (2.3.6)$$

$$k = \frac{\omega}{c}$$

The arguments of the sinusoidal functions ωt and kr are dimensionless. That is, ω has the dimensions of $(\text{time})^{-1}$ and t has the dimension $(\text{time})^{1}$. The product ωt is called the temporal phase. The same is true for kR; k has the dimension $(\text{length})^{-1}$, R has $(\text{length})^{1}$, and the product kR, which is dimensionless, is called the spatial phase.

2.3.2. Traveling Waves

At a fixed position, the signal has temporal dependence $\sin(\omega t)$. At a fixed time, the spatial dependence is $\sin(kR)$. As shown later when solutions of general problems are given, these two concepts can be combined in a form such as

$$\Delta\rho = a \sin(\omega t - kR) \qquad (2.3.7)$$

Recalling $k = \omega/c$ and $c = f\lambda$, $\Delta\rho$ can be written

$$\Delta\rho = a \sin\left[\omega\left(t - \frac{R}{c}\right)\right] \qquad (2.3.8)$$

$$\Delta\rho = a \sin\left[2\pi\left(\frac{t}{T} - \frac{R}{\lambda}\right)\right] \qquad (2.3.9)$$

That this is a wave having the speed c can be demonstrated as follows. Pick an arbitrary phase such as $\omega(t - R/c)$. At later time $t + \Delta t$, the wave will be at $R + \Delta R$. If the phase is the same, we write

$$\omega\left(t + \Delta t - \frac{R + \Delta R}{c}\right) = \omega\left(t - \frac{R}{c}\right) \quad \therefore \quad \frac{\Delta R}{\Delta t} = c$$

The combination $t - R/c$ is indicative of a wave traveling in the positive R direction.

2.4. MULTIPLE CW SOURCES

2.4.1. Two Sources

Suppose that two sources transmit CW signals at frequency ω. To keep the analysis simple, the two amplitudes at the receiver are the same. The distance from source 1 is R_1 and from source 2 is R_2 (Fig. 2.4.1). Both sources have the same phase. The signal at point Q is the sum of the two. The graphical method of adding the signals shows the relationships of the signals and their phases at the observation point in a physically intuitive

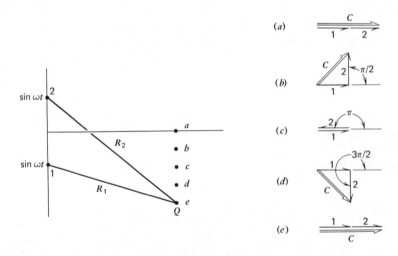

Figure 2.4.1. Graphical combination of the signals at an observation point located at different distances R_1 and R_2 from two sources. The amplitude a at the observation point is assumed to be the same for each signal. The vector sum $|C|$ of the two waves depends on the path difference $R_2 - R_1$. (a) $R_2 - R_1 = 0$, $|C| = 2a$; (b) $R_2 - R_1 = \lambda/4$, $|C| = \sqrt{2}a$; (c) $R_2 - R_1 = \lambda/2$, $|C| = 0$; (d) $R_2 - R_1 = 3\lambda/4$, $|C| = \sqrt{2}a$; (e) $R_2 - R_1 = \lambda$, $|C| = 2a$.

manner. In the graphical technique we let the signal from each source be represented by a phase vector of the amplitude a and spatial phase kR. Since the time dependence is common to both sources, it is ignored.

Assuming that $R_1 \gg \lambda$, the phase changes by 2π many times in distance R_1 and only the residual phase needs to be shown. Similarly for R_2. The phase difference at Q is $kR_2 - kR_1$. Examples of the phase difference are given in Fig. 2.4.1.

The amplitudes at $a, b, \ldots, e$ are the sums of the two vectors. The resultant amplitudes are labeled C.

2.4.2. Several Sources

We combine algebra, geometry, and the phase vectors to develop a means of calculating the amplitude due to many sources. To calculate the relative phase, we need a simple relationship between R_1 and the direction ϕ to the observation point Q. Referring to Fig. 2.4.2, we write

$$R_1^2 = (y - y_1)^2 + x^2 \tag{2.4.1}$$

$$R_1^2 = (R \sin \phi - y_1)^2 + R^2 \cos^2 \phi$$
$$= R^2 - 2Ry_1 \sin \phi + y_1^2 \tag{2.4.2}$$

where $R^2 \cos^2 \phi + R^2 \sin^2 \phi = R^2$

$$R_1 = (R^2 - 2Ry_1 \sin \phi + y_1^2)^{1/2} \tag{2.4.3}$$

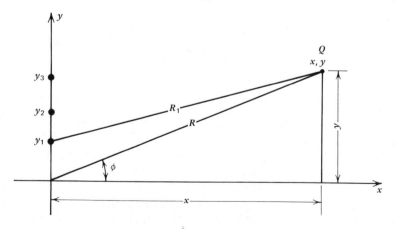

Figure 2.4.2.   Geometry for distances from sources to listening point Q.

Factoring R out of the square root gives

$$R_1 = R \left(1 - \frac{2y_1 \sin \phi}{R} + \frac{y_1^2}{R^2} \right)^{1/2} \tag{2.4.4}$$

where $y_1/R \ll 1$.

The binomial expansion is

$$(1+q)^{1/2} = 1 + \frac{q}{2} - \frac{q^2}{8} + \frac{q^3}{16} + \cdots \tag{2.4.5}$$

where q is a small number. Letting q be $(y_1^2/R^2 - 2y_1 \sin \phi/R)$, (2.4.4) becomes

$$R_1 \simeq R \left[1 - \frac{y_1}{R} \sin \phi + \frac{y_1^2}{2R^2} (1 - \sin^2 \phi) + \cdots \right] \tag{2.4.6}$$

where $\simeq$ means "approximately equal to."

For simplicity, and limiting the values to very small y_1/R, we drop the terms involving y_1^2/R^2 and higher power to obtain

$$R_1 \simeq R - y_1 \sin \phi \tag{2.4.7}$$

This approximation is sometimes called the plane wave or Fraunhofer approximation. Similarly, the distances from sources at y_2, y_3, ..., to listening point Q are

$$R_2 \simeq R - y_2 \sin \phi$$
$$R_3 \simeq R - y_3 \sin \phi \tag{2.4.8}$$

Compared to the spatial phase kR of the reference distance R, the phase of the signal from each source is decreased by $ky_1 \sin \phi$, $ky_2 \sin \phi$, and so on.

For an initial example, assume the three sources are at b, 0, and $-b$ (Fig. 2.4.3). Increments of ϕ cause the path length to increase and the phase difference to increase for each vector. The phase shift between waves from two adjacent sources is

$$\Delta\Phi = kb \sin \phi \tag{2.4.9}$$

The reference vector labeled a_o is the contribution from the center source.

Equal maximum amplitudes occur when contributions from adjacent sources show phase differences given by $kb \sin \phi = 0$, 2π, ... Nulls occur at $2\pi/3$, $4\pi/3$, ... The central maximum is at $\phi = 0$. In the plane wave approximation, the phase differences between equally spaced sources are equal. Eight easily calculable phase differences have been arbitrarily selected for the construction of the amplitude of the interference pattern.

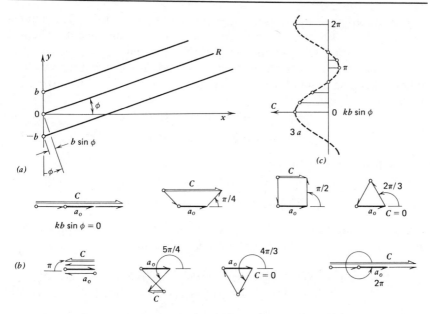

Figure 2.4.3. Construction of the interference pattern of three point sources at very large range. Each source has the same amplitude, frequency, and temporal phase. With reference to the central source, the relative spatial phase shifts of the signals are $-kb \sin \phi$, 0, and $kb \sin \phi$.

The interference pattern is often called the "directional response" because it is a function of angle ϕ.

2.5. MULTIPLE SOURCE ANALYSIS USING COMPLEX NUMBERS

2.5.1. Two Sources

Many operations involve the sums and differences of angles and the products of trigonometric functions. These operations are simplified by using the relations between trigonometric functions and complex exponential functions.

$$e^{i\Phi} = \cos \Phi + i \sin \Phi$$
$$\cos \Phi = \frac{e^{i\Phi} + e^{-i\Phi}}{2}, \qquad \sin \Phi = \frac{e^{i\Phi} - e^{-i\Phi}}{2i} \tag{2.5.1}$$

A review of complex numbers and variables is given in Section A2.1.

The preceding sections used the expression $\sin(\omega t - kR_1)$. The expression $\exp[i(\omega t - kR_1)]$ gives both the sine and cosine components. The real part is the cosine and the imaginary part is the sine. Using the complex exponential, the density fluctuations $\Delta\rho_1$ and $\Delta\rho_2$ can be expressed as follows, keeping in mind that the imaginary part, the sine, will correspond to (2.3.7):

$$\Delta\rho_1 = a \exp[i(\omega t - kR_1)] \tag{2.5.2}$$

$$\Delta\rho_2 = a \exp[i(\omega t - kR_2)] \tag{2.5.3}$$

The sum of the two signals is

$$\Delta\rho = \Delta\rho_1 + \Delta\rho_2 = a\{\exp[i(\omega t - kR_1)] + \exp[i(\omega t - kR_2)]\}$$
$$\Delta\rho = a \exp(i\omega t)[\exp(-ikR_1) + \exp(-ikR_2)] \tag{2.5.4}$$

Let us compute $|\Delta\rho|^2$, that is, $\Delta\rho\,\Delta\rho^*$, where * denotes complex conjugate. The product $\exp(i\omega t)\exp(-i\omega t)$ is unity, and

$$|\Delta\rho|^2 = a^2[\exp(-ikR_1) + \exp(-ikR_2)][\exp(ikR_1) + \exp(ikR_2)]$$
$$= a^2\{2 + \exp[ik(R_1 - R_2)] + \exp[-ik(R_1 - R_2)]\} \tag{2.5.5}$$

and with the aid of (2.5.1) we write

$$|\Delta\rho|^2 = 2a^2[1 + \cos k(R_1 - R_2)] \tag{2.5.6}$$

As in Fig. 2.4.1, the interference maxima occur at $k(R_2 - R_1) = 0$, 2π, $4\pi, \ldots$, and minima occur at π, $3\pi, \ldots$.

To indicate the power of the complex exponential method, we again calculate the response for the three sources in Fig. 2.4.3. At very large R/b, the distances from the sources to the receiver are (2.4.7)

$$R_1 \simeq R - b\sin\phi$$
$$R_2 \simeq R \tag{2.5.7}$$
$$R_3 \simeq R + b\sin\phi$$

Using the exponential notation, the resultant is

$$\Delta\rho = a \exp[i(\omega t - kR + kb\sin\phi)] + a \exp[i(\omega t - kR)]$$
$$+ a \exp[i(\omega t - kR - kb\sin\phi)] \tag{2.5.8}$$

The $\Delta\rho$ is phase dependent, a and $\exp[i(\omega t - kR)]$ can be factored, and the result is

$$\Delta\rho = a \exp[i(\omega t - kR)][1 + \exp(ikb\sin\phi) + \exp(-ikb\sin\phi)]$$

or

$$\Delta\rho = a \exp[i(\omega t - kR)][1 + 2\cos(kb\sin\phi)] \tag{2.5.9}$$

To demonstrate the dependence of the amplitude on ϕ, rewrite $\Delta\rho$ as

$$\Delta\rho \equiv C \exp[i(\omega t - kR)] \tag{2.5.10}$$

where

$$C = [1 + 2\cos(kb\sin\phi)]a \tag{2.5.11}$$

The dependence of C on $kb\sin\phi$ is shown in Fig. 2.4.3c. The nulls occur at $kb\sin\phi = \pm 2\pi/3, 4\pi/3, \ldots$.

2.5.2. Multiple Sources

The calculation of the signal for a large number of sources is simple except for the manipulations that change the answer to a more convenient form. For N sources evenly spaced over a distance W,

$$b = \frac{W}{N-1} \tag{2.5.12}$$

and the density fluctuation $\Delta\rho_n$ of the signal from the nth source, relative to the source at distance R, is

$$\Delta\rho_n = a\exp\left[i\left(\omega t - kR + \frac{nkW\sin\phi}{N-1}\right)\right] \tag{2.5.13}$$

where a is a constant

Since $\omega t - kR$ is common to all the signals, we factor it at the beginning and then suppress it by calculating C as follows:

$$\Delta\rho = a\exp[i(\omega t - kR)]\sum_{n=0}^{N-1}\exp\left(\frac{inkW\sin\phi}{N-1}\right)$$

$$C = a\sum_{n=0}^{N-1}\exp\left(\frac{inkW\sin\phi}{N-1}\right) \tag{2.5.14}$$

This expression has the form of the sum of a geometric series.

$$\sum_{n=0}^{N-1} r^n = \frac{1-r^N}{1-r} \tag{2.5.15}$$

In our case, we recognize that $\exp(iny) = [\exp(iy)]^n$ and

$$r = \exp\left(\frac{ikW\sin\phi}{N-1}\right) \tag{2.5.16}$$

$$C = a\frac{1 - \exp\left(\dfrac{iNkW\sin\phi}{N-1}\right)}{1 - \exp\left(\dfrac{ikW\sin\phi}{N-1}\right)} \tag{2.5.17}$$

The form can be changed to a more convenient one by factoring the exponents using

$$1 - e^{ix} = -e^{ix/2}(e^{ix/2} - e^{-ix/2}) \qquad (2.5.18)$$

Recalling (2.5.1), (2.5.18) simplifies to

$$1 - e^{ix} = -2ie^{ix/2} \sin \frac{x}{2}.$$

Thus

$$C = Na \left\{ \frac{\exp\left[iNk(W/2)\dfrac{\sin \phi}{N-1}\right]}{\exp\left[ik(W/2)\dfrac{\sin \phi}{N-1}\right]} \right\} \frac{\sin\left[Nk(W/2)\dfrac{\sin \phi}{N-1}\right]}{N \sin\left[k(W/2)\dfrac{\sin \phi}{N-1}\right]}$$

$$(2.5.19)$$

The expression in braces has an absolute value of 1 and specifies a phase shift that depends on the choice of origin. For large N, the phase shift is $(kW \sin \phi)/2$ and the phase $i(\omega t - kR)$ becomes $i(\omega t - kR + (kW \sin \phi)/2)$.

The remaining factor in (2.5.19) is known as the "directional response" D. For an N equal source array, D is

$$D = \frac{\sin\left(\dfrac{N}{N-1}\dfrac{kW}{2}\sin \phi\right)}{N \sin\left(\dfrac{1}{N-1}\dfrac{kW}{2}\sin \phi\right)} \qquad (2.5.20)$$

When N is large, and using $\sin[k(W/2)(\sin \phi/(N-1)] \approx k(W/2)$ $(\sin \phi)/(N-1)$, D becomes

$$D = \frac{\sin \dfrac{kW \sin \phi}{2}}{\dfrac{kW \sin \phi}{2}} \qquad (2.5.21)$$

The latter expression has the form $(\sin x)/x$. This function is very important not only in specifying the directional response of a rectangular transducer but also in describing the signals scattered by small plane objects. The function is shown in Fig. 2.5.1. As x tends to zero, $(\sin x)/x$ is indeterminate

$$\lim_{x \to 0} \frac{\sin x}{x} = \frac{0}{0}$$

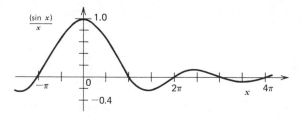

Figure 2.5.1. $(\sin x)/x$.

It is evaluated by forming

$$\lim_{x\to 0}\frac{(d/dx)\sin x}{(d/dx)(x)}=\lim_{x\to 0}\frac{\cos x}{1}=1$$

2.5.3. Continuously Distributed Source

The same result as (2.5.21) can be obtained with less manipulation by integration for a continuous set of sources. In this calculation we are doing an integral over Huygens wavelets and should have the obliquity factor $(1+\cos\phi)/2$ in the expression. It is omitted for simplicity. We normalize the expression by letting the source strength per unit length be W^{-1}. For an element of length dy at position y, the amplitude at the field position is

$$d\rho = a\frac{dy}{W}\exp\left[i(\omega t - kR + ky\sin\phi)\right] \tag{2.5.22}$$

With the aid of Fig. 2.5.2, the signal for the array is the integral of (2.5.22) between $-W/2$ and $W/2$:

$$\Delta\rho = \frac{a}{W}\int_{-W/2}^{W/2}\exp\left[i(\omega t - kR + ky\sin\phi)\right]dy \tag{2.5.23}$$

After factoring $a\exp\left[i(\omega t - kR)\right]$ and again designating the directional response as D, (2.5.23) *becomes*

$$D=\frac{1}{W}\int_{-W/2}^{W/2}\exp\left[iky\sin\phi\right]dy \tag{2.5.24}$$

$$D=\frac{1}{W}\frac{\exp(iky\sin\phi)}{ik\sin\phi}\Bigg]_{-W/2}^{W/2} \tag{2.5.25}$$

$$D=\frac{\exp\left[\dfrac{i(kW\sin\phi)}{2}\right]-\exp\left[\dfrac{-i(kW\sin\phi)}{2}\right]}{ikW\sin\phi} \tag{2.5.26}$$

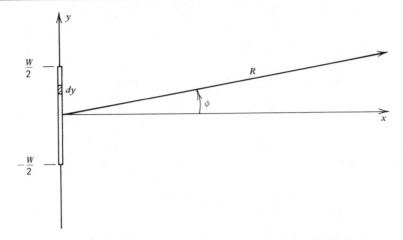

Figure 2.5.2.   Geometry for continuously distributed source.

With the aid of (2.5.1)

$$D = \frac{\sin \dfrac{kW \sin \phi}{2}}{\dfrac{kW \sin \phi}{2}} \tag{2.5.27}$$

This is identical to (2.5.21) obtained by summation. Its graph was plotted in Fig. 2.5.1. An alternative graph of this directional response is in polar coordinates, where the length of the radius vector in the polar pattern is proportional to the directional response at that angle (Fig. 2.5.3).

We have given this analysis for sources aligned in the y direction. The

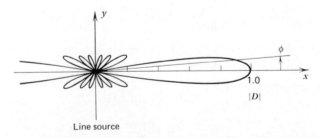

Line source

Figure 2.5.3.   Polar diagram of the directional response of the radiation from a continuous line source along the y axis. The directional response is symmetric about the source. The source is 4λ long. The radial distance from the center of the diagram to the pattern is proportional to the directional response D. Measurements are made at great distances from the source.

same analysis can be used for sources in the z direction. If each of the dy sources on the y axis were a line of sources along the z axis, a rectangular source would be formed. The resulting directional response would be the product of the directional responses for the z and y directions.

The source is regarded as being a "piston source" when the individual elements are very close together or continuous, have the same amplitude, and have the same phase. The piston transducer is usually baffled to ensure that signals from the back side do not interfere with the signals from the front side. Chapter 5 deals with transducers.

2.6. REFLECTION OF SPHERICAL WAVES: FRESNEL ZONES

Our purpose is to discuss reflection of spherical waves from a plane surface.

Many sonar systems have the source and receiver at the same position. In operation, the source transmits a signal for a short time, then the receiver listens for the returning signals. Thus the source is quiet when the reflected and scattered signals come back. For our analysis we assume the source is turned on and transmits a sinusoidal signal for a short time, so that there are no interferences at the receiver position.

The point source is at height h above a plane disk. The calculation consists of summing (or integrating) the contributions of all wavelets from the reflecting disk. We calculate the dependence of the sum on the radius of the disk. Referring to Fig. 2.6.1, the phase of the signal wave front that

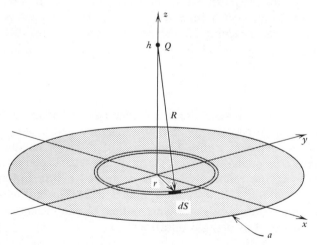

Figure 2.6.1. Reflection geometry for a plane area πa^2. Receiver and source are at the same place Q; dS is an element of area at the radius r.

travels from the source at Q to the typical scattering area element dS, then back to the receiver, a total distance of $2R$ is

$$\Phi = \omega t - 2kR \tag{2.6.1}$$

$$R = (h^2 + r^2)^{1/2} \tag{2.6.2}$$

In summing the contributions from different elements as r increases from zero, the quantity of interest is the change of the phase $2kR$. On letting the smallest value $2kh$ be the reference phase, the relative phase difference, $\Delta\Phi$ for the element dS is

$$\Delta\Phi = 2kR - 2kh \tag{2.6.3}$$

Recalling $k = 2\pi/\lambda$ and $R^2 = (r^2 + h^2)$, (2.6.3) is solved for R and then for r as a function of $\Delta\Phi$.

$$R = (r^2 + h^2)^{1/2} = \frac{\lambda \, \Delta\Phi}{4\pi} + h \tag{2.6.4}$$

$$r^2 = \frac{h\lambda \, \Delta\Phi}{2\pi} + \left(\frac{\lambda \, \Delta\Phi}{4\pi}\right)^2 \tag{2.6.5}$$

We are interested in the case where the reflector is many wavelengths from the source ($\lambda \ll h$) so that (2.6.5) can be approximated by the first term only. Then

$$\Delta\Phi \approx \frac{2\pi r^2}{h\lambda} \tag{2.6.6}$$

The contributions are positive for elements at radii such that $\Delta\Phi$ given by (2.6.6) is in the range $0 \le \Delta\Phi \le \pi$. This range is the first phase zone in Fig. 2.6.2. In the range $\pi \le \Delta\Phi \le 2\pi$, the contributions are negative; this is the first shaded ring. Letting $\Delta\Phi$ be $n\pi$, the phase changes sign at radii

$$r_n \approx \left(\frac{\lambda h}{2}\right)^{1/2} n^{1/2} \tag{2.6.7}$$

Each ring having the same phase sign is called a "Fresnel zone" or "phase zone." The central circle is the first Fresnel zone with $n = 1$.

A signal reflected at a disk having radius r_1 has a maximum value. When the radius of the disk is r_2 the first and second phase zones both contribute, and the reflected signal is decreased nearly to zero. Other maxima and minima follow as the radius is increased and as additional zones contribute. For a disk of infinite radius, an analytical solution gives the reflected signal as being proportional to

$$\left(\frac{1}{2h}\right) \exp\left[i(\omega t - 2kh)\right]$$

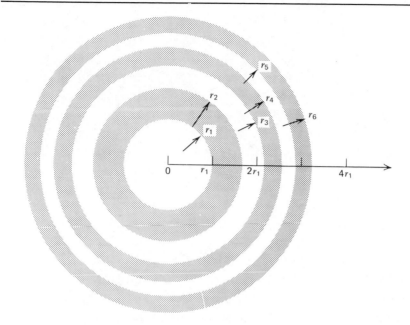

Figure 2.6.2. Fresnel or phase zones for reflection. Relative to the origin, the white areas are the positive phases of the contributions to the signal; shaded areas are the negative phases. The radius of the first zone r_1 is given by (2.6.7) with $n = 1$.

Analytical solutions require a mathematical expression of Huygens' principle and these are in Chapter 10 and Section A10.

2.7. WAVE PROPAGATION

So far our development has been phenomenological. We started at a disturbance and followed the expanding wave fronts using Huygen's principle. We introduced the expression for a traveling wave and then used it to calculate the phase of a signal as a function of time and space.

The usual way to develop the relationship between the properties of the medium and the waves is to derive the wave equation. We do this in the course of giving a physical description of the propagation of waves in a medium having constant sound speed. The many details omitted in this heuristic discussion can be found in Section A2.2.

We make a disturbance in the medium by causing a small spherical source to expand. This causes the local density and pressure to increase

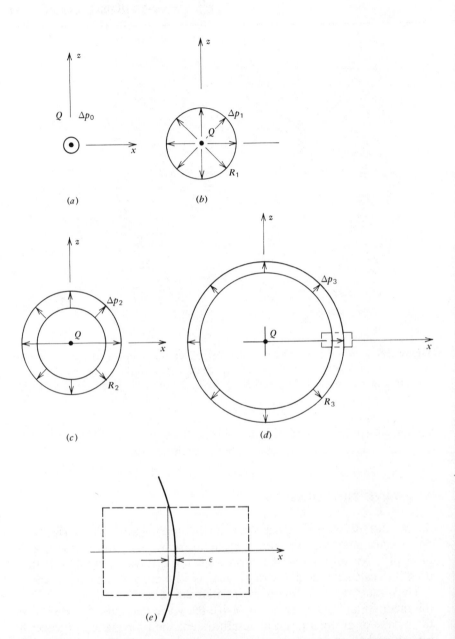

Figure 2.7.1. Disturbance due to expansion of a source at Q. (a), (b), (c), (d) Pressures and motions at successive times; (d) small region for a "plane wave approximation" (dashed area); (e) enlargement of that region: R_3 is so large that the curvature of the wave front is negligible in this region; $\epsilon < \lambda/8$.

because the rest of the medium does not instantaneously move to allow space for the expansion. As sketched in Fig. 2.7.1, the excess pressure Δp_0 at Q causes an acceleration and motion of the medium to R_1. At R_1 the excess pressure Δp_1 causes an acceleration of the medium and motion to R_2. The excess pressure Δp_2 causes the next outward acceleration, and so on.

We make a quantitative comparison of the pressures and accelerations in a small region at very large distance from the source. We have two criteria for choosing the region. First, the curvature of the wave front ε is very small, that is, the width of the region is small compared to the diameter of the first Fresnel zone. Second, the width of the region is very much less than the distance to the source, and spherical spreading of the wave front within the region is negligible. In this region the pressure, velocity, and acceleration of a fluid particle are approximately functions of x.

Assume that the pressure decreases across our region (Fig. 2.7.2). There is a large ambient pressure P_A, which we ignore because it does not change over our region. The net change of pressure acting in the plus x direction is $-(\partial p/\partial x)\,\Delta x$ where ∂ indicates differentiation with respect to one variable, only. Using $p = P_A + \Delta p$ the net force on the volume is

$$F_x = -\left(\frac{\partial \,\Delta p}{\partial x}\,\Delta x\right)\Delta y\,\Delta z \qquad (2.7.1)$$

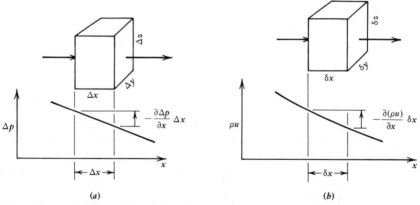

(a)   *(b)*

Figure 2.7.2. Pressure differential across a small volume. a) The pressure differential causes the volume to move to the right because $\partial p/\partial x$ is negative, as shown. (b) Mass flow in a small region: $\delta x\,\delta y\,\delta z$ are the dimensions of a fixed cage; u is the velocity of flow along the x axis, ρ is the density, and ρu is the mass flow.

where $\Delta y\, \Delta z$ is the area. The mass of the volume is $\rho_A\, \Delta x\, \Delta y\, \Delta z$, where ρ_A is the ambient density. Defining u as the velocity component along the x direction and $\partial u/\partial t$ as the acceleration, Newton's law $F = ma$ gives

$$-\left(\frac{\partial\, \Delta p}{\partial x}\, \Delta x\right) \Delta y\, \Delta z = \rho_A \frac{\partial u}{\partial t}\, \Delta x\, \Delta y\, \Delta z \qquad (2.7.2)$$

$$-\frac{\partial\, \Delta p}{\partial x} = \rho_A \frac{\partial u}{\partial t} \qquad \textbf{(2.7.3)}$$

This is the acoustical form of Newton's second law.

The expansion of the source causes an outward mass flow. Within a small cage that is fixed in space, fluid flows in one face and out the other. Since the fluid is compressible, more fluid may flow in than flows out, and the density within the cage may increase. The velocity of the fluid flow along the x direction is sketched in Fig. 2.7.2b. The net mass flow per unit time into the cage is $-[\partial(\rho u)/\partial x]\,\delta x\, \delta y\, \delta z$. This causes a rate of density increase $(\partial\rho/\partial t)$. Equating the two gives

$$-\left(\frac{\partial(\rho u)}{\partial x}\right) \delta x\, \delta y\, \delta z = \left(\frac{\partial\rho}{\partial t}\right) \delta x\, \delta y\, \delta z \qquad (2.7.4)$$

The density is the sum of the ambient value ρ_A and a small change $\Delta\rho \ll \rho_A$. Using this in (2.7.4) and neglecting $\partial(\Delta\rho u)/\partial x$ relative to $\rho_A\, \partial u/\partial x$, (2.7.4) is approximately

$$-\rho_A \frac{\partial u}{\partial x} = \frac{\partial(\Delta\rho)}{\partial t} \qquad \textbf{(2.7.5)}$$

This is the acoustical equation of conservation of mass.

We combine (2.7.3) and (2.7.5) by taking $\partial/\partial x$ of (2.7.3) and $(\partial/\partial t)$ of (2.7.5) and eliminating $\rho_A\, \partial^2 u/(\partial x\, \partial t)$.

$$\frac{\partial^2\, \Delta p}{\partial x^2} = \frac{\partial^2(\Delta\rho)}{\partial t^2} \qquad (2.7.6)$$

The dependence of Δp on $\Delta\rho$ is the result of experiment. It is approximately linear, and is called the acoustic equation of state,

$$\Delta p \simeq \frac{E}{\rho_A}\, \Delta\rho \qquad \textbf{(2.7.7)}$$

where E is the bulk modulus of elasticity. This is Hooke's law. It assumes that an instantaneous applied pressure Δp causes an instantaneous proportional increase of density $\Delta\rho$. Actually there is a slight time lag in the response, which is discussed in Chapter 3.

We use (2.7.7) to eliminate either Δp or $\Delta \rho$ from (2.7.6) and obtain the wave equations for pressure and density.

$$\frac{\partial^2 \Delta p}{\partial x^2} = \frac{\rho_A}{E} \frac{\partial^2 \Delta p}{\partial t^2} \tag{2.7.8}$$

$$\frac{\partial^2 \Delta \rho}{\partial x^2} = \frac{\rho_A}{E} \frac{\partial^2 \Delta \rho}{\partial t^2} \tag{2.7.9}$$

It is customary to write the wave equations without the Δ's.

$$\frac{\partial^2 p}{\partial x^2} = \frac{\rho_A}{E} \frac{\partial^2 p}{\partial t^2}$$

$$\frac{\partial^2 \rho}{\partial x^2} = \frac{\rho_A}{E} \frac{\partial^2 \rho}{\partial t^2}$$

$$\tag{2.7.10}$$

These equations relate the spatial dependence of the disturbance to the temporal dependence of the disturbance. Later we show that E/ρ_A is c^2. In rectangular coordinates there are identical equations for particle velocity and particle displacement.

Wave propagation is an important mechanism for transferring energy. The disturbance is propagated without transporting mass. For example, if we were to dye the fluid around the source, we would find that although the dyed fluid remained close to the source, the energy of this disturbance was passed along from volume to volume as the wave moved outward.

2.8. ON SOLUTIONS OF THE WAVE EQUATION

2.8.1. Acoustic pressure waves

The wave equation is one of the fundamental equations in physics. Most advanced texts on geophysics either start with the wave equation or derive it in the first chapter. In the rest of the book, the authors demonstrate their skills in solving that equation. It is possible to derive the wave equation in much more generality than it can be solved.

We derived the wave equation for a one-dimensional plane wave disturbance. As long as the disturbance is a plane wave, it does not matter what direction it is going in because the coordinate axis can always be chosen in the direction of propagation. In a homogeneous medium, as shown later, the spherical wave solution to the wave equation can be reduced to an equivalent plane wave problem. The mathematical techniques used to rotate coordinates are given in texts on analytic geometry and calculus.

Recall that partial differentiation means that the function is to be differentiated only with respect to that variable. For example,

$$F = x^2 + y^2 + z^2 \qquad (2.8.1)$$

$$\frac{\partial F}{\partial x} = 2x \qquad (2.8.2)$$

where all the other variables are treated as constants. Sometimes the variables are combined such that partial differentiation with respect to one variable is proportional to differentiation with respect to the other. Let us suppose that in the function $F(x, a)$ the variables x and a always enter as $(x + a)$. Letting $q = x + a$, we have

$$F(x + a) = F(q) \qquad (2.8.3)$$

Partial differentiations of $F(a + x)$ with respect to a and x are then

$$\frac{\partial F}{\partial x} = \frac{\partial F}{\partial q}\frac{\partial q}{\partial x}$$
$$\frac{\partial F}{\partial a} = \frac{\partial F}{\partial q}\frac{\partial q}{\partial a} \qquad (2.8.4)$$

thus

$$\frac{\partial F}{\partial x}\left(\frac{\partial q}{\partial x}\right)^{-1} = \frac{\partial F}{\partial a}\left(\frac{\partial q}{\partial a}\right)^{-1} \qquad (2.8.5)$$

To solve the wave equation we assume a solution, test the solution, and evaluate the constants. As a solution to (2.7.10), we propose

$$p_+ = AF_+\left(t - \frac{x}{c}\right) \qquad (2.8.6)$$

where A and c are constants. Note that c must have the dimensions of speed. We will show that p_+ satisfies the wave equation.

Let

$$q = t - \frac{x}{c}, \qquad \frac{\partial q}{\partial t} = 1, \qquad \frac{\partial q}{\partial x} = -\frac{1}{c} \qquad (2.8.6a)$$

form

$$\frac{\partial^2 p_\pm}{\partial x^2} = A\frac{\partial^2 F_\pm}{\partial x^2} = A\frac{\partial^2 F_\pm}{\partial q^2}\left(\frac{\partial q}{\partial x}\right)^2 = \frac{A}{c^2}\frac{\partial^2 F_\pm}{\partial q^2}$$

and

$$\frac{1}{c^2}\frac{\partial^2 p_+}{\partial t^2} = \frac{A}{c^2}\frac{\partial^2 F_+}{\partial t^2} = \frac{A}{c^2}\frac{\partial^2 F_+}{\partial q^2}\left(\frac{\partial q}{\partial t}\right)^2$$

$$\frac{1}{c^2}\frac{\partial^2 p_+}{\partial t^2} = \frac{A}{c^2}\frac{\partial^2 F_+}{\partial q^2}$$

Thus

$$\frac{\partial^2 p_+}{\partial x^2} = \frac{1}{c^2}\frac{\partial^2 p_+}{\partial t^2} \tag{2.8.7}$$

Comparison of (2.8.7) and (2.7.10) shows that (2.8.6) is a solution of the wave equation, provided

$$c^2 = \frac{E}{\rho_A} \tag{2.8.8}$$

An equation such as (2.8.8) with c^2 proportional to medium elasticity (or stiffness or restoring force and so on) and inversely proportional to density (or similar inertial parameter) is common to *all* mechanical waves.

Similarly, we test a second solution

$$p_- = BF_-\left(t+\frac{x}{c}\right) \tag{2.8.9}$$

and let q be redefined as follows for F_-:

$$q = t + \frac{x}{c}, \qquad \frac{\partial q}{\partial x} = \frac{1}{c}, \qquad \frac{\partial q}{\partial t} = 1 \tag{2.8.9a}$$

Repeating the operations, we obtain

$$\frac{\partial^2 p_-}{\partial x^2} = \frac{1}{c^2}\frac{\partial^2 p_-}{\partial t^2} \tag{2.8.10}$$

Since both p_+ and p_- satisfy the wave equation, their sum also satisfies it. Therefore a more general solution to the wave equation is

$$p = AF_+\left(t-\frac{x}{c}\right) + BF_-\left(t+\frac{x}{c}\right) \tag{2.8.11}$$

Functions having the form $F_+(t-x/c)$ and $F_-(t+x/c)$ are traveling waves propagating with the speed c; F_+ goes in the $+x$ direction and F_- goes in the $-x$ direction; $F_+(q_+)$ represents a wave traveling in the $+x$

direction because

$$q_+ = t - \frac{x}{c}$$

$$x = (t - q_+)c \qquad (2.8.12)$$

For any given pressure represented by the value of the function $F_+(q_+)$, x must reach larger positive values as t increases. Hence the function $F_+(q_+)$ is moving in the $+x$ direction. An example of the transformation is sketched in Fig. 2.8.1.

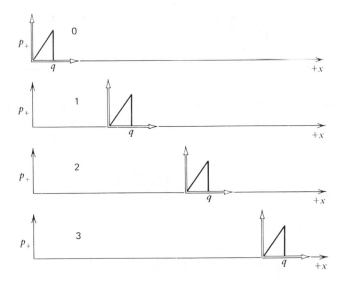

Figure 2.8.1. Traveling wave. The pressure along the x direction is observed at a sequence of times. The q coordinate system moves at speed c along the plus x direction.

Similarly, for the negative direction

$$q_- = t + \frac{x}{c}$$

$$x = (q_- - t)c \qquad (2.8.13)$$

For constant q_-, as t grows, x moves further in the negative x direction. Ordinarily we know from the context of the problem when to use the positive traveling solution, negative traveling solution, or both.

2.8.2. Relations Between Wave Parameters

Throughout this section we have studied the pressure wave principally because most hydrophones are pressure-sensitive devices. But there are sensors of other types.

The relation between acoustic particle velocity and acoustic pressure is quite important. Recalling (2.7.3) and dropping the Δ, we find

$$-\frac{\partial p}{\partial x} = \rho_A \frac{\partial u}{\partial t} \qquad (2.8.14)$$

where u and p are traveling waves. We integrate this expression by expressing $\partial u/\partial t$ in terms of $\partial u/\partial x$. We let $F = u$, $a = t$; and $q = t - x/c$ in (2.8.5) and obtain

$$-c\frac{\partial u}{\partial x} = \frac{\partial u}{\partial t}$$

Substitution in (2.8.14) gives

$$\frac{\partial p}{\partial x} = \rho_A c \frac{\partial u}{\partial x} \qquad (2.8.15)$$

The use of (2.8.9a) for a negative x traveling wave gives the same expression with a minus sign.

Integration follows directly because both sides are $\partial/\partial x$, and, ignoring the constant of integration, and restoring the Δ

$$\Delta p = \pm(\rho_A c)u \qquad \textbf{(2.8.16)}$$

where the plus sign is for waves traveling in the positive x direction and minus sign is for the negative x direction. To people trained in electricity this resembles Ohm's law, with the acoustic pressure analogous to voltage, acoustic particle velocity replacing electric current, and $(\rho_A c)$ taking the place of resistance in dc circuits, or impedance in ac circuits. The analogue is used frequently, and the $\rho_A c$, "rho-c," of a material is probably its most common acoustical characterization.

Similarly, the conservation of mass equation (2.7.5) leads to the relation [by letting $F = \Delta\rho$, using (2.8.6a) to obtain $\partial \Delta\rho/\partial t = -c\, \partial \Delta\rho/\partial x$, then integrating over x]

$$\frac{u}{c} = \frac{\Delta\rho}{\rho_A} \qquad \textbf{(2.8.17)}$$

Since $\Delta\rho \ll \rho_A$, (2.8.17) shows that the particle velocity is very much less than the speed of sound. The ratio u/c is sometimes called the acoustic Mach number.

Finally, from (2.8.16) and (2.8.17) we can readily derive the valuable relation between the acoustic pressure and the acoustic density

$$\Delta p = c^2 \Delta \rho \qquad \text{(2.8.18)}$$

where we include the Δ to facilitate comparison with (2.7.7). If we had known this relation earlier, we could have moved directly from (2.7.6) to the wave equation for pressure (2.8.7) and (2.8.10) or to the equivalent form for acoustic density.

Equations (2.8.18) can be used to calculate c by forming $dp/d\rho$ if the equation of state $p = p(\rho)$ is known.

2.8.3. Harmonic Signals

In marine geophysics and ocean acoustics, experiments are often made at single frequencies. The source transmits a sinusoidal wave, a tone, a CW signal. In practice the CW signal is achieved by turning the source on, letting it run for many cycles, then turning it off. Such a signal is called a "gated wave train" or a "ping." When measurements are made in a steady state condition, the ping can be treated as a CW signal.

For a CW disturbance traveling in the $+x$ direction, several expressions for the pressure signal are

$$p = a_1 \cos(\omega t - kx)$$
$$p = a_2 \sin(\omega t - kx)$$
$$p = a_3 \exp[i(\omega t - kx)]$$

where a_1, a_2, and a_3 are constants. For plane waves, we choose the constants by defining P^2 as being the mean (absolute) square sound pressure or P as being the root mean square (rms) sound pressure. The average is taken over one wave period, $T = 2\pi/\omega$. For each of the expressions, P^2 is as follows:

$$P^2 = \frac{1}{T} \int_0^T a_1^2 \cos^2(\omega t - kx)\, dt = \frac{a_1^2}{2}$$

$$P^2 = \frac{1}{T} \int_0^T a_2^2 \sin^2(\omega t - kx)\, dt = \frac{a_2^2}{2}$$

$$P^2 = \frac{1}{T} \int_0^T |a_3 \exp[i(\omega t - kx)]|^2\, dt = a_3^2$$

The replacement of a_1 and a_2 by $\sqrt{2}P$ and a_3 by P gives the following expressions for p:

$$p = \sqrt{2}P \cos k(ct - x) = \sqrt{2}P \cos (\omega t - kx)$$

$$p = \sqrt{2}P \sin k(ct - x) = \sqrt{2}P \sin (\omega t - kx) \qquad (2.8.19)$$

$$p = P \exp [i(\omega t - kx)]$$

Note. We are not taking the real part of the complex p but are using both components of p. *The order* $(kx - \omega t)$ *or* $(\omega t - kx)$ *does not change the character of the positive traveling wave.* All these forms are equivalent. We choose the form of p to suit the problem.

We replace ρ_A/E in (2.7.10) by $1/c^2$. The second partial time derivative of any of the expressions in (2.8.19) is

$$\frac{\partial^2 p}{\partial t^2} = -\omega^2 p \qquad (2.8.20)$$

Therefore, for harmonic time dependence the wave equation is

$$\frac{\partial^2 p}{\partial x^2} = -\frac{\omega^2}{c^2} p \qquad (2.8.21)$$

$$\frac{\partial^2 p}{\partial x^2} + k^2 p = 0 \qquad \mathbf{(2.8.22)}$$

When harmonic time dependence is assumed, the sound pressure $p(x, t)$ is sometimes written as $p_s \exp (i\omega t)$. In this form p_s is a space-dependent function only, and the wave equation is used in the form (2.8.22).

2.9. REFLECTION AND TRANSMISSION COEFFICIENTS FOR PLANE WAVE SIGNALS

Procedures for calculating reflection and transmission use the solutions for the signals in the upper and lower media and fit them together at the boundary. The fitting conditions are called boundary conditions. We use two assumptions:

1. The interface does not have an excess pressure on one side or the other.

2. The two media maintain contact at the interface as the signal reflects and passes through the interface.

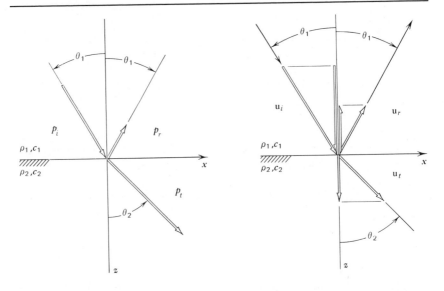

Figure 2.9.1. Reflection geometry: particle velocities (boldface type) are vectors.

The geometry is in Fig. 2.9.1. Labeling p_i, p_r, and p_t as the instantaneous incident, reflected, and transmitted pressures, at the interface, respectively the first condition yields

$$p_i + p_r = p_t \qquad (2.9.1)$$

The second condition requires that the *vertical components* of the particle displacements or velocities be equal on each side. We consider the particle velocities. Using $\mathbf{u}_i$, $\mathbf{u}_r$, and $\mathbf{u}_t$ (bold face type for vectors) for the incident, reflected, and transmitted particle velocities, respectively, the vertical components are

$$u_i \cos \theta_1 + u_r \cos \theta_1 = u_t \cos \theta_2 \qquad (2.9.2)$$

where θ_1 is the incident angle and θ_2 is the refracted angle. The θ_1 and θ_2 are related by Snell's law $(\sin \theta_1)/c_1 = (\sin \theta_2)/c_2$; the magnitude of $\mathbf{u}$, u, is related to the pressure by (2.8.16). We use $\rho_1 c_1 u_i = p_i$ for the downgoing wave ($+z$ direction) and $-\rho_1 c_1 u_r = p_r$ for the upgoing reflected wave.

$$p_i = \rho_1 c_1 u_i$$
$$p_r = -\rho_1 c_1 u_r \qquad (2.9.3)$$
$$p_t = = \rho_2 c_2 u_t$$

Substitution of (2.9.3) in (2.9.2) gives the second condition

$$p_i \frac{\cos \theta_1}{\rho_1 c_1} - p_r \frac{\cos \theta_1}{\rho_1 c_1} = p_t \frac{\cos \theta_2}{\rho_2 c_2} \qquad (2.9.4)$$

Simultaneous solution for $\mathcal{R}_{12} \equiv p_r/p_i$ and $\mathcal{T}_{12} \equiv p_t/p_i$ gives the *pressure reflection and transmission coefficients*

$$\mathcal{R}_{12} = \frac{\rho_2 c_2 \cos \theta_1 - \rho_1 c_1 \cos \theta_2}{\rho_2 c_2 \cos \theta_1 + \rho_1 c_1 \cos \theta_2} \qquad (2.9.5)$$

$$\mathcal{T}_{12} = \frac{2\rho_2 c_2 \cos \theta_1}{\rho_2 c_2 \cos \theta_1 + \rho_1 c_1 \cos \theta_2} \qquad (2.9.6)$$

where

$$\theta_2 = \arcsin \left(\frac{c_2}{c_1} \sin \theta_1 \right) \qquad (2.9.7)$$

Snell's law (2.2.3) and (2.9.7) applies for all angles of incidence and all ratios c_2/c_1. For $c_2 < c_1$, $(c_2/c_1) \sin \theta_1$ is less than 1 for all angles of incidence.

The case $c_2 > c_1$ leads to the important condition of "total reflection." As $(c_2/c_1) \sin \theta_1$ tends to 1, θ_2 tends to 90° and becomes 90° at $\sin \theta_1 = c_1/c_2$. This is the critical angle θ_c.

$$\sin \theta_c = \frac{c_1}{c_2} \qquad (2.9.8)$$

Incident angles greater than critical are handled by using Snell's law and the relation $\cos^2 a + \sin^2 a = 1$, to calculate $\cos \theta_2$.

$$\sin \theta_2 = \frac{c_2}{c_1} \sin \theta_1$$

$$\cos \theta_2 = \left[1 - \left(\frac{c_2}{c_1} \right)^2 \sin^2 \theta_1 \right]^{1/2} \qquad (2.9.9)$$

$$\cos \theta_2 = \pm i b_2$$

where $b_2 = [(c_2/c_1)^2 \sin^2 \theta_1 - 1]^{1/2}$. We choose the minus sign to cause the signal to attenuate exponentially in the lower medium (Section A2.3.3). Using (2.9.5), $\mathcal{R}_{12}$ becomes

$$\mathcal{R}_{12} = \frac{\rho_2 c_2 \cos \theta_1 + i \rho_1 c_1 b_2}{\rho_2 c_2 \cos \theta_1 - i \rho_1 c_1 b_2} \qquad \text{for} \quad \theta_1 > \theta_c \qquad (2.9.10)$$

The numerator is the complex conjugate of the denominator and the absolute square of $\mathcal{R}_{12}$ is 1. For angles greater than critical, we write

$$\mathcal{R}_{12} = e^{+2i\Phi} \tag{2.9.11}$$

$$\Phi = \arctan \frac{\rho_1 c_1 b_2}{\rho_2 c_2 \cos \theta_1} \tag{2.9.12}$$

At total reflection, the signal has a phase lag 2Φ.

Most acoustic experiments are made with signals from finite sources, and the actual wave fronts at the interfaces are spherical. Since it has long been a practice to use the *plane wave* reflection coefficients, we need to examine the approximation. We recall from Section 2.2 and Fig. 2.2.2c that the reflected wave front appears to spread spherically outward from an image source. The image strength is $\mathcal{R}I$. Using a result from Brekhovskikh (1960), p. 255), the spherical reflection coefficient is written

$$\mathcal{R} = \mathcal{R}_{12} - \frac{iN}{kR} \tag{2.9.13}$$

$$N = \frac{\mathcal{R}_{12}'' + \mathcal{R}_{12}' \cot \theta_1}{2} \tag{2.9.14}$$

where R is the distance from the image source to the receiver and the derivatives $\mathcal{R}_{12}'$ and $\mathcal{R}_{12}''$ are taken with respect to θ_1. The value of N is negligible for θ_1 *not too close to the critical angle*, large kR, and $\theta_1 > 0$. We interpret (2.9.13) and (2.9.14) to mean that simple plane wave concepts of reflection and transmission will fail near the critical angle in most of our experiments.

Near the critical angle, part of the energy of the sound wave travels along the interface in the lower medium. As it travels in the lower medium it loses energy into the upper medium. This process is discussed in detail in Section 8.3.1.

2.10. REFLECTION AND TRANSMISSION AT THIN LAYERS

The physical model for this discussion is the sequence of horizontal layers of sediment on lake and sea floors (Fig. 2.10.1). If the layers are thick enough, or if the duration of the signal is short enough, the signals from the interfaces come back at different times and we can separate them. For signals that are longer than the travel times in the layers, the multiple reflections overlap.

The simplified model is a horizontally stratified half space. The incident signal is a plane wave having angular frequency ω. The CW reflection and transmission coefficients are useful when the signal is a long ping.

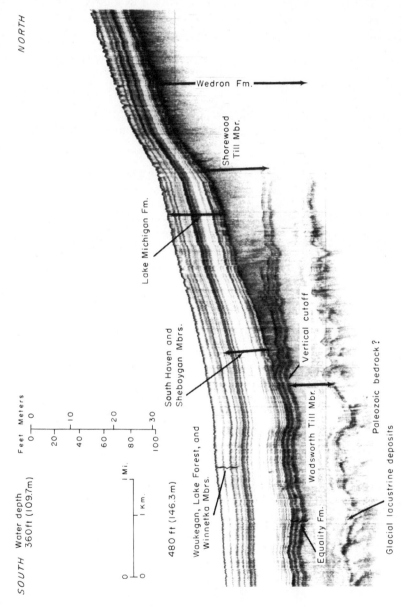

Figure 2.10.1. Subbottom layers in Lake Michigan. "Enlarged seismic reflection profile from an area near the northern end of the southern lake basin showing the southward pinchout of the Shorewood Member of the Wedron Formation. Note that the Equality Formation is separated from the Shorewood Till Member by a vertical cutoff at the pinch-out of the till in the Shorewood." The transmitted signal is 1 cycle of a 7 kHz sinusoidal pulse. From Lineback *et al.* (1974, p. 18). Courtesy of the Illinois State Geological Survey.

65

A second application is in the design of sonar windows. The sound signals pass from the transducer, through the protective window, and into the water. The transmission coefficient through the window depends on the thickness, c, and ρ. The parameters are chosen to maximize the transmission of sound signals through the window.

2.10.1. Derivation for a Layered Half Space

The geometry and notation for the derivation of a layered half space are in Fig. 2.10.2. For the pressure 1, 2 reflection and transmission coefficients (2.9.5) and (2.9.6) are

$$\mathcal{R}_{12} = \frac{\rho_2 c_2 \cos \theta_1 - \rho_1 c_1 \cos \theta_2}{\rho_2 c_2 \cos \theta_1 + \rho_1 c_1 \cos \theta_2} \tag{2.10.1}$$

$$\mathcal{T}_{12} = \frac{2\rho_2 c_2 \cos \theta_1}{\rho_2 c_2 \cos \theta_1 + \rho_1 c_1 \cos \theta_2} \tag{2.10.2}$$

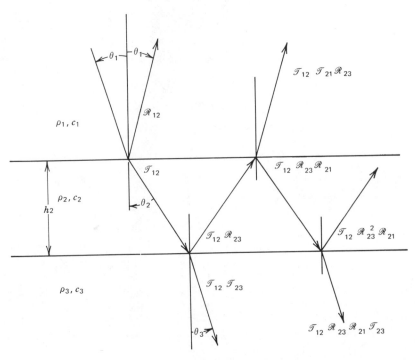

Figure 2.10.2.   Reflection and transmission at a thin layer.

with similar expressions for $\mathcal{R}_{23}$ and $\mathcal{T}_{23}$. From the figure, the total up-traveling signal is the sum of an infinite number of partial transmissions and reflections. Each path within the layer has a phase delay $2k_2h_2 \cos \theta_2$, where $k_2 \cos \theta_2$ is the vertical component of the wave number in the layer. By letting the incident signal have unit amplitude, the total reflection $\mathcal{R}_{13}$ is

$$\mathcal{R}_{13} = \mathcal{R}_{12} + \mathcal{T}_{12}\mathcal{T}_{21}\mathcal{R}_{23} \exp(-2i\Phi_2) + \mathcal{T}_{12}\mathcal{T}_{21}\mathcal{R}_{23}^2\mathcal{R}_{21} \exp(-4i\Phi_2) + \cdots$$
$$(2.10.3)$$

$$\Phi_2 \equiv k_2h_2 \cos \theta_2 \qquad (2.10.4)$$

After $\mathcal{R}_{12}$, terms in (2.10.3) have the form of a geometric series

$$S = \sum_{n=0}^{\infty} r^n = (1-r)^{-1} \qquad \text{for} \quad r < 1$$
$$(2.10.5)$$
$$\mathcal{R}_{13} = \mathcal{R}_{12} + \mathcal{T}_{12}\mathcal{T}_{21}\mathcal{R}_{23} \exp(-2i\Phi_2) \sum_{0}^{\infty} [\mathcal{R}_{23}\mathcal{R}_{21} \exp(-2i\Phi_2)]^n$$

Our reduction requires manipulation, and we use the following relations, which the reader can verify,

$$\mathcal{R}_{12} = -\mathcal{R}_{21}$$
$$\mathcal{T}_{12}\mathcal{T}_{21} = 1 - \mathcal{R}_{12}^2 \qquad (2.10.6)$$

to express $\mathcal{R}_{13}$

$$\mathcal{R}_{13} = \frac{\mathcal{R}_{12} + \mathcal{R}_{23} \exp(-2i\Phi_2)}{1 + \mathcal{R}_{12}\mathcal{R}_{23} \exp(-2i\Phi_2)} \qquad (2.10.7)$$

The transmission through the layer for a unit incident signal is

$$\mathcal{T}_{13} = \mathcal{T}_{12}\mathcal{T}_{23} \exp(-i\Phi_2) + \mathcal{T}_{12}\mathcal{T}_{23}\mathcal{R}_{23}\mathcal{R}_{21} \exp(-3i\Phi_2) + \cdots$$
$$(2.10.8)$$

Again, this is a geometric series, and the sum is

$$\mathcal{T}_{13} = \frac{\mathcal{T}_{12}\mathcal{T}_{23} \exp(-i\Phi_2)}{1 + \mathcal{R}_{12}\mathcal{R}_{23} \exp(-2i\Phi_2)} \qquad (2.10.9)$$

Both the reflection and transmission coefficients are oscillatory functions and depend on $\Phi_2 = k_2h_2 \cos \theta_2$. They are functions of frequency and angle of incidence for a given layer.

We derive the pressure reflection coefficient for multiple layers by repeated applications of the single layer coefficient, (2.10.7). As in Fig. 2.10.3 the reflection from the lower half space is $\mathcal{R}_{(n-1)n}$. Using (2.10.7),

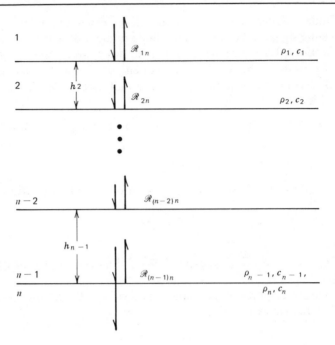

Figure 2.10.3.   Reflection from a layered half space.

the reflection coefficient at the top of the $n-1$ layer is found to be

$$\mathcal{R}_{(n-2)n} = \frac{\mathcal{R}_{(n-2)(n-1)} + \mathcal{R}_{(n-1)n} \exp\left(-2i\Phi_{n-1}\right)}{1 + \mathcal{R}_{(n-2)(n-1)}\mathcal{R}_{(n-1)n} \exp\left(-2i\Phi_{n-1}\right)} \qquad (2.10.10)$$

$\mathcal{R}_{(n-2)n}$ is the reflection coefficient for the layers beneath the interface. On repeating the calculation, $\mathcal{R}_{(n-3)n}$ is found to be

$$\mathcal{R}_{(n-3)n} = \frac{\mathcal{R}_{(n-3)(n-2)} + \mathcal{R}_{(n-2)n} \exp\left(-2i\Phi_{n-2}\right)}{1 + \mathcal{R}_{(n-3)(n-2)}\mathcal{R}_{(n-2)n} \exp\left(-2i\Phi_{n-2}\right)} \qquad (2.10.11)$$

The calculation continues upward to the top and

$$\mathcal{R}_{1n} = \frac{\mathcal{R}_{12} + \mathcal{R}_{2n} \exp\left(-2i\Phi_{2}\right)}{1 + \mathcal{R}_{12}\mathcal{R}_{2n} \exp\left(-2i\Phi_{2}\right)} \qquad (2.10.12)$$

We conceptually simplify the formulation of the reflection from a multiple layered half space by letting the reflection coefficient $\mathcal{R}_{1n}$ represent all the frequency and angle dependence. Then for large source distance and the plane wave approximation, the acoustic problem is stated as being the reflection from a half space that has the composite pressure reflection coefficient, $\mathcal{R}_{1n}$.

For example, we assume that the layer represents a sonar window. The coupling medium behind the window, usually castor oil, is c_1, ρ_1, the window is c_2, ρ_2, and the water is c_3, ρ_3. Since the phase is unimportant here, we calculate the absolute square of the transmission coefficient.

$$|\mathcal{T}_{13}|^2 = (\mathcal{T}_{12}\mathcal{T}_{23})^2 [1 + (\mathcal{R}_{12}\mathcal{R}_{23})^2 + 2\mathcal{R}_{12}\mathcal{R}_{23} \cos (2\Phi_2)]^{-1}$$

$$(2.10.13)$$

The transmission is largest when the expression in brackets is the smallest. The values of ρc for the oil and water are nearly the same; thus $\mathcal{R}_{23} \approx -\mathcal{R}_{21}$ and

$$[1 + (\mathcal{R}_{12}\mathcal{R}_{23})^2 + 2\mathcal{R}_{12}\mathcal{R}_{23} \cos (2\Phi_2)] \approx 1 + \mathcal{R}_{12}^4 - 2\mathcal{R}_{12}^2 \cos 2\Phi_2$$

The transmission is best when

$$2\Phi_2 = 0, 2\pi, \ldots$$

$$k_2 h_2 \cos \theta_2 = n\pi \qquad \text{where} \quad n = 0, 1, 2, \ldots$$

$$(2.10.14)$$

At vertical incidence h_2 is

$$h_2 = \frac{n\lambda_2}{2}$$

$$(2.10.15)$$

where λ_2 is the wavelength in the window. An example of the calculation of the reflection and transmission at a thin layer (sonar window) appears in Fig. 2.10.4. One maximum of the transmission is at 215 kHz. This is the frequency for $\lambda/2 = 0.25$ cm in the silicone rubber. The maxima of the transmission coefficient correspond to the minima of the reflection coefficient.

2.10.2.   Total Reflection at a Layered Half Space

When the pressure wave is totally reflected at the nth layer in a multiple-layered half space, the reflection coefficient for the half space reduces to a unit amplitude and a phase shift. The geometry is shown in Fig. 2.10.3. The reflection coefficient at the nth layer is

$$\mathcal{R}_{(n-1)n} = \exp (2i\Phi_n)$$

$$(2.10.16)$$

$$\Phi_n = \arctan \left(\frac{\rho_{n-1}c_{n-1}b_n}{\rho_n c_n \cos \theta_{n-1}} \right)$$

$$(2.10.17)$$

$$b_n = \left(\frac{c_n^2}{c_{n-1}^2} \sin^2 \theta_{n-1} - 1 \right)^{1/2} = \left[\left(\frac{c_n}{c_1} \right)^2 \sin^2 \theta_1 - 1 \right]^{1/2}$$

$$(2.10.18)$$

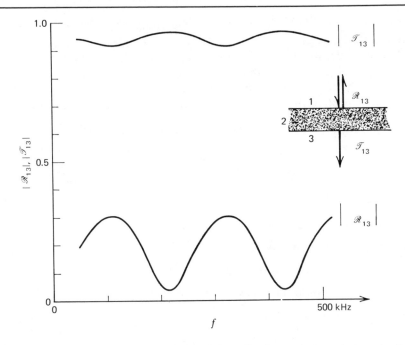

Figure 2.10.4. Frequency dependence of the reflection and transmission at a thin layer. Layer 2 is RTV silicone rubber 602, $\rho = 0.99$ g/cm^3, $c = 1076$ m/s. (Folds, 1974.) Medium 1 is water $\rho_w = 1$, g/cm^3, $c = 1500$ m/s. Medium 3 is oil $\rho_w = 0.95$ g/cm^3, $c = 1450$ m/s. Layer thickness $= 0.25$ cm.

where the incident angle θ_n is greater than the critical angle. The assumption that the nth layer can be treated as a half space is based on the thickness of the layer being considerably greater than the penetration of the pressure field into the layer. From Section A2.3.3, the thickness is greater than $c_n/\omega b_n$ in this case.

Using (2.10.10) the reflection at the next layer is found to be

$$\mathcal{R}_{(n-2)n} = \frac{\mathcal{R}_{(n-2)(n-1)} + \exp\left[2i(\Phi_n - \Phi_{n-1})\right]}{1 + \mathcal{R}_{(n-2)(n-1)} \exp\left[2i(\Phi_n - \Phi_{n-1})\right]} \qquad (2.10.19)$$

$$\Phi_{n-1} = k_{n-1}h_{n-1}\cos\theta_{n-1} \qquad (2.10.20)$$

To write (2.10.19) in the form of a unit amplitude and phase shift, we factor $\exp\left[i(\Phi_n - \Phi_{n-1})\right]$ out of the numerator and denominator. Then the numerator is the complex conjugate of the denominator. The expansion of the exponentials and collection of the real and imaginary parts gives

$$\mathcal{R}_{(n-2)n} = \exp(2i\Phi_{n-2}) \tag{2.10.21}$$

$$\Phi_{n-2} = \arctan\left[\frac{c_{n-2}\rho_{n-2}\cos\theta_{n-1}}{c_{n-1}\rho_{n-1}\cos\theta_{n-2}}\tan(\Phi_n - \Phi_{n-1})\right] \tag{2.10.22}$$

We proceed upward layer by layer, and at the bottom of the first layer we write

$$\mathcal{R}_{1n} = \exp(2i\Phi_1)$$

$$\Phi_1 = \arctan\left[\frac{\rho_1 c_1 \cos\theta_2}{\rho_2 c_2 \cos\theta_1}\tan(\Phi_3 - \Phi_2)\right]$$

$$\Phi_2 = k_2 h_2 \cos\theta_2$$

The total reflection from a half space has unit amplitude and a phase shift.

PROBLEMS

Problems 2.2

2.2.1. Use an incident plane wave and Huygens' wavelets to prove that the angle of incidence equals the angle of reflection.

2.2.2. Find or improvise a ripple tank (a shallow pan of water or perhaps a quiet pond) and experimentally study the reflection and diffraction of wave fronts.

2.2.3. In the ripple tank, compare the amplitude of the wave originating at a small pole relative to the one originating at the edge of a plane reflector.

Problems 2.3

These are experimental problems. The reader can try them in a ripple tank.

2.3.1. Impulsively excite a ripple (touch the surface). Observe and compare the wave with the waves from a harmonically disturbed surface (when you wiggle your finger up and down). Sketch what you see.

2.3.2. Harmonically excite the surface at a low and high frequency by wiggling your finger at different rates. Observe and sketch the wave pattern.

2.3.3. Repeat Problem 2.3.2 using two fingers separated a few centimeters. Sketch.

2.3.4. Repeat Problem 2.3.2 when your finger is near a wall of the tank. Sketch.

In the following problems use scales such that all the functions can be plotted on the same sheet of graph paper. This will help the reader to see relationships. Ordinate scale of ± 1 cm (or $\frac{1}{2}$ in.) $= \pm 1$ and abscissa of 2 cm (or 1 in.) $= \pi$ are suggested.

2.3.5. Plot sin x over a range of x from -2π to $+4\pi$.

2.3.6. Plot sin $(x + \pi/2)$

$$-2\pi \leq x \leq 4\pi$$

2.3.7. Plot cos x for $-2\pi \leq x \leq 4\pi$.

2.3.8. Graphically add sin x and sin $(x + \pi/2)$.

2.3.9. Plot sin $(x + \pi)$.

2.3.10 Graphically add sin x + sin $(x + \pi)$.

2.3.11. The purpose is to demonstrate the traveling wave character of the function sin $(\omega t - kR)$. An ordinate of ± 0.05 cm $= a$ and an abscissa of 2 cm $= \pi$ are suggested. In sketching sine waves it is usually sufficient to plot values at 0, $\pi/2$, π, and so on, and sketch the shape in between.

(a) For $\Delta\rho = a \sin(\omega t - kR)$

$\omega = 30\pi$ rad/s

$k = \pi/50$ per meter

Using a common abscissa, sketch $\Delta\rho$ as a function of R, for the following times: $t = 0$, $t = 1/180$ s, $t = 2/180$ s, $t = 3/180$ s, $t = 4/180$ s, $t = 5/180$ s, $t = 6/180$ s.

(b) Identify a crest and calculate the speed at which it appears to move out in R.

Problems 2.4

2.4.1. We gave an example of two harmonic sources (same frequency) disturbing the medium. Describe physical situations that would correspond to the two-source problem.

2.4.2. Suppose two sources have different frequencies. Would you observe interference patterns? If you were to listen to the disturbance, what would you hear?

2.4.3. A harmonic plane wave disturbance illuminates a small object. Use Huygens' principle to show how a source at the position of the object can be used to describe the scattering of waves by the object.

2.4.4. A line array has four sources of equal strength a. Use the graphical construction technique to show that there are maxima at $\Phi = kb \sin \Phi = 0$, 2π, 4π, ...

2.4.5. Calculate the interference pattern for a two-source array. Source 1 has amplitude a and source 2 has amplitude $1.5a$.

Problems 2.5

In all the multiple-source problems we assume that the observer is very far from the sources and that the plane wave approximation can be used. The coordinate system shown below applies to the problems. The sources are assumed to be in very deep water and at large distance from the bottom and surface. The sound speed in all problems is 1500 m/s, and the depth dependence of the speed is ignored.

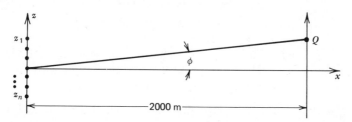

2.5.1. A pair of sources are on the z axis at $z_1 = 5$ m, $z_2 = -5$ m. The source amplitudes are a and signal frequency f is 1500 Hz. The speed of sound is 1500 m/s. Calculated and graph the interference $|C|$ at Q as functions of $\sin \phi$ and ϕ $(-0.1 \le \sin \phi \le 0.1)$.

2.5.2. Four sources like those in Problem 2.5.1 are placed at -7.5, -2.5, 2.5, and 7.5 m. Calculate the graph $|C|$ for the same range of parameters as in Problem 2.5.1.

2.5.3. Two sources are placed along the x axis and in line with the direction to the receiver. The sources have frequencies $f = 3000$ Hz and amplitudes a. The positions are $x_1 = 0$, $x_2 = 10.5$ m.

Calculate and graph the interference function $|C|$ over the range $-0.3 \le \phi \le 0.3$ and plot it versus ϕ, where ϕ is measured relative to the x axis.

2.5.4. Twenty sources are equally spaced over 1.9 m along the z axis. The sources have equal amplitudes $a/20$ and signal frequency $f = 15{,}000$ Hz. Calculate and graph $|C|$ over the range of $-0.2 \le \sin \phi \le 0.2$.

2.5.5. A continuous source along the z axis extends from -1 to $+1$ m. The frequency is $f = 15{,}000$ Hz and the amplitude is $a/2$ m. Calculate and graph $|C|$ over the range $-0.2 \le \phi \le 0.2$.

2.5.6. Two sources on the z axis are as follows: at $z_1 = -5$ m, $a_1 = -a$; at $z_2 = 5$ m, $a_2 = a$. For each, $f = 1500$ Hz. Compute and graph the interference $|C|$ as in Problem 2.5.1.

2.5.7. Two sources, having frequency 1500 Hz have the following positions and amplitudes:

$$a_1 = (1+i)/2^{1/2}, \quad z_1 = -5 \text{ m}; \quad a_2 = (1-i)/2^{1/2}, \quad z_2 = 5 \text{ m}.$$

Compute and graph $|C|$. How would this result compare with $|C|$ for $a_1 = i$ and $a_2 = 1$?

Problems 2.6

Problems 2.6.1 and 2.6.2 use the same information, (a) 3.5 kHz, (b) 12.5 kHz, and (c) 45 kHz and represent the signal frequencies f of many echo sounders in use today. Use sound speed $c = 1500$ m/s.

2.6.1. Estimate the radius of the first phase zone at the bottom for these instruments in the following water depths: 100, 1000, and 3200 m.

2.6.2. To find a reflector having about 10 m radius, how would you expect each of the systems to perform at 100, 1000, and 3200 m?

2.6.3. A ship makes a track over a 40 m radius reflector at 3200 m depth. The reflector is in the horizontal plane. Use Figure 2.6.2 to estimate and draw a curve of the relative amplitude of the signal received by the sonar. The signal frequency is $f = 1500$ Hz. *Hint.* Cut a circle the size of the reflector, slide it over the drawing of the phase zones, and estimate the sum of the positive and negative wavelets.

Problems 2.8

2.8.1. Prove that F, F_1, and F_2 are all solutions of the plane wave equation where $F = F_1(x - ct) + F_2(x + ct)$.

2.8.2. Show that the following are solutions of the wave equation

(a) $X_+ = a \sin(kx - \omega t)$, $X_+ = a \sin\left[\omega\left(t - \dfrac{x}{x}\right)\right]$

  $X_- = a \sin(kx + \omega t)$, $X_- = a \sin\left[\omega\left(t + \dfrac{x}{c}\right)\right]$

(b) $X_+ = a \cos(kx - \omega t)$, $X_+ = a \cos\left[\omega\left(t - \dfrac{x}{c}\right)\right]$

  $X_- = a \cos(kx + \omega t)$, $X_- = a \cos\left[\omega\left(t + \dfrac{x}{c}\right)\right]$

(c) $X_+ = a\, e^{i(kx - \omega t)}$, $i = \sqrt{-1}$, $X_+ = a \exp[i(\omega t - kx)]$

(d) $X_+ = a \sin(kx - \omega t + \theta)$, θ constant

(e) $X = \exp\left[-\dfrac{(x - ct)^2}{L^2}\right]$

2.8.3. Solve the following.
  (a) Derive (2.8.16) from (2.7.3).
  (b) Compare the relative acoustic particle velocity to acoustic density for air (assume $\rho_0 = 1\ \text{kg/m}^3$ and $c = 3.4 \times 10^2\ \text{m/s}$) with the ratio for water (assume $\rho_0 = 10^3\ \text{kg/m}^3$, and $c = 1.5 \times 10^3\ \text{m/s}$).

2.8.4. The speed of sound in air can be readily calculated because the equation of state is known. The propagation is known to be close to adiabatic for all audible frequencies, therefore $p\rho^{-\gamma} = $ constant, where γ is the ratio of specific heats for the gas.
  (a) *Form* $dp/d\rho$ and show that the speed of sound in air is given by

$$c = \left(\frac{\gamma P_A}{\rho_A}\right)^{1/2}$$

  (b) Assume that $\gamma = 1.4$, calculate the speed of sound in air under standard conditions (assume $P_A = 10^5\ \text{N/m}^2$ and $\rho_A = 1.29\ \text{kg/m}^3$).

Problems 2.9

2.9.1. Calculate and graph $\mathcal{R}_{12}$ and $\mathcal{T}_{12}$ versus θ_1 for water sediment interface, $\rho_1 = 1$ g/cm^3, $c_1 = 1.5 \times 10^3$ m/s, and $\rho_2 = 1.4$ g/cm^3, $c_2 = 1.48 \times 10^3$ m/s. Determine whether $\mathcal{R}_{12} = 0$ for some angle of incidence.

2.9.2. Calculate and graph $\mathcal{R}_{12}$ and $\mathcal{T}_{12}$ versus θ_1 for water sediment interference, $\rho_1 = 1$ g/cm^3, $c_1 = 1.5 \times 10^3$ m/s, $\rho_2 = 2$ g/cm^3, and $c_2 = 2 \times 10^3$ m/s. Determine θ_c and graph the phase of $\mathcal{R}_{12}$.

2.9.3. Calculate and graph $\mathcal{R}_{12}$ and $\mathcal{T}_{12}$ for a source above the air-water interface.

2.9.4. Calculate and graph $\mathcal{R}_{12}$ and $\mathcal{T}_{12}$ for reflection and transmission from a source below the water-air interface.

2.9.5. Show that if $X(x)$, $Z_1(z)$, and $T(t)$ are functions of x, z, and t, respectively, then $p = XZ_1 \, e^{i\omega t}$ gives

$$\frac{\partial^2 X}{\partial x^2} + k_x^2 X = 0$$

$$\frac{\partial^2 Z_1}{\partial z^2} + k_z^2 Z_1 = 0$$

$$k_x^2 + k_z^2 = k^2 = \frac{\omega^2}{c^2}$$

Hint. If the sum of independent functions is a constant, each of the functions is equal to a constant.

2.9.6. Show that the solution p is

$$p = P \exp\left[i(\omega t - k_x x - k_z z)\right]$$

where P is the pressure amplitude.

2.9.7. Use the boundary conditions and the form of p in Problem 2.9.6 to calculate $\mathcal{R}_{12}$ and $\mathcal{T}_{12}$. Let $k_1 = \omega/c_1$, $k_2 = \omega/c_2$; $k_{1x} = k_1 \sin \theta_1$, $k_{1z} = k_1 \cos \theta_1$; and $k_{2x} = k_2 \sin \theta_2$, $k_{2z} = k_2 \cos \theta_2$.
Hint. Equation (2.8.15) and k_{1z} change sign for the reflected signal.

2.9.8. Letting p_1 be the signal in medium 1 and p_2 in medium 2, show that p_2 is exponentially attenuated in medium 2 for $c_2 > c_1$ and $\theta_1 > \theta_c$.

2.9.9. For a source in region 1, $c_2 = 2c_1$ and $\rho_2 = 2\rho_1$, calculate and graph $|\mathcal{R}_{12}|$ and Φ for $\theta_1 = 0$ to $\pi/2$. Use the graph to estimate $\mathcal{R}'_{12}$ and $\mathcal{R}''_{12}$. ($\mathcal{R}''_{12} \to -\infty$ at critical angle). For $f = 1$ kHz and $R = 1000$ m, estimate the range of θ for which N in (2.9.13) is negligible.

2.9.10. Calculate the reflection and transmission coefficients for particle velocity rather than for pressure.

2.9.11 Compare the reflection coefficients for particle velocity and acoustic pressure for a signal incident on a free surface.

2.9.12. Calculate and compare the reflection and transmission coefficients for pressure squared and for pressure times velocity.

THREE

SOUND TRANSMISSION IN THE OCEAN

Wave theory and ray theory are two methods of calculating sound transmission in an inhomogeneous medium such as the ocean. We employ an approximation that is a combination of both. We use ray methods to follow the acoustic signal from the source to the vicinity of the object or surface. Wave theory is the method by which we calculate the actual scattering process. Finally, we use ray theory to follow the scattered signal back to the receiver. This chapter gives a description of the sound field in an inhomogeneous ocean and a method of calculating the attenuation and transmission loss of the signal. The scattering characteristics of objects and the sea floor are discussed in later chapters.

Geophysicists and underwater acousticians have different customs in the use of the word attenuation. From the Encylopedic Dictionary of Exploration Geophysicists (R. E. Sheriff ed, Society of Exploration Geophysics, Tulsa, Okla., 1973), we give the first four definitions: "**Attenuation:** 1. A reduction in amplitude or energy, such as might be produced by passage through a filter. 2. A reduction in the amplitude of seismic waves, such as produced by divergence, reflection and scattering, and absorption. 3. That portion of the decrease in seismic or sonar signal strength with distance not dependent on geometrical spreading. This decrease depends on the physical characteristics of the transmitting media, involving reflection, scattering, and absorption. 4. If the amplitude of a plane wave is reduced by the factor $\exp[-\alpha(f)x]$ in traveling a distance of x meters, then the attenuation factor is $\alpha(f)$. Often thought to be linear with frequency, sometimes thought to be quadratic with frequency."

Geophysicists usually use attenuation in the sense of 1 and 2. When they use attenuation in the sense of 3 and 4 they usually, and explicitly, explain it as being due to reflection, scattering, and absorption. Underwater acousticians usually use the more restricted definition 3.

3.1. THE RADIATION OF POWER

Assume that the source is a small sphere. It expands or contracts uniformly and the signal travels outward. In a constant speed medium, the signal is the same in all directions. The signal satisfies the wave equation in spherical coordinates and in this case is only a function of the radial distance R. From (A2.2.24) and (A2.2.19) the wave equation is

$$\frac{1}{R^2}\frac{\partial}{\partial R}\left[R^2\frac{\partial}{\partial R}p(R,t)\right]=\frac{1}{c^2}\frac{\partial^2}{\partial t^2}p(R,t) \qquad (3.1.1)$$

where $p(R, t)$ is the outward traveling pressure signal. "Guessing" the answer, we write

$$p(R,t)=\frac{b}{R}g(R,t) \qquad (3.1.2)$$

where b is a constant. Substitute (3.1.2) into (3.1.1). The wave equation for $g(R, t)$ is

$$\frac{\partial^2}{\partial R^2}g(R,t)=\frac{1}{c^2}\frac{\partial^2}{\partial t^2}g(R,t) \qquad (3.1.3)$$

which has the same form as the one-dimensional plane wave equation. Its solutions are $g(R, t) = g(t - R/c)$ and $g(t + R/c)$.

We choose the outgoing signal $p(R, t) = (b_c/R)g(t - R/c)$, in complex form with time dependence $\exp(i\omega t)$, and write

$$p(R,t)=\frac{b_c}{R}\exp\left[i\omega\left(t-\frac{R}{c}\right)\right]$$

Next we evaluate b_c. The temporal average of the absolute square of $p(R, t)$ over one period T is

$$\langle|p(R,t)|^2\rangle=P^2=\frac{1}{T}\int_0^T\frac{b_c^2}{R^2}\,dt=\frac{b_c^2}{R^2}$$

where P is the rms sound pressure. We assume that the rms source pressure is P_0 at distance R_0, therefore $b_c = P_0 R_0$.

Alternatively, the real expression can be used.

$$p(R, t) = \frac{b_r}{R} \cos\left[\omega\left(t - \frac{R}{c}\right)\right]$$

The mean square value of the cosine function is $\frac{1}{2}$, again the source is P_0 at R_0, therefore $b_r = \sqrt{2}\, P_0 R_0$. Using these results, three expressions for $p(R, t)$ are

$$p(R, t) = P_0 \frac{R_0}{R} \exp\left[i\omega\left(t - \frac{R}{c}\right)\right]$$

$$p(R, t) = \sqrt{2} P_0 \frac{R_0}{R} \cos \omega\left(t - \frac{R}{c}\right) \qquad \textbf{(3.1.4)}$$

$$p(R, t) = \sqrt{2} P_0 \frac{R_0}{R} \sin \omega\left(t - \frac{R}{c}\right)$$

For a relation between the acoustic pressure and particle velocity, we rewrite (2.8.14) for the radial component

$$\frac{\partial p(R, t)}{\partial R} = -\rho_A \frac{\partial u(R, t)}{\partial t} \qquad (3.1.5)$$

where $u(R, t)$ is the radial component of the particle velocity. Substituting (3.1.4) into (3.1.5) and using $\omega = kc$ yields

$$\frac{ikb}{R} \exp\left[i(\omega t - kR)\right]\left(1 + \frac{1}{ikR}\right) = \rho_A \frac{\partial u(R, t)}{\partial t} \qquad (3.1.6)$$

Since u must have time dependence $\exp(i\omega t)$, we use $\partial u/\partial t = i\omega u(R, t)$. Then (3.1.6) becomes

$$p(R, t)\left(1 + \frac{1}{ikR}\right) = (\rho_A c) u(R, t) \qquad (3.1.7)$$

For large kR (i.e., at many wavelengths from the source), u is in phase with p and the relation is the same as for plane waves

$$p(R, t) \simeq (\rho_A c) u(R, t) \qquad kR \gg 1 \qquad \textbf{(3.1.8)}$$

The source transmits power into the water. (Power is force times the velocity in the same direction.) The instantaneous intensity, the power that passes through a unit area, is $p(R, t)u(R, t)$ when both quantities are real. Alternatively we use the real part of $p(R, t)u^*(R, t)$ when the quantities are complex. For a sound pressure P_0 at R_0, the complex pressure and particle velocity are

$$p(R, t) = P_0 \frac{R_0}{R} \exp\left[i\omega\left(t - \frac{R}{c} \right) \right]$$

(3.1.9)

$$u(R, t) = P_0 \frac{R_0}{R\rho_A c}\left(1 + \frac{1}{ikR} \right) \exp\left[i\omega\left(t - \frac{R}{c} \right) \right]$$

The average intensity $\langle I \rangle$, or average power passing through a unit area, at distance R is

$$\langle I \rangle = \mathrm{Real}\, \frac{1}{T} \int_0^T p(R, t) u^*(R, t)\, dt$$

The substitution of (3.1.9) and integration yields

$$\langle I \rangle = P_0^2 \frac{R_0^2}{R^2 \rho_A c}$$

(3.1.10)

Although the result is valid at any range, we suggest making source calibration measurements at larger range $(kR \gg 1)$ to avoid possible complications near the source. See section 5.3.

The total power from the source Π is the integral of $\langle I \rangle$ that passes through the surface of the sphere of radius R.

$$\Pi = \int_S \langle I \rangle\, dS = \int_{4\pi} \langle I \rangle R^2\, d\Omega$$

where $d\Omega = dS/R^2$ is the element of solid angle.

Since I and R are constant over the spherical surface

$$\Pi = \langle I \rangle R^2 \int d\Omega$$
$$= 4\pi P_0^2 \frac{R_0^2}{\rho_A c}$$

(3.1.11)

This is the power radiated through the spherical surface at any radius, provided there is no energy loss in the medium.

The rms pressure at the reference distance R_0 is

$$P_0 = \left(\frac{\Pi \rho_A c}{4\pi R_0^2} \right)^{1/2}$$

(3.1.12)

In mks units, pressure is in newtons per square meter $(N/m^2) =$ pascals (Pa), power is in watts, distance in meters, and ρ_A in kilograms per cubic meter. Generally 1 m serves as the reference R_0 because it is convenient in computations.

From (3.1.9) the rms pressure P at distance R is inversely proportional to R.

$$P = \frac{P_0 R_0}{R} = \left(\frac{\Pi \rho_A c}{4 \pi R^2}\right)^{1/2} \tag{3.1.13}$$

3.1.1. Sound Intensity Level, Sound Pressure Level, and the Decibel

In many measurements we are interested in the ratio of the sound intensity I relative to a reference intensity I_r. Since the ratio may be very large or very small, it is convenient to use the logarithm of the ratio. In decibels (dB), the ratio is called the sound intensity level, (SIL)

$$SIL \equiv 10 \log_{10} \frac{I}{I_r} \text{ dB re } I_r \tag{3.1.14}$$

Since the sound intensity is proportional to the square of the sound pressure, we can rewrite (3.1.14) as follows:

$$SPL \equiv 20 \log_{10} \frac{P}{P_r} \text{ dB re } P_r \tag{3.1.15}$$

where P_r is the reference sound pressure. This is called the *sound pressure level* (*SPL*). The reference must be stated or implied.

We express (3.1.13) in decibels by taking $20 \log_{10}$ of P/P_r

$$SPL = SL - 20 \log_{10} \frac{R}{R_0} \tag{3.1.16}$$

where $SL \equiv 20 \log_{10}(P_0/P_r)$ dB re P_r is the *source level* at the distance R_0. The reference sound pressure is conventionally 1 Pa or 1 μPa. The older literature uses 1 μ bar. Eq. (3.1.16) is the "sonar equation" for a spherically spreading sound pressure. For each doubling of the distance the *SPL* decreases by 6 dB (6 dB $\approx 20 \log_{10} 2$).

The difference between the source level and the *SPL* is called the transmission loss (*TL*). The transmission loss due to spherical spreading is

$$TL = SL - SPL = 20 \log_{10}(R/R_0) \text{ dB.} \tag{3.1.17}$$

3.2. RAY TRACING

We use Snell's law to trace the path of a small sector of a wave front as it travels within an inhomogeneous medium because it gives the travel time and direction of the wave front. This ray method is valid for high frequency signals—that is, when λ is much less than the water depth and much less than the distances from the source to object and object to receiver. A strong condition is that the change of sound speed is negligible over several wavelengths. Sound is partially reflected at sharp changes of the sound speed.

Comparisons of theory and experiment since the 1940s have shown that ray approximation errors are small when diffraction effects are absent. The troublesome diffraction effects are caused by inhomogeneities of sound speed in the ocean. In typical situations the sound rays turn upward or downward and create a shadow zone or the sound rays cross and come to a focus. Wave theory is used to calculate the sound field in such regions.

3.2.1. Application of Snell's Law

The fluidity of water and the stability of less dense over more dense water cause oceans and lakes to be nearly horizontally stratified in local regions. In the open ocean the dimensions of a stratified region may be several hundred kilometers. The horizontal stratification assumption is poor where there are currents and lateral changes of the type of water mass. We use the simplification that the sound speed $c(z)$ is only a function of depth.

We use Snell's law, $(\sin \theta_1)/c_1 = (\sin \theta_2)/c_2$, to trace rays. It is easy to show that Snell's law also applies in a continuously stratified medium. The directions of the signal as it goes from medium 1 to 2, or 2 to 3, and so on, in a stratified medium are as follows:

$$\frac{\sin \theta_1}{c_1} = \frac{\sin \theta_2}{c_2} = \frac{\sin \theta_3}{c_3} \cdots = \frac{\sin \theta_n}{c_n} \quad \text{or} \quad \frac{\sin \theta_1}{c_1} = \frac{\sin \theta_n}{c_n}$$

(3.2.1)

where the angles are measured between the ray and the vertical (Fig. 3.2.1). As we let the layer thicknesses become infinitesimal, c becomes a continuous function of depth, labeled $c(z)$. For an incident angle θ_i at the initial depth z_i, we find θ at depth z by Snell's law:

$$\frac{\sin \theta}{c(z)} = \frac{\sin \theta_i}{c(z_i)} = a$$

(3.2.2)

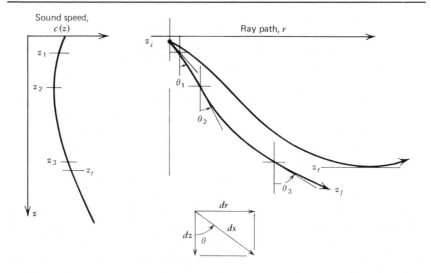

Figure 3.2.1.   Ray paths in a continuously stratified medium. *Inset,* differential distance along the path *ds* in terms of the horizontal and vertical displacements *dr* and *dz.*

where a is the Snell's law constant of the ray. The piecewise application of Snell's law to a continuous $c(z)$ is sketched in Fig. 3.2.1.

The determinations of the ray paths and travel times require an integration involving Snell's law for the particular $c(z)$ in the medium. The integral is basically dependent on z because the sound speed is a function of z. The differential distance along the ray ds and the travel time dt are

$$ds = \frac{dz}{\cos \theta} \tag{3.2.3}$$

$$dt = \frac{ds}{c(z)} = \frac{dz}{c(z) \cos \theta} \tag{3.2.4}$$

The differential horizontal displacement is

$$dr = \tan \theta \, dz \tag{3.2.5}$$

For more convenient expressions, we use Snell's law (3.2.2) to replace the $\sin \theta$, $\cos \theta$, and $\tan \theta$ with the following:

$$\sin \theta = ac(z)$$
$$\cos \theta = [1 - a^2 c^2(z)]^{1/2} \tag{3.2.6}$$
$$\tan \theta = ac(z)[1 - a^2 c^2(z)]^{-1/2}$$

Using these, the total travel time and horizontal displacement are the integrals of dt and dr from the initial depths z_i to the final depths z_f.

$$t_f - t_i = \int_{z_i}^{z_f} dt$$

$$= \int_{z_i}^{z_f} \frac{dz}{c(z)[1 - a^2c^2(z)]^{1/2}} \qquad (3.2.7)$$

$$r_f - r_i = \int_{z_i}^{z_f} dr$$

$$= \int_{z_i}^{z_f} \frac{ac(z)\,dz}{[1 - a^2c^2(z)]^{1/2}} \qquad (3.2.8)$$

where the integrations are along the ray. These equations can be evaluated for $c(z)$ when the initial position z_i and initial angle θ_i are given.

Since the ray follows the Snell's law path, it becomes horizontal at depth z if $\sin \theta_i = c(z_i)/c(z)$. The ray path then turns upward, still obeying Snell's law. The signal is totally reflected when the medium has an increasing sound speed below the turning depth z_t (Fig. 3.2.1). When the ray turns, the limits of the integral are z_i to z_t for the downward ray, and a new integration is initiated for the upward moving ray.

It is customary to use approximations to an actual $c(z)$ in the evaluation of the integrals. The linear approximation is most often used because of its simplicity (Fig. 3.2.2). More complicated sound speed–depth functions can be broken into several linear segments.

In applying the linear approximation to the integration of (3.2.7) and (3.2.8), we replace the actual $c(z)$ by a series of layers and evaluate the

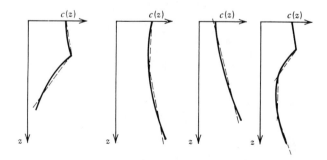

Figure 3.2.2. Linear approximations to $c(z)$. Approximations are the dashed lines.

integrals for the ray path within each layer. The total ray path is the sum of the pieces. We use the layer between z_1 and z_2 as an example. Within this layer

$$c(z) = c(z_1) + b(z - z_1) \qquad \text{for} \quad z_1 \le z \le z_2 \qquad \textbf{(3.2.9)}$$

$$b = \frac{d[c(z)]}{dz} \qquad \textbf{(3.2.10)}$$

where $c(z_1)$ is the sound speed at depth z_1 and b is the constant gradient of the sound speed within the layer. The direct substitution of (3.2.9) in (3.2.7) and (3.2.8) gives messy expressions, and we change variables to simplify the algebra.

$$w \equiv z - z_1 + \frac{c(z_1)}{b} \qquad \textbf{(3.2.11)}$$

$$dw = dz \qquad (3.2.12)$$

$$c(w) = bw \qquad (3.2.13)$$

In evaluating the integrals, both the initial and final depths must be within or on the boundaries of the layer. On replacing $c(z)$ by $c(w)$ and changing variables, (3.2.7) and (3.2.8) become

$$t_f - t_i = \int_{w_i}^{w_f} \frac{dw}{bw(1 - a^2 b^2 w^2)^{1/2}} \qquad (3.2.14)$$

$$r_f - r_i = \int_{w_i}^{w_f} \frac{abw \, dw}{(1 - a^2 b^2 w^2)^{1/2}} \qquad (3.2.15)$$

where the limits follow by substituting z_i and z_f in (3.2.11). If the ray is being traced through the layer, w_i is w_1, w_f is w_2, and θ_2, r_2, and t_2 are the initial conditions for the next layer. If the ray turns or becomes horizontal within the layer, w_f is the turning depth. The integrations follow with the aid of a table of integrals (Appendix A1):

$$t_f - t_i = \frac{1}{b} \log_e \frac{w_f[1 + (1 - a^2 b^2 w_i^2)^{1/2}]}{w_i[1 + (1 - a^2 b^2 w_f^2)^{1/2}]} \qquad \textbf{(3.2.16)}$$

or

$$t_f - t_i = \frac{1}{b} \log_e \frac{w_f(1 + \cos \theta_i)}{w_i(1 + \cos \theta_f)} \qquad (3.2.17)$$

and

$$r_f - r_i = \frac{1}{ab} [(1 - a^2 b^2 w_i^2)^{1/2} - (1 - a^2 b^2 w_f^2)^{1/2}] \qquad \textbf{(3.2.18)}$$

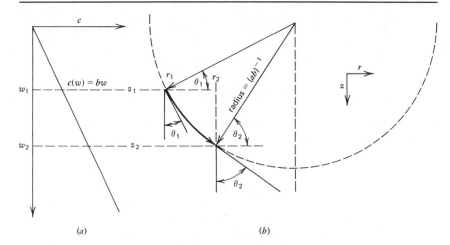

Figure 3.2.3. (*a*) Sound speed profile. (*b*) Circular ray path for linear dependence of speed on depth between z_1 and z_2. Radius is $(ab)^{-1} = w/(\sin \theta)$, where $a = (\sin \theta)/c$ and b is the gradient of the speed.

or

$$r_f - r_i = \frac{1}{ab} (\cos \theta_i - \cos \theta_f)$$

$$= (\text{radius})(\cos \theta_i - \cos \theta_f) \qquad (3.2.19)$$

where radius $= (ab)^{-1}$ is the radius of the circular path followed by a ray in the region where the linear gradient is b (Fig. 3.2.3).

The computation of the travel time along a curved ray path and the r displacement is easy when we know the gradient of sound speed b. We should enter some words of caution: the gradient b is usually subject to error because the whole change of $c(z)$ is only a few percentage points.

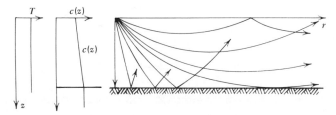

Figure 3.2.4. Rays in isothermal, isohaline water. The sound speed increases slightly with depth because of the pressure dependence, causing an upward curvature of the sound paths.

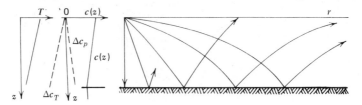

Figure 3.2.5. Negative sound speed gradient. Incremental change of speed caused by pressure is Δc_p, greater change caused by temperature is Δc_T, and resultant sound speed is $c(z)$.

When the speed tends to a constant, b tends to zero and (3.2.11) to (3.2.19) are unsatisfactory. In shallow water an average sound speed can be chosen for travel time–distance computations. When b is too small, we use an average constant speed and assume straight ray paths.

3.2.2. Speed of Sound and Ray Paths in the Ocean

The type of ray path is associated with the dependence of the sound speed on depth. Typical situations are illustrated in Figs. 3.2.4 to 3.2.7. We show temperature as well as speed of sound because these are the quantities most often measured. Simplifying the results of the Naval Research Laboratory studies (Medwin, 1975), the dependence of sound speed on temperature, salinity, and pressure within 1 km of the surface is found to be approximately

$$c = 1449.2 + 4.6T - 0.055T^2 + 0.00029T^3$$
$$+ (1.34 - 0.010T)(S - 35) + 1.58 \times 10^{-6} P_a \quad \textbf{(3.2.20)}$$

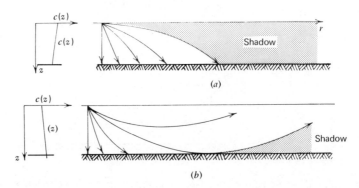

Figure 3.2.6. Shadow zones, hatched regions. Only the direct ray paths are shown at the shadow region; bottom is nonreflecting for illustration.

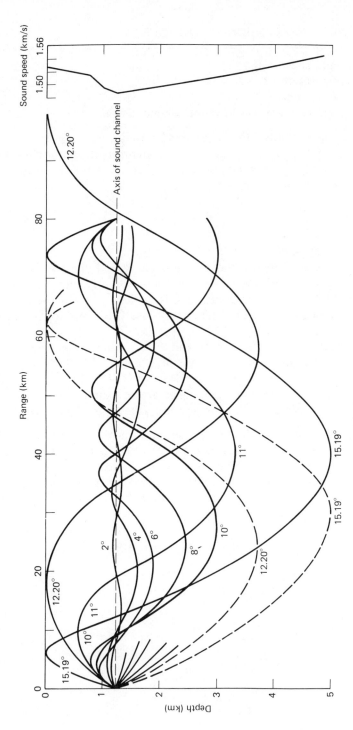

Figure 3.2.7. Ray diagram for typical Atlantic Ocean sound channel, depicting channeled rays and refracted-surface-reflected (RSR) rays; sound speed profile is at the right. The angles are grazing angles at the axis of the sound channel. (Ewing and Worzel, 1948.)

where c = sound speed (m/s)

T = temperature (°C)

S = salinity (ppt)

P_a = gauge pressure due to water column, (N/m²)

Ignoring compressibility $P_a' = \rho_A g z$ and using $\rho_A \simeq (1 + S \times 10^{-3}) \times 10^3$ kg/m³, $g = 9.8$ m/s², $S = 35$ppt, and z = depth (m), the last term is written $0.016z$. The dependence of sound speed on depth in isothermal isohaline water is therefore, typically

$$b = \frac{dc}{dz} \simeq 0.016 \text{ s}^{-1} \tag{3.2.21}$$

When the temperature has a large decrease with increasing depth, the temperature effect overrides the pressure effect and the sound speed gradient is negative. Both cases, negative and positive gradients of speed, have the possibility of shadow zones, that is, regions where the sound ray does not penetrate. The shadow zones for these two cases are shown in Fig. 3.2.6.

The stratification of the deep ocean and the resulting ray paths depends on location, season, and time of day. As an example, Fig. 3.2.7 displays, the ray traces computed by Ewing and Worzel (1948). The steepest ray shown just grazes the bottom. The angles are measured relative to the horizontal in long-range ray tracing and are called "grazing angles."

Illustrative Problem. The water temperature in the Arctic is nearly isothermal. The sound speed increases with depth because of the pressure effect. Assume a surface temperature of 0°C and salinity of 35 ppt, determine the initial angle, range, and travel time for the ray that starts at the surface (Fig. 3.2.4), turns at 2 km depth, and returns to the surface.

Solution. Using eq. (3.2.20)

$$c(z) = 1449 + 0.016z \text{ m/s}$$

$$c(0) = 1449 \text{ m/s}$$

$$c(2000 \text{ m}) = 1481 \text{ m/s}$$

Applying Snell's law

$$\sin \theta_1 = \frac{1449}{1481} = 0.9784$$

$$\cos \theta_1 = 0.2068$$

$$a = \frac{\sin \theta_1}{c(z_1)} = 6.752 \times 10^{-4} \text{ s/m}$$

at the turning depth, 2000 m

$$\sin \theta_2 = 1 \qquad \cos \theta_2 = 0$$

Calculate $w = z + \dfrac{c(0)}{b}$ where $b = 0.016 \text{ s}^{-1}$ and $w = 9.056 \times 10^{+4} + z$ (m).

The ray is traced from

$$w_1 = 9.056 \times 10^4 \text{ m}$$

to

$$w_2 = 9.256 \times 10^4 \text{ m}$$

With the aid of (3.2.17) and (3.2.19), substitution of the numbers yields

$$t_2 - t_1 = 13.11 \text{ s}$$

$$r_2 - r_1 = 19.14 \times 10^3 \text{ m}$$

The surface-to-surface travel time is

$$t = 26.22 \text{ s}$$

and surface-to-surface distance is

$$r = 38.28 \times 10^3 \text{ m}$$

The path is the arc of a circle.

3.2.3. Divergence and Convergence in a Horizontally Stratified Ocean

Section 3.1 presented the sound transmission for a source in a constant speed medium. When sound rays converge or diverge, the change of separation of closely spaced rays can be used to estimate the sound transmission loss.

Within a few wavelengths of an omnidirectional source, the curvature of the ray path is small and can be ignored. The power radiates uniformly outward, and we choose a small increment of angle $\Delta\theta$ and bound it by straight rays above and below (Fig. 3.2.8). In the ray approximation, no power flows across the rays. For this case the flux of power at a large range can be estimated by computing the area perpendicular to the rays at the horizontal range r. The sound power so bounded is constant in the absence of absorption.

Referring to Fig. 3.2.8 for the geometry, at the reference spherical radius R_0 the power $\Delta\Pi$ within the ring of width $\Delta\theta$ radians is found to be

$$\Delta\Pi = \frac{P_0^2}{\rho_0 c_0} (2\pi R_0 \sin \theta_0) R_0 \, \Delta\theta \qquad (3.2.22)$$

where the impedance is $\rho_0 c_0$ at the reference radius R_0.

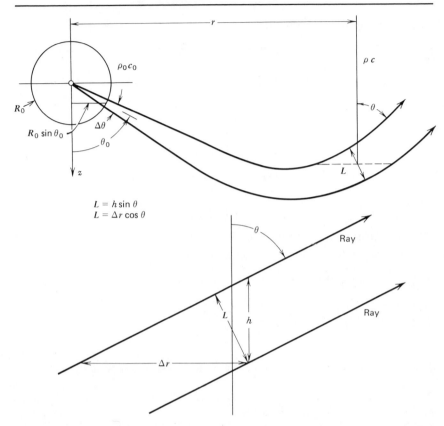

Figure 3.2.8.   Spread of power within $\Delta\theta$ in a horizontally stratified medium. The figure can be rotated about the z axis. *Lower inset*, geometry for ray separation L, incremental range Δr, and vertical ray separation h.

The area intercepted by the sound power at horizontal range r is $2\pi rL$, where L is the distance between the rays as measured along the wave front. Since the sound power is constant, $\Delta\Pi$ also equals the power passing through the area $2\pi rL$; that is,

$$\Delta\Pi = \frac{2\pi R_0^2}{\rho_0 c_0} P_0^2 \sin\theta_0 \Delta\theta = \frac{2\pi rLP^2}{\rho c}$$

Therefore

$$P^2 = \frac{P_0^2 R_0^2 \rho c \sin\theta_0 \, \Delta\theta}{\rho_0 c_0 rL} \tag{3.2.23}$$

where ρc is the impedance at range r.

Usually it is more convenient to calculate the vertical separation of the rays h than the wave front spreading L. If the upper and lower bounding rays are not converging or diverging rapidly, we can estimate L from numerical values of h by the formula (Fig. 3.2.8, inset)

$$L = |h \sin \theta| \tag{3.2.24}$$

Alternatively we can use a single depth and the horizontal separation of a pair of rays Δr to compute L.

$$L = |\Delta r \cos \theta| \tag{3.2.25}$$

The transmission equations for a non-absorbing medium [using (3.2.24) in (3.2.23)] are written as follows:

$$P^2 = P_0^2 \frac{R_0^2 \Delta \theta}{rh} \frac{\rho c}{\rho_0 c_0} \frac{|\sin \theta_0|}{|\sin \theta|} \tag{3.2.26}$$

where

$$P_0^2 = \frac{\rho_0 c_0 \Pi}{4 \pi R_0^2} \tag{3.2.27}$$

see (3.1.11).

The transmission loss caused by a stratified medium is obtained by forming

$$
\begin{aligned}
TL \text{ (dB)} &= -20 \log_{10} \frac{P}{P_0} \\
&= 10 \log_{10} \frac{r}{R_0} + 10 \log_{10} \frac{h}{R_0 \Delta \theta} \\
&\quad - 10 \log_{10} \frac{\rho c}{\rho_0 c_0} - 10 \log_{10} \frac{\sin \theta_0}{\sin \theta}
\end{aligned} \tag{3.2.28}
$$

In a constant speed medium, (3.2.26) reduces to

$$P^2 = \frac{P_0^2 R_0^2}{R^2} \tag{3.2.29}$$

and (3.2.28) reduces to $TL = 20 \log_{10}(R/R_0)$ dB for a spherically spreading wave.

Nearly everyone who does ray tracing discovers that sound rays can cross (Fig. 3.2.9) and that (3.2.26) becomes infinite at the crossing point. The crossing regions are focusing or convergence zones. In fact, although the intensity can be large, it is finite. As in optics, the intensity depends on the sound wavelength, the amount of energy focused, and how nearly

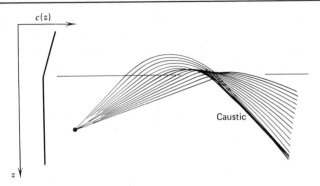

Figure 3.2.9. Crossing ray paths. The caustic is the envelope of the family of rays.

perfect the focusing is. Here we are considering rays that have the same travel times from the source to the crossing point. If the travel times are sufficiently different, short pings or impulsive signals do not interfere and there is no focusing of sound.

Qualitative studies are made by using equally spaced rays at the source and doing simple things like counting the number of rays that pass through a small region at the field position. For illustration, consider the 3° of sound source that is bounded by the downward starting ray at 12.20°, which later grazes the surface, and the downward starting ray at 15.19°, which grazes the bottom, in Fig. 3.2.7. Redrawing the pair of rays gives Fig. 3.2.10. The convergence of the ray paths in the "convergence

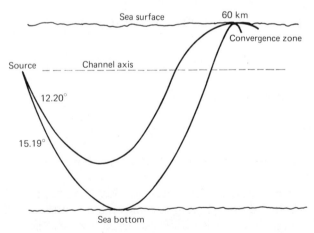

Figure 3.2.10. Convergence zone for the surface-limited and bottom-limited rays. Redrawn from the 12.20° and 15.19° downward-starting rays in Fig. 3.2.7.

zone" at about 62 km is labeled. Many experiments have verified the existence of convergence zones and very high sound intensities within the zones. The focal region is circular around the z axis because of circular symmetry.

Quantitative calculations in the focusing region use solutions of the wave equation. We give a simplified calculation in Section A3.1.

3.2.4. Reflections at the Sea Floor

Now we include reflection at the bottom. The wave front interacts at a *plane interface* and is reflected at each local area. Since the angle of *reflection* equals the angle of incidence, the reflected wave front has the same curvature and appears to come from an image source beneath the interface, as in Fig. 2.2.2.

The ray path construction (Fig. 3.2.11a) shows the reflection as occurring at a point on the bottom. Actually, from Huygens' principle and Fig. 2.2.2, we know that the reflection process involves a fairly large area and includes many Fresnel zones. It is customary to use the *plane wave reflection* coefficient and θ_i for computations. The latter is a good approximation as long as θ_i is not too close to the critical angle (2.9.13).

In the ray-wave approximation, we follow the expanding wave front from the source to its interaction at the interface. The spreading of the wave front is calculated by tracing a pair of closely spaced rays. At the interface the rays are reflected into the new directions and continue. The pressure amplitude is multiplied by the pressure reflection coefficient at the bottom $\mathcal{R}_{12}$.

For the image construction, we replace the medium below the interface by a mirror image. The image source pressure becomes $\mathcal{R}_{12}P_0$. For constant sound speed and the geometry in Fig. 3.2.11b, the sound pressure at the receiver, ignoring the direct arrival, is

$$P = \frac{\mathcal{R}_{12}P_0R_0}{R_1 + R_2} \qquad (3.2.30)$$

where the source pressure is P_0 at distance R_0.

When curvature of the ray paths is important, the ray is traced from the image source to the receiver. The image source pressure is $\mathcal{R}_{12}P_0$, and this replaces P_0 in (3.2.26).

$$P^2 = (\mathcal{R}_{12}P_0)^2 \frac{R_0^2\Delta\theta}{rh} \frac{\rho c}{\rho_0 c_0} \frac{|\sin\theta_0|}{|\sin\theta|} \qquad (3.2.31)$$

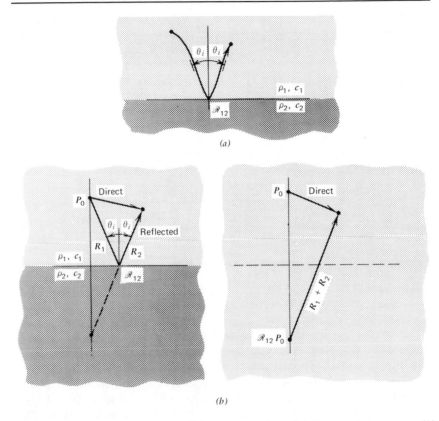

Figure 3.2.11. Ray paths and image construction. (*a*) Ray path in ocean. (*b*) Image construction.

In using this method it is important to remember that we assumed the interface to be a large plane "mirror" and that no scattering processes are involved. Chapter 10 demonstrates that a rough interface causes the sound waves to be scattered into other directions in which case the image method is not applicable.

3.3. ATTENUATION DUE TO SOUND ABSORPTION

Sound is attenuated also by conversion of its acoustic energy into the random form, heat. The earliest realistic studies of sound absorption were conducted by Stokes in England and Kirchhoff in Germany in the mid-nineteenth century. The Stokes-Kirchhoff "classical" theory of attenuation attributed propagation losses to the thermal conductivity and the viscosity of the medium. The thermal conductivity loss for water turns out to be negligible.

The significant losses in water are caused by shear viscosity and bulk viscosity. The shear viscosity is due to frictional forces during relative motion between adjacent layers of the liquid. The bulk viscosity, due to molecular rearrangements that take place during a sound wave cycle, is even more important than the shear viscosity, in water.

The time required for molecular reordering in response to changing pressure is called the relaxation time. The loss of acoustic energy depends on the relaxation time of the process, compared to the period of the sound wave. The loss per cycle is small when the relaxation time is very different from the sound period, but when the two are approximately equal, the loss per cycle is a maximum. The relaxation times for the three known molecular processes in seawater are approximately 10^{-11} s for the freshwater component, 10^{-5} s for the magnesium sulfate, and 10^{-3} s for the boric acid in seawater.

The mechanisms for energy loss cause the sound pressure in a plane wave to have a rate of decrease with distance that is proportional to the original pressure:

$$\frac{dP}{dx} = -(\alpha_e)P \qquad (3.3.1)$$

Integrating, we find

$$P = P_i \exp\left(-\alpha_e x\right) \qquad \textbf{(3.3.2)}$$

where P_i is the rms pressure at $x_i = 0$, P is the rms pressure at x, and α_e is the exponential pressure attenuation rate. The units of α_e are nepers (Np) per unit distance.

The plane wave attenuation in decibels is obtained by taking $20 \log_{10}$ of the pressure ratio in (3.3.2).

$$\text{dB loss} = 20 \log_{10} \frac{P_i}{P} = (\alpha_e)x \, (20 \log_{10} e)$$

$$= 8.686(\alpha_e)x \qquad \textbf{(3.3.3)}$$

Generally people use the logarithmic rate of pressure attenuation, or attenuation coefficient,

$$\alpha = \frac{\text{dB loss}}{\text{distance}} = 8.686 \, \alpha_e \qquad \textbf{(3.3.4)}$$

Then dB loss $= \alpha x$.

The physical origins of the absorption in water are described in Section A3.2. The attenuation rate for fresh water at sea level is

$$\alpha_F = \frac{4.34}{\rho_F c_F^3}\left(\frac{4\mu_F}{3} + \mu_F'\right)\omega^2 \text{ dB/m} \qquad (3.3.5)$$

where $\rho_F \approx 1000 \text{ kg/m}^3$

$c_F =$ m/s from (3.2.20) with $S = 0$ (≈ 1461 m/s for $T = 14°C$)

$\omega =$ rad/s

$\mu_F =$ dynamic (or absolute) coefficient of shear viscosity for freshwater ($\approx 1.2 \times 10^{-3}$ N·s/m^2 for $T = 14°C$)

$\mu_F' =$ dynamic (or absolute) coefficient of bulk viscosity for freshwater ($\approx 3.3 \times 10^{-3}$ N·s/m^2 for $T = 14°C$)

The attenuation rate is calculated by using the values of the physical constants at the appropriate water temperature. It is plotted in Fig. 3.3.1 for freshwater at 14°C. See Fig. A3.2.2 for viscosities at other temperatures and eqs. (A3.2.24) and (A3.2.25) for the small corrections at other depths.

Seawater has two additional relaxation processes. One is due principally to the magnesium sulfate (MgSO$_4$) component and the other is caused mainly by the boric acid in seawater. After rewriting (3.3.5) in terms of frequency in kilohertz and adding the seawater relaxation effects from Sections A3.2.3 and A3.2.4 we have the attenuation rate in seawater.

$$\alpha_S = \underbrace{\frac{1.71 \times 10^8 (4\mu_F/3 + \mu_F')f^2}{\rho_F c_F^3}}_{\text{freshwater}} + \underbrace{\left(\frac{SA'f_{rm}f^2}{f^2 + f_{rm}^2}\right)(1 - 1.23 \times 10^{-3}\,P_a)}_{\text{magnesium sulfate relaxation}}$$

$$+ \underbrace{\frac{A''f_{rb}f^2}{f^2 + f_{rb}^2}\text{ dB/m}}_{\text{boric acid relaxation}} \qquad (3.3.6)$$

where $f_{rm} =$ relaxation frequency (kHz) for MgSO$_4$; see Fig. A3.2.3 for temperature dependence

$f_{rb} =$ relaxation frequency (kHz) for boric acid; see Fig. A3.2.4 for temperature dependence

$A' = 2.03 \times 10^{-5}$ dB/[(kHz)(ppt)(m)]

$A'' = 1.2 \times 10^{-4}$ dB(kHz)$^{-1}$(m)$^{-1}$

$S =$ salinity (ppt) by weight

$f =$ frequency (kHz)

$P_a =$ gauge pressure due to water column (atm)

Equation (3.3.6) is plotted in Fig. 3.3.1 for $S = 35$ ppt and $T = 14°C$ at sea level. The magnesium sulfate and boric acid relaxation effects, drawn in dashed lines, show the typical relaxation behavior, with α proportional to f^2 for $f \ll f_r$ and α = constant for $f \gg f_r$. For this example, $f_{rm} = 111$ kHz [from (A3.2.27) or Fig. A3.2.3], and $f_{rb} = 1.23$ kHz [from (A3.2.29) or Fig. A3.2.4]. The attenuation rates in freshwater and seawater (sum of the three components) are indicated by solid lines.

The absorption has been treated rather generally in this section. However in special cases a water volume can change its attenuation rate orders of magnitude in a few minutes. For example, the tiny bubbles caused by the wake of a passing boat or a high sea state can create a very high, time-varying, attenuation rate.

Knowing the attenuation rate for plane waves, we calculate the change in pressure for spherical waves going from R_0 to R in a constant speed medium by replacing P_i by P_0R_0/R (3.1.13) and range x by $(R - R_0)$ in (3.3.2)

$$P = P_0 \frac{R_0}{R} \exp\left[-(\alpha_e)(R - R_0)\right] \qquad (3.3.7)$$

Taking $20 \log_{10}(P_0/P)$ gives the signal change in decibels. This defines the transmission loss TL for spherical spreading in a homogeneous absorbing medium,

$$TL = 20 \log_{10} \frac{R}{R_0} + \alpha(R - R_0) \text{ dB} \qquad \textbf{(3.3.8)}$$

where the α (dB/m) is α_F for freshwater or α_S for seawater and R and R_0 are in meters. Generally $R \gg R_0$; therefore $R - R_0 \approx R$.

Equation (3.3.8) shows the difference in range dependence between the logarithmic transmission loss due to spreading (first term), and the linear transmission loss due to absorption. Inserting a few numbers into (3.3.8) makes it clear that the spreading loss is dominant near the source, whereas the absorption loss is the major term for longer ranges. Consider $f = 10$ kHz where $\alpha_S \approx 10^{-3}$ dB/m. For the double distance $R = 2$ m and $R_0 = 1$ m there is a 6 dB divergence loss and only 10^{-3} dB absorption loss. However in propagating from 10 to 20 km, the 6 dB divergence loss is exceeded by an absorption loss of $(10^{-3}$ dB/m$)(10^4$ m$) = 10$ dB.

For each frequency there is a transition range R_t at which the two *rates* of loss are equal.

$$\frac{d}{dR}\left(20 \log_{10} \frac{R}{R_0}\right) = \frac{d}{dR}[\alpha(R - R_0)] \qquad (3.3.9)$$

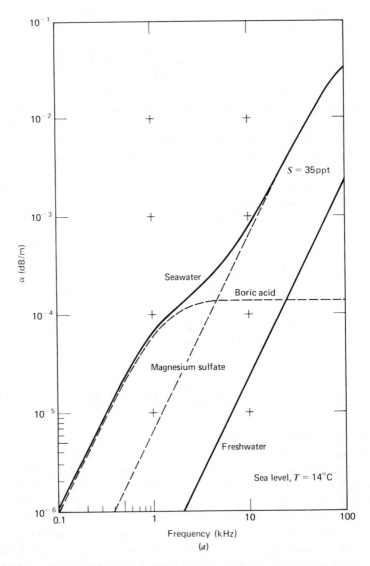

Figure 3.3.1. Attenuation rate due to absorption in freshwater and seawater: (*a*) 0.1 to 100 kHz; (*b*) 10 to 10,000 kHz. Dashed lines indicate contributing attenuation rates due to relaxation processes.

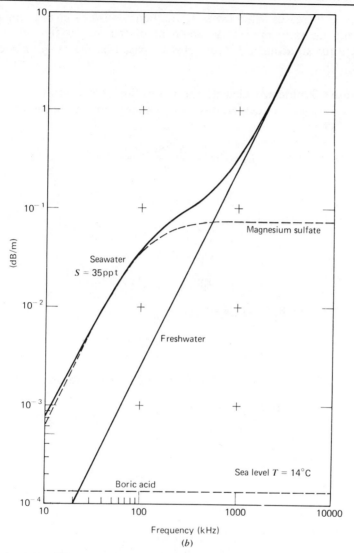

Frequency (kHz)

(b)

Performing the differentiation

$$R_t = \frac{8.68}{\alpha} \qquad (3.3.10)$$

For example, for 10 kHz sound, we have $R_t = 8.68/10^{-3} = 8680$ m. The range at which $\alpha(R - R_0)$ begins to exceed $20 \log_{10}(R/R_0)$ is greater than R_t.

Since α increases with increasing frequency, *high frequencies are practical only for short ranges.* A judgment of the proper frequency to be selected for an acoustical study must be based on the range implications of (3.3.10).

Illustrative Problem. Calculate the absorption coefficient α at 10 kHz in sea level water of temperature 0°C, salinity 35 ppt. From (3.3.6) the freshwater part is

$$\alpha_F = \frac{1.71 \times 10^8 (4\mu_F/3 + \mu_F')f^2}{\rho_F c_F^3}$$

From Fig. A3.2.2, $\mu_F\,(0°C) = 1.8 \times 10^{-3}$, $\mu_F' = 5.5 \times 10^{-3}$ mks. From (3.2.20), we have

$$c_F = 1449.2 - (1.34)(35)$$
$$= 1402.3 \text{ m/s}$$
$$\alpha_F = 4.90 \times 10^{-5} \text{ dB/m}$$

From (3.3.6) with $P_a = 0$, we write

$$\alpha_{rm} = \frac{SA'f_{rm}f^2}{f^2 + f_{rm}^2}$$

From Fig. A3.2.3 or (A3.2.27), $f_{rm} = 59.2$ kHz. Therefore

$$\alpha_{rm} = \frac{(35)(2.03 \times 10^{-5})(59.2)(10)^2}{10^2 + (59.2)^2} = 1.17 \times 10^{-3} \text{ dB/m}$$

From (3.3.6) we have

$$\alpha_{rb} = \frac{A''f_{rb}f^2}{f^2 + f_{rb}^2}$$

From Fig. A3.2.4 or (A3.2.29), $f_{rb} = 0.9$ kHz. Therefore

$$\alpha_{rb} = \frac{(1.2 \times 10^{-4})(0.9)(10)^2}{10^2 + (0.9)^2} = 1.07 \times 10^{-4} \text{ dB/m}$$

Adding the three components yields

$$\alpha_s = \alpha_F + \alpha_{rm} + \alpha_{rb} = 1.32 \times 10^{-3} \text{ dB/m}$$

Note that the attenuation is higher than for 14°C water (Fig. 3.3.1) principally because of the shift of f_{rm} to a lower frequency.

3.4. THE SONAR EQUATION

All the characteristics of the sonar system, the sound transmission, and the scattering characteristics of objects are combined in the sonar equation. For the present the system factor is simply the source pressure P_0 at R_0 because we assume an omnidirectional source and receiver. The sound transmission includes all the effects of spreading loss and absorption loss. We now calculate the sonar equation for one-way transmission.

On including the absorption loss in (3.2.26) we have

$$P^2 = P_0^2 \frac{R_0^2 \Delta\theta}{rh} \frac{\rho c}{\rho_0 c_0} \frac{|\sin\theta_0|}{|\sin\theta|} 10^{-\alpha(R-R_0)/10} \tag{3.4.1}$$

where α and R are measured along the ray path.

The sonar equation is usually expressed in decibels after all the quantities have been made dimensionless by the use of reference distances and reference sound pressures. Expressed in decibels, the sonar equation for transmission to a distant point ($R \gg R_0$) is,

$$SPL = SL - TL \tag{3.4.2}$$

where

$$SPL = 10 \log_{10} \left(\frac{P}{P_r}\right)^2 \quad \text{dB re } P_r \tag{3.4.3}$$

$$SL = 10 \log_{10} \left(\frac{P_0}{P_r}\right)^2 \quad \text{dB re } P_r \tag{3.4.4}$$

or, using (3.1.11) with Π in water, $\rho_0 = 10^3 \text{ kg/m}^3$, $c_0 = 1500 \text{ m/s}$,

$$SL = 10 \log \Pi + 50.8 \text{ dB re } 1 \text{ Pa}.$$

$$TL = 10 \log_{10} \left(\frac{r}{R_0}\right) + 10 \log_{10} \frac{h}{R_0 \Delta\theta}$$

$$- 10 \log_{10} \frac{\rho c}{\rho_0 c_0} - 10 \log_{10} \frac{\sin\theta_0}{\sin\theta} \tag{3.4.5}$$

$$+ \alpha R \text{ dB}$$

where P_r is the reference pressure, R_0 is the reference distance, SL refers to *source level*, TL to *transmission loss*, and SPL to *sound pressure level* at the field position. It is customary to use 1 m as the reference distance R_0.

As a historical inheritance from psychological acoustics, some of the older underwater sound literature used the threshold of hearing,

0.0002 dyne/cm^2, for the reference pressure. During the 1950s and 1960s, 1 dyne/cm$^2 = 1$ μbar was popular. In mks units the reference pressure is 1 N/m^2, the pascal (or Pa). In the 1970s it appears that the micropascal (μPa: 10^{-6} N/m^2) will be popular.

When the medium is homogeneous and the spreading is spherical, (3.4.5) reduces to (3.3.8) with $R \gg R_0$, $TL = 20 \log_{10} \dfrac{R}{R_0} + \alpha R$.

Equations (3.4.1), (3.4.2) are for ray paths that are completely within the water. When the ray paths reflect at the bottom and surface, the amplitude of the reflected signal is the amplitude of the incident signal multiplied by the reflection coefficient. Presumably the local interface is plane and is much larger than the first few Fresnel zones. It is customary to insert the reflection process into the sonar equation by defining the *bottom loss*

$$BL \equiv -20 \log_{10} \mathcal{R}_{12} \tag{3.4.6}$$

The *SPL* is

$$SPL = SL - TL - BL \tag{3.4.7}$$

If there are several reflections at the boundaries, we have

$$BL = BL_1 + BL_2 + \cdots \tag{3.4.8}$$

where BL_1, BL_2, ..., are for the reflections at each local area.

The scattering characteristics of objects are also specified in decibels. As an example of the use of the sonar equation in a backscattering experiment, we let the transmission loss from the source to the object be TL_1, the loss from the object to the receiver be TL_2, and the scattering characteristic of the object be the target strength *TS*. The sound pressure level at the receiver is then given by the following form of the sonar equation:

$$SPL = SL - TL_1 + TS - TL_2 \tag{3.4.9}$$

Illustrative Problem. We compute the intensity of the signal that returns to the surface for the Arctic condition suggested in the illustrative problem in Section 3.2. The source power is assumed to be 1000 W. Reflections at the free surface are ignored. The frequency is 10 kHz.

Solution. The loss of intensity due to divergence is estimated from the spreading of closely spaced rays. It is more convenient to compute the spreading of the rays as they cross $z = 0$. Hence the sonar equation, with the aid of (3.2.24), (3.2.25), (3.2.27), and (3.4.5), is found to be

$$SPL = SL - TL$$

$$SL = 10 \log_{10} \Pi + 10 \log_{10} (\rho_0 c_0) - 10 \log_{10} (4\pi)$$

$$TL = 10 \log_{10}\left(\frac{r}{R_0}\right) + 10 \log_{10}\left(\frac{\Delta r}{R_0 \Delta \theta}\right)$$

$$- 10 \log_{10} \frac{\sin \theta_0}{\cos \theta} - 10 \log_{10}\left(\frac{\rho c}{\rho_0 c_0}\right) + \alpha R$$

Note $\theta = \theta_0$ and $\rho c = \rho_0 c_0$ because the rays return to the surface; mks units are used throughout. The first ray is θ_1. The second ray is designated as θ_1', where $\theta_1' = \theta_1 - 0.01$ rad

$\sin \theta_1 = 0.9784$	$\theta_1 = 1.3626$ rad $= 78.07°$
$\cos \theta_1 = 0.2067$	$\theta_1' = \theta_1 - \Delta\theta = 1.3526$ rad
$\Delta\theta = 0.01$ rad	$\cos \theta_2 = 0$
$\sin \theta_1' = 0.9763$	$\cos \theta_2' = 0$
$\cos \theta_1' = 0.2165$	$b = 0.016 \text{ s}^{-1}$
$a' = 6.737 \times 10^{-4}$ s/m	$c_0 = 1449$ m/s
$a = 6.752 \times 10^{-4}$ s/m	

With the aid of (3.2.19), the surface to surface distances are found to be

$$r = \frac{2(\cos \theta_1 - \cos \theta_2)}{ab}$$

$$r = 38.28 \times 10^3 \text{ m}$$

$$r' = 40.17 \times 10^3 \text{ m}$$

$$\Delta r = 1.896 \times 10^3 \text{ m}$$

Note. Assume that

$$\rho_0 = 1.035 \times 10^3 \text{ kg/m}^3$$

$$c_0 = 1449 \text{ m/s}$$

Substitution into the sonar equation with $\alpha = 0$ yields

$$SL = 10 \log_{10} \Pi + 50.8 \text{ dB, } \Pi \text{ in watts}$$

$$= 80.8 \text{ dB re 1 Pa} = 200.8 \text{ dB re 1 } \mu\text{ Pa}$$

$$TL = 92.1 \text{ dB (spreading loss only)}$$

$$SPL = -11.3 \text{ dB re 1 Pa} = +108.7 \text{ dB re 1 } \mu\text{Pa}$$

$$P = 0.272 \text{ Pa}$$

This can be compared to the spherical spreading loss in a uniform medium: For a slant range R equal to the horizontal range r, the loss would be

$$TL = 20 \log_{10}(38.28 \times 10^3) = 91.7 \text{ dB re 1 m}$$

For the same horizontal range r, at the same initial angle $\theta_1 = 78.07°$, the slant range in a uniform medium would be

$$R = \frac{r}{\cos \theta} = 1.852 \times 10^5 \text{ m}$$

and the transmission loss would be

$$TL = 20 \log_{10}(1.852 \times 10^5) = 105.4 \text{ dB re 1 m}$$

To calculate the absorption loss αR we need the path length along the arc of the circle that starts and ends at incidence angle $78.1°$. The grazing angles are $11.9°$ and the angle subtending the arc is $\phi = 23.8° = 0.415$ rad. The radius of the arc is

$$\frac{1}{ab} = \frac{1}{(6.75 \times 10^{-4} \text{ s/m})(0.016 \text{ s}^{-1})} = 9.26 \times 10^4 \text{ m}$$

The arc length $s = (\text{radius})(\phi) = (9.26 \times 10^4 \text{ m})(0.415) = 3.85 \times 10^4 \text{ m}$. From the illustrative problem in Section 3.3 we have

$$\alpha(10 \text{ kHz}) = 1.32 \times 10^{-3} \text{ dB/m}$$

The attenuation loss is

$$\alpha R = (1.32 \times 10^{-3})(3.85 \times 10^4) = 50.7 \text{ dB}$$

Adding the losses, we find the total transmission loss,

$$TL = 92.1 + 50.7 = 142.8 \text{ dB}$$

PROBLEMS

Problem 3.1

3.1.1.   A point source radiates 1000 W of acoustic power. Plot a graph of the rms pressure as a function of range from 1 to 10^4 m. A log-log graph is suggested.

Problems 3.2

3.2.1. Ray trace calculation. The speed-depth profile for a point in the South Pacific has been approximated by the three linear segments. The parameters of the four points defining the segments are as follows:

	Speed (m/s)	Depth (m)
z	1495	0
z_1	1495	500
z_2	1485	1000
z_3	1520	4000

The problem is to compute the ray paths and travel times of rays incident at z at angles with the normal of 10°, 30°, 60°, 70°, 80°, and 90°. The travel paths are from z to z_1, z_1 to z_2, and z_2 to upper and lower turning depths.

3.2.2. The source is above an interface in a medium having sound speed c and the receiver is below the interface where the sound speed is c_2. The source and receiver are displaced a horizontal distance x from each other. Use calculus to show that the minimum time for a ray path to go from the source to a point on the interface then to the receiver yields the Snell's law path.

Problems 3.3

3.3.1. Determine the attenuation constants (dB/m) for 0°C seawater at the following frequencies: $f = 1$, 12.5, 25, and 50 kHz.

3.3.2. Determine the attenuation constants (dB/m) for 18°C freshwater at the following frequencies: $f = 1$, 12.5, 25, 50, and 100 kHz.

3.3.3. You have equipment that can transmit omnidirectionally at the power of 1000 W. Assuming a received signal of 1 Pa, over approximately what distance would you transmit in seawater for the frequencies of Problem 3.3.1?

3.3.4. The same equipment as in Problem 3.3.3 is used in freshwater. For a 1 Pa signal, what are the ranges for the frequencies of Problem 3.3.2?

3.3.5. Using the attenuation constants found in Problem 3.3.1, calculate and tabulate for comparison the loss due to spherical divergence from 1 to 10, 100, 1000, and 10,000 m; then compare with the loss due to absorption to the same distances for the frequencies 1, 12.5, 25, and 50 kHz.

3.3.6. Calculate the decibel correction required to convert a signal level from dB re 1 μbar to dB re 1 Pa.

3.3.7. Prove that the loss per wavelength $\alpha\lambda$ is a peak when $f = f_{rm}$ in the $MgSO_4$ relaxation process. Is the situation different for boric acid?

3.3.8. Use (3.2.6) and Figs. A3.2.2, A3.2.3, and A3.2.4 to draw the attenuation curve (dB/m) versus frequency 0.1 kHz to 10,000 kHz for $S = 35$ ppt and $T = 10°C$.

3.3.9. Rewrite (3.2.6) in the generalization of (3.2.5)

$$\alpha = \frac{4.34}{\rho_F c_F^3}\left(\frac{4\mu_F}{3} + \mu_F' + \mu_m' + \mu_b'\right)\omega^2$$

and identify the bulk viscosities due to $MgSO_4$ and boric acid μ_m' and μ_b'.

3.3.10. Calculate and graph the variation of α_S with temperature for the range 0 to 30°C at frequency 10 kHz.

SUGGESTED READING

R. T. Beyer and S. V. Lecher, *Physical Ultrasonics*, Academic Press, New York, 1969. A general text in which acoustics in liquids and nonlinear acoustics are of particular interest to marine acousticians.

M. Born and E. Wolf, *Principles of Optics*, Pergamon Press, Oxford, rev. 1965. Chapter III gives foundation of geometrical optics and ray trace methods. Section 8.8 gives the three-dimensional intensity near a focus. This is a basic reference book in theoretical physics.

L. M. Brekhovskikh, *Waves in Layered Media*, Academic Press, New York, 1960. Detailed calculations of the field near caustics, Section 38.4–38.6. Section 41 contains shadows and transmission in a thin surface layer. This is a basic reference. The combination of ray calculations and plane wave reflection coefficients is commonly used. Section 19 gives the applicability of the plane wave reflection coefficients when the wave front is spherical.

K. F. Herzfeld and T. A. Litovitz, *Absorption and Dispersion of Ultrasonic Waves*, Academic Press, New York, 1959. The relaxation processes are discussed in detail for a large variety of materials.

C. C. Leroy, "Sound propagation in the Mediterranean Sea," Chapter 11, in V. M. Albers, Ed., *Underwater Acoustics*, Vol. 2, Plenum Press, New York, 1967. A description of experimental measurements of low frequency sound attenuation and pulse deformations for turning rays. Finds a $\pi/2$ relative phase shift for rays that are turned by strong gradients of the sound speed. The pulse deformation or phase shift of pulses turned at weak gradients is small. Evidence of a low frequency relaxation process $f_r \simeq 1$ kHz.

Physics of Sound in the Sea. Originally 1946, reprinted by U.S. Navy, Government Printing Office, 1969. Excellent discussions of basic underwater sound theory in the first three chapters. The mathematics are fairly simple and they emphasize the physical phenomena. Comparisons of theory and experiment appear in later chapters. Theoretical explanations of the attenuations are now known to be inadequate.

I. Tolstoy, "Phase changes and pulse deformation in acoustics," *J. Acoust. Soc. Am.* **44,** 675–683 (1968). Shows the process of caustic formation and the turning of the ray path by a strong gradient of the sound speed give phase shifts of $-\pi/2$. The interpretation of theory and experimental data is controversial; see Tolstoy, *J. Acoust. Soc. Am.* **43,** 380 (1968) and R. M. Barash, *J. Acoust. Soc. Am.* **43,** 378–380 (1968).

I. Tolstoy, *Wave Propagation,* McGraw-Hill, New York, 1973. Minimizes the level of his mathematics and uses much physical insight. Section 3.4 contains careful discussions of the use of the WKB approximation for wave ray calculations in media having discrete and continuous variation of the sound speed.

FOUR

SIGNALS, FILTERS, AND RANDOM FUNCTIONS

We need ways to express the characteristics of signals, noise, and the sonar system when our signal is more complicated than the sinusoidal signal. Whether we do it analytically or conceptually, the representation of any signal as the sum of a set of sinusoidal waves having different frequencies is extremely useful.

4.1. FOURIER REPRESENTATION

The signal $x(t)$ can be written

$$x(t) = \sum_m X'_m \cos (2\pi m f_1 t) + X''_m \sin (2\pi m f_1 t)$$

or recalling $\exp (i\theta) = \cos \theta + i \sin \theta$

$$x(t) = \sum_m X_m \exp (i2\pi m f_1 t)$$

$$m = 0, \pm 1, \pm 2, \ldots$$

where X'_m, X''_m, and X_m are complex amplitude coefficients and f_1 is the fundamental frequency. This is the Fourier series representation of the signal. The components X_m comprise the "spectrum" of the signal. An

110

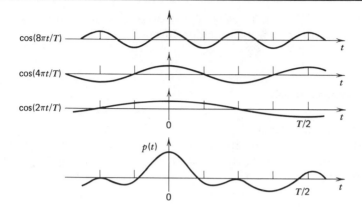

Figure 4.1.1. Synthesis of $p(t)$ using three cosine waves of frequency $1/T$, $2/T$, $4/T$, $f_1 = 1/T$.

example of using three cosine waves to synthesize a signal appears in Fig. 4.1.1. If we were to carry this example further, we would find that $x(t)$ consists of segments that repeat at the fundamental frequency f_1 and that *it repeats periodically forever.* The simplest periodic signals are the sinusoids $\sin(2\pi ft)$, $\cos(2\pi ft)$, and $\exp(i2\pi ft)$. Transmitted and received (pressure) waves that have this time dependence are often called *carrier waves* and are abbreviated CW.

We use the Fourier representation because it is a convenient way to measure the characteristics of a sonar system and to do analytical calculations. To simplify our discussion, we describe analogue signals in this chapter and put the corresponding description of digital signals in Appendix A4. We use some results from the appendix and refer to equations there.

The pressure wave is converted to an electrical signal at the receiving hydrophone. The electrical signal has the same time variations as the pressure wave and is called the *analogue* signal. The electrical signal is processed by electrical circuits in the sonar system. For digital processing the electrical signal is sampled and converted into a string of numbers for input to the digital computer. The sampling and digitizing process is called *analogue* to *digital conversion* (A/D conversion).

Whether your system is primarily analogue or digital, at some time you will sample data. The rule for the data sampling operation is: *the sampling frequency must be greater than twice the highest frequency component in the signal.* This minimum sampling frequency is called the *Nyquist rate.* This rule is very general and also applies to spatial sampling of the ocean environment. When it is obeyed, the original signal can be recovered from the sample by an interpolation formula.

4.1.1. Periodic Signals

We designate the signal as $g(t)$. The periodic signal repeats itself every T_1 seconds, $g(t) = g(t + T_1) = g(t + 2T_1) = \cdots$. When we measure a periodic signal, we often average $|g(t)|^2$ over one period. Denoting the averaging operation by $\langle \ \rangle$, we write

$$\langle |g(t)|^2 \rangle = \frac{1}{T_1} \int_0^{T_1} |g(t)|^2 \, dt \qquad (4.1.1)$$

where $|g(t)|^2 = g^*(t)g(t)$ and the symbol $*$ represents the complex conjugate.

If $g(t)$ is an analogue signal in volts, the mean square signal has the units volts squared.

The frequency representation of $g(t)$ is

$$g(t) = \sum_{m=-\infty}^{\infty} G_m \exp(i2\pi mf_1 t) \qquad (4.1.2)$$

where

$$G_m = \frac{1}{T_1} \int_0^{T_1} g(t) \exp(-i2\pi mf_1 t) \, dt$$

$$f_1 = \frac{1}{T_1} \qquad (4.1.3)$$

The G_m's are the complex amplitude coefficients of the sinusoidal waves $\exp(i2\pi mf_1 t)$. We use the capital to denote the coefficient and the subscript m for the harmonic number of the sinusoidal wave. When $g(t)$ is in volts, the units of G_m also are volts.

For a simple example, let $g(t) = \cos(2\pi f_1 t)$. Evaluation of (4.1.2) using $\cos(2\pi f_1 t) = [\exp(i2\pi f_1 t) + \exp(-i2\pi f_1 t)]/2$ gives $G_1 = G_{-1} = \frac{1}{2}$.

The introduction of negative frequencies can be considered as an artifact of the Fourier expansion. They are the consequence of allowing m to have positive and negative integer values. In this context, the frequency f_1 is positive.

Illustrative Problem. Calculate the amplitude coefficients for a signal that repeats at the period T_1. For the first period,

$$g(t) = 1 \qquad \frac{-\Delta t}{2} \le t \le \frac{\Delta t}{2}$$

$$g(t) = 0 \qquad \frac{\Delta t}{2} \le t \le T_1 - \frac{\Delta t}{2}$$

Using (4.1.3) and shifting the integration limits to simplify the calculation

$$G_m = \frac{1}{T_1} \int_{-\Delta t/2}^{\Delta t/2} (1) \exp\left(-i2\pi m f_1 t\right) dt$$

$$= \frac{\exp\left(-i2\pi m f_1 t\right)}{-i2\pi m f_1 T_1} \Bigg|_{-\Delta t/2}^{\Delta t/2}$$

$$= \frac{\exp\left[i\pi m f_1 \Delta t\right] - \exp\left(-i\pi m f_1 \Delta t\right)]}{i2\pi m f_1 T_1}$$

Using $f_1 = 1/T_1$, we write

$$G_m = \frac{1}{m\pi} \sin \frac{m\pi \Delta t}{T_1}$$

4.1.2 Transient Signals

Most sonar signals are transients. They occur for a brief time, then are quiet. Designating the signal as $g(t)$, the integral of the absolute square of $g(t)$ is finite.

$$I_{gg} = \int_{-\infty}^{\infty} |g(t)|^2 \, dt = \text{finite value} \qquad (4.1.4)$$

where the double subscript indicates the integral of the squared signal. The integral I_{gg} has the units of volts squared · seconds, when $g(t)$ is in volts. The representation of $g(t)$ as the sum of sinusoidal waves is the same as for a periodic signal (4.1.2) except that the sum is replaced by an integral. For transient functions we follow the convention that $G(f)$ is the spectrum of $g(t)$, $X(f)$ is the spectrum of $x(t)$, and so on. This is analogous to our notation of G_m being the amplitude coefficient of $g(t)$ when $g(t)$ is periodic. The integral expression for $g(t)$ is

$$g(t) = \int_{-\infty}^{\infty} G(f) \exp\left(i2\pi f t\right) df \qquad \textbf{(4.1.5)}$$

where the amplitude coefficients G_m have been replaced by $G(f)$, the spectrum of the signal. For $g(t)$ in volts, $G(f)$ has the units of volts per hertz or volt $\cdot$ seconds. The companion expression is

$$G(f) = \int_{-\infty}^{\infty} g(t) \exp(-i2\pi ft)\, dt \qquad \textbf{(4.1.6)}$$

These two integrals, called a "transform pair," show that given a $g(t)$, we can calculate a spectrum, or vice versa. The interrelationship of $G(f)$ and G_m is given in (A4.2.5) and (A4.2.6), and is $T_1 G_m \rightarrow G(f)$ and $mf_1 \rightarrow f$.

You may have noticed that the frequencies go from $-\infty$ to $+\infty$ in (4.1.2) and (4.1.5). This comes directly from the analytical representation of $g(t)$. In analogue systems, the frequencies of signals from oscillators are assumed to be positive. We only display the spectra for positive frequencies in the figures in this chapter. We ignore the phases of the sinusoidal components and show the absolute values of $|G_m|$ and $|G(f)|$.

Examples of several $g(t)$ and their spectra are given in Fig. 4.1.2. The short or gated sine wave, called a "ping," is the signal for many sonar systems. The maximum of $G(f)$ occurs at the frequency of the sine wave. Long pings have narrow $G(f)$ and short pings have wide $G(f)$.

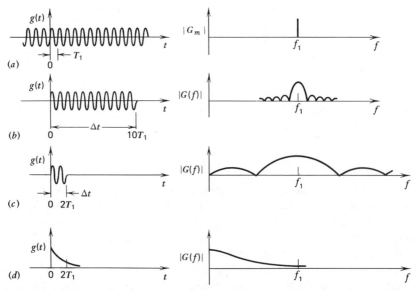

Figure 4.1.2. Examples of signals and their spectra. The same time or frequency scales are used for all $g(t)$ or their spectra. The vertical units are arbitrary. Only positive frequencies are shown. (a) CW; (b), (c) short pings; (d) explosive source, $T_e = T_1$.

Illustrative Problem. The transient signal is on from $-\Delta t/2$ to $\Delta t/2$ and is zero all the rest of the time.

$$g(t) = 1 \qquad \frac{-\Delta t}{2} \le t \le \frac{\Delta t}{2}$$

$$= 0 \qquad \text{otherwise}$$

Using (4.1.4) we write

$$\int_{-\Delta t/2}^{\Delta t/2} |g(t)|^2 \, dt = \Delta t$$

and the condition (4.1.4) is verified. Now we can calculate $G(f)$ using (4.1.6).

$$G(f) = \int_{-\Delta t/2}^{\Delta t/2} (1) \exp(-i2\pi ft) \, dt$$

$$G(f) = \frac{\exp(i\pi f \Delta t) - \exp(-i\pi f \Delta t)}{i2\pi f}$$

$$G(f) = \frac{\sin(\pi f \Delta t)}{\pi f}$$

Sometimes we wish to compare $G(f)$ and the corresponding values of g_m when $f = mf_1$. From (A4.2.5) we find

$$G(mf_1) \equiv T_1 G_m \qquad\qquad \textbf{(4.1.7)}$$

Thus

$$G_m = \frac{G(mf_1)}{T_1} = \frac{\sin(\pi mf_1 \Delta t)}{\pi mf_1 T_1}$$

$$G_m = \frac{\sin(m\pi \Delta t/T_1)}{m\pi}$$

Comparing with the illustrative problem in Section 4.1.1, we conclude that the evaluation of the infinite integral for the single transient gives the envelope of the G_m for the same transient when it repeats periodically.

4.2. FILTERS AND NOISE

4.2.1. Input and Output of Filters

Electronic and other analogue filters are used in sonar systems. To avoid a discussion of circuit analysis, we treat them as "black boxes." The filters are linear; that is doubling the input voltage doubles the output voltage and the sum of two signals at the input gives the sum at the output.

It is customary to describe the operation of a filter in terms of its frequency response. A direct way of measuring the response is to use an oscillator at the input and a meter or oscilloscope to measure the input and output voltages (Fig. 4.2.1), where e_i is the input voltage, e_o is the output, and the ratio e_o/e_i is the response $F(f)$. The analogue filter operations are sketched in the figure; f_{co} is the so-called cutoff frequency. For a high pass filter, the output of a CW signal is relatively small for frequencies less than f_{co}. Commonly, f_{co} is the frequency at which $(e_o/e_i)^2$ is $\frac{1}{2}$. Then f_{co} is the "half power" frequency. Other definitions of f_{co}, such as $e_o/e_i = \frac{1}{2}$, are used less often. Usually $F(f)$ is dimensionless.

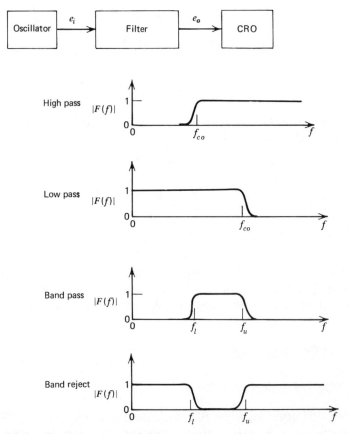

Figure 4.2.1. Common types of filter response, plotted versus the oscillator frequency $f > 0$, where f is the frequency of the oscillator. The output is measured by a cathode ray oscilloscope (CRO).

Considering the signal to be the sum of sinusoidal components, the amplitude and phase of each sinusoidal component is altered by the filter response. For a filter response $F(f)$ and signal spectrum $G(f)$, the spectrum of the output is

$$H(f) = G(f)F(f) \qquad \textbf{(4.2.1)}$$

Just as $g(t)$ and $G(f)$ are the transform pair for the input (4.1.5) and (4.1.6), $h(t)$ and $H(f)$ are the transform pair for the output

$$h(t) = \int_{-\infty}^{\infty} G(f)F(f) \exp(i2\pi ft)\, df$$

$$\textbf{(4.2.2)}$$

$$h(t) = \int_{-\infty}^{\infty} H(f) \exp(i2\pi ft)\, df$$

Examples of the effect of filters on $G(f)$ and $g(t)$ are presented in Fig. 4.2.2. We deal with analogue and digital filter input-output relations in Section A4.4.

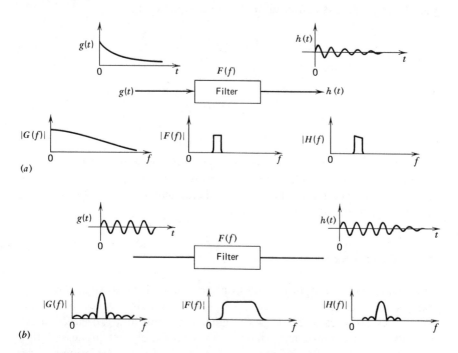

Figure 4.2.2. Examples of input and output signals for filters. (a) Wide band signal and narrow filter. (b) Narrow band signal and wide filter.

The integral squared signal at the output of a filter is

$$I_{hh} = \int_{-\infty}^{\infty} |h(t)|^2 \, dt$$

Using (A4.6.12) I_{hh} is

$$I_{hh} = \int_{-\infty}^{\infty} |G(f)|^2 |F(f)|^2 \, df \qquad \textbf{(4.2.3)}$$

$$= 2 \int_{0}^{\infty} |G(f)|^2 |F(f)|^2 \, df$$

When $F(f)$ is very narrow and $|G(f)|^2$ is nearly constant over the pass band of $F(f)$ we replace $|G(f)|^2$ by its average value and remove it from the integral.

$$I_{hh} = 2\langle |G(f)|^2 \rangle \int_{0}^{\infty} |F(f)|^2 \, df \qquad (4.2.4)$$

For narrow $F(f)$ the integral of $|F(f)|^2$ can be replaced by an ideal band pass filter F_I that is centered at the midfrequency f_o.

$$F_I = 1 \qquad \text{for} \quad f_o - \frac{\Delta f}{2} \le f \le f_o + \frac{\Delta f}{2} \qquad (4.2.5)$$

$$= 0 \qquad \text{otherwise}$$

$$\int_{0}^{\infty} |F(f)|^2 \, df = \Delta f$$

$$I_{hh} = 2 \Delta f \langle |G(f)|^2 \rangle \qquad \textbf{(4.2.6)}$$

This equation is used to analyze analogue measurements of signals. The signal is passed through a set of narrow band pass filters and I_{hh} is measured for each filter. Then $\langle |G(f)|^2 \rangle$ is calculated from (4.2.6) by using Δf and the result of measurements.

4.2.2 Random Signals

If we listen to the output of a hydrophone, in addition to signals we hear extraneous sounds from sources such as fish, breaking waves, boats, and rain. These sounds interfere with signal reception and are called *noise*. Usually the noise pressures fluctuate unpredictably. A process that is unpredictable is said to be *random*. Both desired and interfering sources may transmit random pressure signals.

All measurements of random signals are made by similar procedures. Since the random or noiselike signal is of long duration, we time average the noise signal squared. The mean squared noise signal is

$$\langle |n(t)|^2 \rangle \equiv \frac{1}{T} \int_0^T |n(t)|^2 \, dt \qquad (4.2.7)$$

where T is very much greater than the period of the lowest frequency component of the noise.

The output of the filter for noise input is $h_n(t)$. Analogous to (4.2.3) we write the mean square output for noise

$$\langle |h_n(t)|^2 \rangle = 2 \int_0^\infty S_{nn}(f) \, |F(f)|^2 \, df \qquad (4.2.8)$$

where $S_{nn}(f)$ is the spectral density of the noise. $S_{nn}(f)$ has the units of volts squared per hertz when the signal is in volts. Acoustic noise pressures are in pascals or micropascals, and $S_{nn}(f)$ has the units of pascals squared per hertz or micropascals squared per hertz. The mean squared pressure in a 1 Hz bandwidth is $2S_{nn}(f)$, positive frequencies.

An analogue system for measuring the spectral density of noise (random) signals is shown in Fig. 4.2.3. In analogue measurements, we can regard the spectral components for negative frequencies as being folded on top of those for positive frequencies and that both positive and negative frequency components are measured at the same time. The output of the mth filter is $2\Delta f S_{nn}(f_m)$ where we include the functional dependence f_m. Sometimes, and particularly in analogue measurements, the factor 2 is absorbed and the spectral density for positive frequencies only is written as (Section A4.6)

$$P_{nn}(f_m) = 2 S_{nn}(f_m)$$

For simplicity in comparing results of digital analysis with analogue measurements, we use $2S_{nn}(f)$ to report the spectral density of noise. Noise measurements are also reported as spectrum levels in dB with reference to $1 \ \mu Pa^2/Hz$

$$\text{Spectrum level} \equiv 10 \log_{10} \left[\frac{2 S_{nn}(f)}{\mu Pa^2/Hz} \right] dB \qquad (4.2.9)$$

In the ordinary sense, a random signal cannot be expanded in a Fourier series because it is not periodic. We can use the following artifice to make an expansion of a random signal: Suppose a portion of a random signal $n(t)$ is recorded between the times $t = 0$ and $t = T$. Assuming that T is much larger than any other time of interest, we use the measured values of $n(t)$ between 0 and T to *construct* a *periodic signal* $n(t) = n(t + mT)$ where m is an integer. This signal can be expanded as a Fourier series. This procedure has become

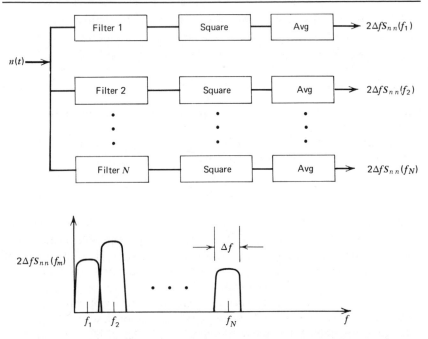

Figure 4.2.3. Analogue measurements of the spectral density of a noise input $n(t)$; f_m is the center frequency. The factor 2 is for consistency with standard digital computational procedure. Most analogue measurements yield $P_{nn}(f_m) = 2S_{nn}(f_m)$. The averaging time is much much greater than $1/\Delta f$.

the basis of digital techniques for the analysis of random signals. Generally the digital techniques use the fast Fourier transform FFT.

4.2.3 Noise

The ocean is noisy. Our equipment is noisy. The ships that carry us are noisy. Except for the few who like to measure the background noise in the ocean, all this noise interferes with signal transmissions. Gross classifications of noise are (*a*) electrical self-noise of the sonar system including thermal or *Johnson noise*, (*b*) ambient or background noise in the ocean (including ships), and (*c*) noise voltages generated by the ship and transducer when they are moving through the water (mechanical and hydrodynamic noise).

The system self-noise—the noise output when there is no input—comes from the amplifiers and filters. In a well-designed sonar, most of the system noise comes from the first amplifier in the receiver. The lower limit of the self-noise is thermal-resistor noise caused by spontaneous fluctuations of voltage in a resistor. The mean square electrical noise power, or Johnson noise power, is proportional to the absolute temperature and bandwidth. Often the Johnson noise level is used as a reference for specifying the performance of amplifiers and transducers. At low frequencies (i.e., <5 Hz), there are fluctuations of the current through transistors that give apparent noise levels that may be larger than the

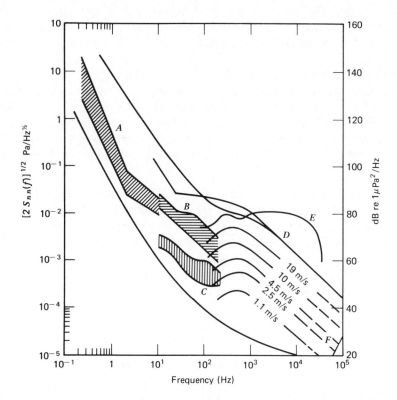

Figure 4.2.4. Spectral density of ocean noise, $f > 0$. The analogue measurements are for positive frequency; the measurements are reduced to 1 Hz bandwidths; the ordinates are square root of spectral density $[2S_{nn}(f)]^{1/2}$ and spectrum level $10 \log_{10}[2S_{nn}(f)/(\mu Pa^2/Hz)]$ in decibels. Heavy solid lines indicate empirical maximum and minimum prevailing noise curves: A, seismic; B, ships; C, quiet lake; D, wind; E, rain; F, thermal noise. From *Ocean Acoustics* by Tolstoy and Clay. Copyright 1966, McGraw-Hill Book Company. Used with permission of McGraw-Hill Book Company.

Johnson noise. Careful choice of components and design can reduce input transistor noise.

Ambient or background noise in the ocean is caused by a large number of sources. Every source of sound contributes to the noise level. The amount of contribution depends on the transmission loss between the source and the receiver. Since low frequency sounds usually have smaller transmission losses, we would expect all the sources of low frequency sound in the ocean to be important. In Fig. 4.2.4 the noise in the band less than 10 Hz is attributed to seismic activity (earthquakes). It is believed that the noise in the 10–150 Hz band is due largely to the machinery of distant ships. The noise due to waves, rain, and wind would be expected to depend on the local weather. The dependence on wind speed is important for frequencies above 100 Hz. Ocean thermal noise becomes significant at frequencies above 60 kHz.

Acoustic measurements made from moving ships show three additional kinds of noise: the acoustic noise from the ship, the mechanical vibrations of the receiving transducer, and flow noise. Flow noise is due to the hydrophone moving through local pressure fluctuations in the water. The design and mounting of the transducer to reduce the effects of these kinds of noise is a highly skilled art.

Numerical Example. The system is intended to receive signals having a bandwidth of 1 kHz and a center frequency of 10 kHz. Use the ambient noise data (4.5 m/s wind speed, Fig. 4.2.4) to estimate the noise signal. First we plot $2S_{nn}(f)$ versus frequency on linear scales (Fig. 4.2.5). The mean noise pressure squared is given by (4.2.8). Assume that $F(f) = 1$ for the band 9.5–10.5 kHz and is zero outside. The mean square filter output is the shaded area on the inset.

$$\langle |h_n(t)|^2 \rangle = 2 \int_{f_1}^{f_2} S_{nn}(f)\, df \tag{4.2.10}$$

$$= \left(\frac{1.35 + 1.05}{2} \times 10^{-8} \, \text{Pa}^2/\text{Hz} \right)(1000 \, \text{Hz})$$

$$= 1.2 \times 10^{-5} \, \text{Pa}^2$$

The rms noise is 3.5×10^{-3} Pa.

4.2.4. Filters, Signals, and Noise

Filters are used for a variety of reasons, perhaps the most important being to reduce the effect of noise and unwanted signals. The desired signal and

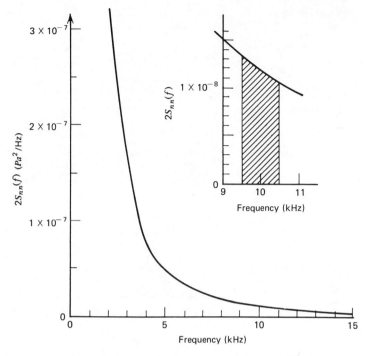

Figure 4.2.5. Spectral density of ambient noise on linear scale, $f > 0$; data from Fig. 4.2.4, 4.5 m/s. *Inset,* enlargement for 10 kHz.

noise having spectral densities $|G(f)|^2$ and $S_{nn}(f)$ respectively are received (Fig. 4.2.6a). The purpose of the filter $F(f)$ is to enhance the signal relative to the noise. This is possible when the signal and noise have different spectral densities, as in Fig. 4.2.6b, c. At the input to the filter, the mean square noise and the mean square signal are proportional to the shaded areas. As drawn, the mean square signal is about two-thirds the mean square noise. Since the spectral density of the signal is largely between the frequencies f_L and f_U, it is reasonable to pass the signal and noise through a filter having these cutoffs. This eliminates contributions of $S_{nn}(f)$ outside the pass band. Comparison of the shaded areas within the pass band indicates that the mean square signal at the output is approximately 2.5 times the mean square noise. From our experience, it is very difficult to identify a signal correctly when the signal is less than the noise. Identifications are more reliable when the mean square signal is more than twice the noise, within the pass band.

The choice of an *approximate filter* to pass the signal and reduce the

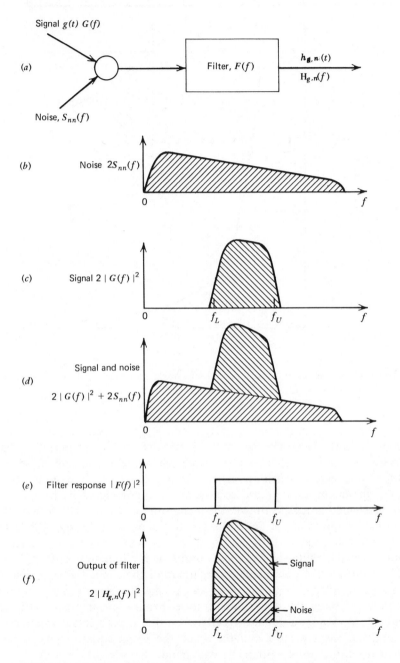

Figure 4.2.6. Signal, noise, and filter action (positive frequencies).

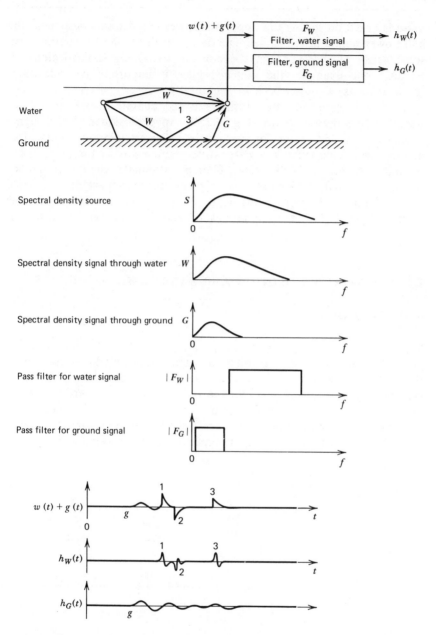

Figure 4.2.7. Filter system for separating signals: $w(t)$ represents the waterborne signals and $g(t)$ the ground path signal labeled G; output of F_W is $h_W(t)$; output of F_G is $h_G(t)$. F_G has a small response to the impulsive nature of $w(t)$.

noise is based on setting the upper and lower cutoff frequencies near the half power points of the signal. The design of filters that are optimum in some sense of the word has become an extremely sophisticated art. Because one example, the *matched filter*, is important in underwater sound systems, we describe it in Section 4.4.

A second important use of filters is to separate signals that have travelled by different paths (Fig. 4.2.7). An explosion makes a signal having a very wide range of frequencies. The attenuation of high frequency signal components in water is very much smaller than it is in the sediment. As Fig. 4.2.7 illustrates, filter F_W attenuates the low frequency subbottom signals and passes the higher frequency part of the signal transmission having paths in the water, F_G passes the low frequency part of both signals. The signals appear as a function of time at the bottom of the figure.

4.3 TEMPORAL RESOLUTION AND BANDWIDTH OF SIGNALS

4.3.1. Temporal Resolution

Suppose we are using a sonar system to count the density of a fish population and we wish to resolve the signals from individual fish. To do this, we must shorten the pulse duration to prevent the signals from running together (Fig. 4.3.1), but the rise and decay times of the transducer, the filter, and the receiver impose a limitation on pulse shortening. We use the error function (also called the Gaussian function) $\exp\left[-t^2/(\Delta t/2)^2\right]$ as a simple approximation to the envelope of the very

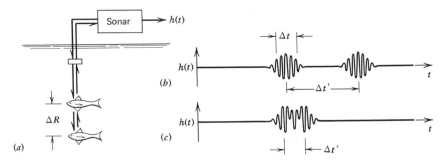

Figure 4.3.1.   Echoes from closely spaced objects. (*a*) The physical situation. (*b*) Separated returns. (*c*) Echoes are beginning to merge and the two signals are barely resolved, $\Delta t = \Delta t'$.

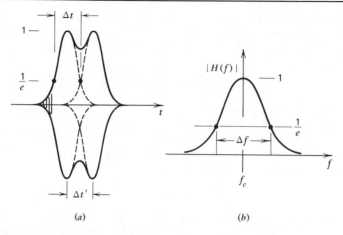

Figure 4.3.2. Resolution of a pair of signals. (*a*) The envelopes of a pair of overlapping signals. The carrier frequency is suppressed for simplicity. (*b*) The spectrum of a single signal.

short ping. The output of the sonar system is

$$h(t) = \exp\left[\frac{-t^2}{(\Delta t/2)^2}\right] \exp(i2\pi f_c t) \qquad (4.3.1)$$

where f_c is the carrier frequency. The nominal ping width is Δt. We find that the resolution does not depend on f_c.

For the range difference ΔR (Fig. 4.3.1) the difference of travel times of the sound scattered at fish a and fish b is

$$\Delta t' = \frac{2\Delta R}{c} \qquad (4.3.2)$$

If $\Delta t \ll \Delta t'$ the scattered pings are resolved. We assume that the signals are just barely resolvable when they are separated by $\Delta t' = \Delta t$ (Fig. 4.3.2).

Applying (4.1.6) to $h(t)$, the spectrum is found to be

$$H(f) = \int_{-\infty}^{\infty} h(t) \exp(-i2\pi ft) \, dt \qquad (4.3.3)$$

The substitution of $h(t)$ and the evaluation of the infinite integral with the aid of an integral table, yields

$$H(f) \sim \exp\left[\frac{-\pi^2(f-f_c)^2 \Delta t^2}{4}\right] \qquad (4.3.4)$$

The bandwidth of a sonar system Δf depends on a rather arbitrary definition of the bandwidth. We use the $1/e$ to $1/e$ specification for Δf as in Fig. 4.3.2b, and

$$H(f) = \exp\left[\frac{-(f-f_c)^2}{(\Delta f/2)^2}\right]$$ **(4.3.5)**

Equating the two expressions for $H(f)$ gives $\Delta f \Delta t = 4/\pi$. This is the result for a "good" signal. Signals having the same bandwidth may have much longer duration. Arbitrarily we use $\Delta f \Delta t = 1$ as a lower limit and write the criterion for resolution

$$\Delta f \Delta t \geq 1$$ **(4.3.6)**

This defines the minimum bandwidth Δf that is needed to permit the identification of objects separated by Δt. This condition, sometimes called the "classical uncertainty principle," is a consequence of the Fourier transformations (4.1.5) and (4.1.6).

Numerical Example. The fish in Fig. 4.3.1 are 0.3 m apart. What minimum bandwidth is required to resolve them? From (4.3.2) we have

$$\Delta t' = \frac{2(0.3)}{1500}$$

$$\Delta t' = 4 \times 10^{-4}\,\text{s}$$

Applying (4.3.6), the minimum bandwidth is $(\Delta t' = \Delta t)$

$$\Delta f \geq (\Delta t)^{-1} = 2500\,\text{Hz}$$

4.3.2. Bandshifting or Heterodyning Operations

We have shown that the temporal resolution of the system depends on the *bandwidth* Δf. When recording sonar signals, only the required bandwidth need be recorded; the carrier frequency does not have to be retained. Bandshifting or heterodyning consists of multiplying the signal by a CW (heterodyning) signal f_H, then filtering the information to pass the desired frequency band, as in Fig. 4.3.3.

For simplicity we use $g(t) = \cos(2\pi f_c t)$ and remember that the spectrum of the signal really extends from f_c by $\pm\Delta f/2$. The multiplication by the local oscillator signal yields the heterodyned signal.

$$h_H(t) = \cos(2\pi f_H t) \cos(2\pi f_c t)$$ **(4.3.7)**

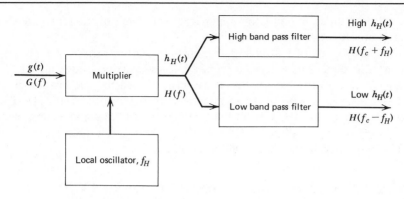

Figure 4.3.3. Bandshift of heterodyning operation.

The product of the cosine functions is rewritten as the sum of two cosines.

$$h_H(t) = 0.5\{\cos\left[2\pi(f_H - f_c)t\right] + \cos\left[2\pi(f_H + f_c)t\right]\} \qquad \textbf{(4.3.8)}$$

After passing $h_H(t)$ through a low pass filter, the center of the frequency band Δf is at $|f_H - f_c|$, as in Fig. 4.3.4b. The upper band is obtained by using a high pass filter. This simple system is adequate as long as the frequency band of the shifted signal is greater than zero. Its virtue is that recording or processing of low frequency bandshifted information is easier (less costly) than it is at the carrier frequency.

Numerical Examples. Let us assume that a sonar system transmits pings of duration 10^{-3} s and has a 50,000 Hz carrier frequency. We wish to tape

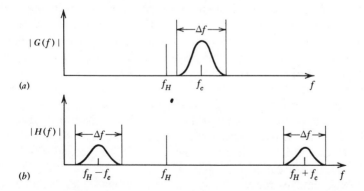

Figure 4.3.4. Bandshifting a signal to different frequency regions (positive frequencies). (a) Original signal. (b) Bandshifted signals. The band pass filter can be adjusted to pass the lower band $f_H - f_c$ or the upper band $f_H + f_c$.

record the echoes. What is the upper frequency limit of the tape recorder that is able to record the echoes?

Obviously a recorder having response to more than 50,000 Hz can record the data. Much less is required if the carrier frequency of the echoes is not of interest, that is, if we wish to measure the envelopes of the signal. A 10^{-3} s ping has an approximate bandwidth of 10^3 Hz. The signal can be heterodyned from the 50,000 Hz band to a convenient band say from 1000 to 2000 Hz. Now the tape recorder requirement is an upper frequency limit of only 2000 or 3000 Hz.

4.4. IMPROVING THE SIGNAL TO NOISE POWER RATIO

Good experiments require signal to noise power ratios that are considerably larger than unity at the output of the receiving system. The signal level is proportional to the amount of energy transmitted. In a simple way, we can improve the signal to noise power ratio by making a bigger "bang" at the source. This is easy when high explosives are chosen for the source. When the source is an electronically driven transducer, the peak power is limited by the amplifier and by cavitation. The only way to increase the transmitted energy is to transmit long pings instead of short ones. For a temporal resolution Δt, we need a bandwidth $\Delta f \geq 1/\Delta t$.

To send a long ping and keep the bandwidth large, we can vary the frequency of the signal during transmission. The swept signal or chirp in Fig. 4.4.1 is an example. The transmission starts at frequency f_1 and sweeps to f_2. The bandwidth is $f_2 - f_1 = \Delta f$. As Fig. 4.4.2a indicates, each short segment of the signal a, b, c, ... is unique and has the travel time T. If we can contrive a circuit or a computer program that adds the peaks at a, b, c, ..., the amplitude of the peak is increased and the temporal resolution becomes the width of the peak. It is difficult to isolate the peaks when the signal level is less than the noise level and we are obliged to operate on the whole signal.

The combination of a tapped delay line and an adding network is a simple way to add the peaks. The delay times are $t_e - t_d$, $t_e - t_c$, and so on (Fig. 4.4.2b). Adding the signal at these time delays aligns the peaks, as in Fig. 4.4.2c. The sum is a very narrow peak having a fixed time delay relative to the initial arrival of t_e. The low amplitude wiggles before and after the peak are called the "side lobes." A particular combination of time delays matches the wave form of a particular signal. This combination is called a "matched filter." If a different signal, such as a "downswept chirp," is passed through an upswept chirp matched filter, the output will be low amplitude wiggles and no peak.

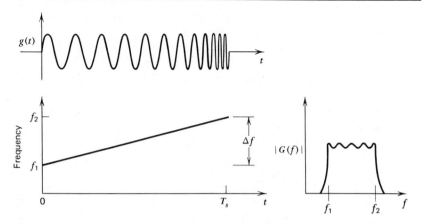

Figure 4.4.1. Swept signal or chirp. (a) The signal $g(t)$. (b) Frequency of the signal as a function of time. (c) Spectrum of the signal: the turn-on and turn-off transients broaden the spectrum outside f_1 and f_2.

Next we calculate the amount of improvement of the signal to noise power ratio due to a matched filter. For simplicity we ignore constant factors such as amplifier gains and attenuations in the filter because they are the same for both the signal and noise wave forms.

The peak output of a matched filter for its matching signal is $(N)(s)$, where s is the amplitude of the chirp signal and N is the number of temporal resolution elements in the signal. Recalling our discussion in Section 4.3.1, the duration of a signal is $\Delta t \geq 1/\Delta f$, where Δf is its frequency bandwidth. Then Δt or $1/\Delta f$ is the minimum duration of the signal and is the temporal resolution element. For a signal duration T_s, the maximum number of temporal resolution elements is $T_s/\Delta t$ or

$$N = T_s \, \Delta f \qquad\qquad (4.4.1)$$

where Δf is $f_2 - f_1$ as in Fig. 4.4.1. The width of the peak is Δt. The matched filter compresses the long transmission T_s into a peak having width Δt. It is a way of achieving a minimum duration signal for a given bandwidth.

When noise is passed into the matched filter, the wave forms at each of the taps are samples of the original noise wave forms at different times. Since the time delays taps were not chosen to match the wave form of the noise, the peaks do not add. For time delays larger than Δt, the samples are sufficiently different that the wave forms add as the sums of squares. (Section A4.6 gives the relationship of the spectral density of the noise and its temporal measure.) Again N is the maximum number of different

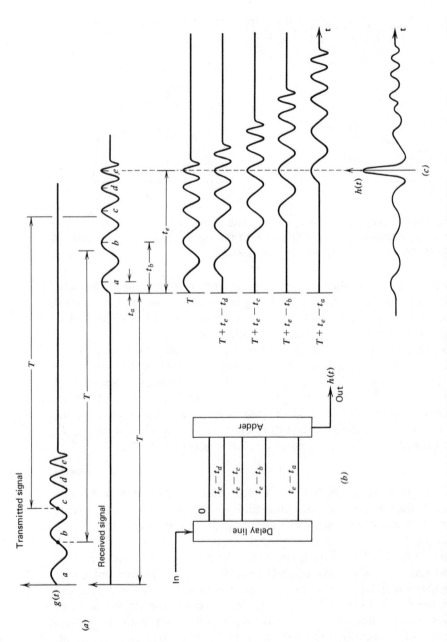

Figure 4.4.2. The chirp transmission. (*a*) Transmitted and received signals. (*b*) Delay line and adder type of filter. (*c*) Addition of the delayed signals forms a peak at *e*.

samples of the wave form of the noise. For rms noise amplitude σ_n, the mean square output of the matched filter is $N\sigma_n^2$.

The peak signal squared is $|N_s|^2$ and the noise is $|N\sigma_n^2|$. The ratio is

$$\frac{|\text{output peak signal}|^2}{\langle\text{noise}^2\rangle} = \frac{|Ns|^2}{N\sigma_n^2} \tag{4.4.2}$$

$$= \frac{N|s|^2}{\sigma_n^2} \tag{4.4.3}$$

where $|s^2|/\sigma_n^2$ is the input signal to noise power ratio. The improvement of the signal to noise power ratio is N or $T_s\,\Delta f$; $T_s\,\Delta f$ is also called the *time bandwidth product*.

We demonstrated in Fig. 4.4.2b, a procedure for matching the combination of a tapped delay line and adder to a signal. This is one way of processing the signal. Other approaches are analytically equivalent, although the hardware or computer programs may differ. The delay line processor and its analytic equivalents are called "matched filters," "correlators," and "time compressors."

One of the equivalent descriptions of the tapped delay line is its frequency response $F_g(f)$. By measurement or much manipulation in Fourier integrals, we find that the "delay line filter" has a frequency response that is proportional to the *complex conjugate* of the *signal spectrum*

$$F_g(f) \sim G^*(f) \tag{4.4.4}$$

Sections A4.5 to A4.7, show that the matched filter $F_g(f)$ is an optimum way to detect the presence of $g(t)$. When the noise spectrum changes over the bandwidth of the signal, $F_g(f)$ is modified to $G^*(f)/S_{nn}(f)$. Even if the phase response cannot be matched, setting the filter $|F(f)| = |G(f)|/S_{nn}(f)$ improves the output signal to noise ratio.

Numerical Example.   The source transmits a chirp signal which has duration 1.6 s and sweeps from 3.3 to 3.7 kHz. For a matched filter, find (*a*) the improvement of the output signal to noise ratio and (*b*) the wave form of the output signal.

(*a*.   The signal to noise improvement is approximately the factor N.

$$N = T_s\,\Delta f$$
$$\Delta f = f_2 - f_1$$
$$N = (1.6)(400) = 640$$

(*b*. The output wave form is given by

$$h(t) = \int_{-\infty}^{\infty} G^*(f)G(f)e^{i2\pi ft}\, df$$

We assume that G is flat within the band f_1 to f_2 and zero outside. For convenience, we write the frequencies as follows:

$$\frac{f_1+f_2}{2} = f_0$$

$$f_1 = f_0 - \frac{\Delta f}{2}$$

$$f_2 = f_0 + \frac{\Delta f}{2}$$

$$|GG^*|^{1/2} = N \qquad \text{for} \quad -f_2 \le f \le -f_1$$
$$f_1 \le f \le f_2$$

$$|GG^*|^{1/2} = 0 \qquad \text{otherwise}$$

$$h(t) = N\int_{-(f_0+\Delta f/2)}^{-(f_0-\Delta f/2)} \exp{(i2\pi ft)}\, df + N\int_{(f_0-\Delta f/2)}^{(f_0+\Delta f/2)} \exp{(i2\pi ft)}\, df$$

Integration and evaluation at the limits gives

$$h(t) = 2N \cos{(2\pi f_0 t)}\, \frac{\sin{(\pi\, \Delta ft)}}{\pi t}$$

The output signal has a carrier frequency f_0 and an envelope of the form $(\sin x)/x$. The null to null width of the envelope is $2/\Delta f$.

PROBLEMS

Problems 4.1

4.1.1. Calculate the first $20 G_m$ for the example in the illustrative problem in Section 4.1.1. Let $\Delta t = 0.01$ s and $T_1 = 0.1$ s. Compare the envelope of the set of G_m with $(\sin x)/x$ Fig. 2.5.1.

4.1.2. Calculate the following.
   (*a*) G_m for

$$g(t) = \exp{\left[\frac{-t}{(\tau)}\right]} \qquad \text{for} \quad 0 \le t \le T_1$$

   $g(t)$ repeats at period T_1.

(b) The $|G_m|^2$ of the G_m for

$$\tau = 0.01 \text{ s} \qquad \text{and} \qquad T_1 = 0.1 \text{ s} \qquad \text{for} \quad m = 0 \text{ to } 10$$

4.1.3. Calculate the spectrum of the explosive shock signal

$$g(t) = = \exp\left(\frac{-t}{\tau}\right) \qquad \text{for} \quad t \geq 0$$

$$g(t) = 0 \qquad \text{for} \quad t < 0$$

(a) Prove that this expression satisfies the condition (4.1.4).
(b) Calculate $G(f)$ and evaluate it for $\tau = 0.01$ s.
(c) Use your values of $G(f)$ and (4.1.7) to calculate and graph the G_m in Problem 4.1.2.

This wave form is called an exponential pulse.

4.1.4. Sometimes signals have a rounded shape that can be described by

$$g(t) = \exp\left[\frac{-t^2}{(\tau/2)^2}\right]$$

(a) Show that $G(f) \sim \exp\left[-\pi^2 \tau^2 f^2 / 4\right]$. This wave form is often called a Gaussian pulse.
(b) Compare graphs of $g(t)$ and $G(f)$ for $\tau = 0.01$ and 0.005 s.

4.1.5. Calculate the following.
(a) The $G(f)$ of

$$g(t) = \exp(2\pi i f_c t) \qquad -\frac{\Delta t}{2} \leq t \leq \frac{\Delta t}{2}$$

$$= 0 \qquad \text{otherwise}$$

Answer. $G(f) = \dfrac{\sin\left[\pi(f_c - f)\,\Delta t\right]}{\pi(f_c - f)}$

This signal is called a ping.

(b) Calculate and graph real $g(t)$ and $G(f)$ for $\Delta t = 0.01$ and 0.005 s, $f_c = 50 \times 10^3$ Hz.

Sketch a few cycles of $g(t)$ and then sketch the envelopes.

Problems 4.2

4.2.1. The signal has an exponential wave form $\exp(-t/\Delta t)$. It is passed through a band pass filter.
(a) Use (4.2.1) to calculate $H(f)$.

(b)   Use (4.2.2) to calculate an *approximate* wave form. (Use both positive and negative frequency components.) Let $\Delta t = 10^{-4}$ s.

$$F(f) = 1 \quad \text{for} \quad f_1 \leq f \leq f_2, \quad -f_2 \leq f \leq -f_1$$
$$F(f) = 0 \quad \text{otherwise,}$$
$$f_1 = 95 \text{ Hz}$$
$$f_2 = 105 \text{ Hz}$$

Graph the output and compare it to the input. Ignore the time shift of one relative to the other.

4.2.2. The apparatus in Fig. 4.2.3 is used to measure a "noise signal." The outputs and filter pass bands are as follows:

Channel	f_L(Hz)	f_H(Hz)	Mean Square Volts (V^2)
1	10	15	3.0
2	25	32	2.5
3	30	40	2.5
4	40	55	2.2

The filters have sharp cutoffs below and above their pass bands and $F(f) = 1$ within the pass band. Estimate the spectral density of the noise.

4.2.3. The output of a commercial spectrum analyzer is in rms volts. Calculate the spectral density when the instrument reads 4.7 V. Assume the frequency bandwidth is 7 Hz centered at 100 Hz.

Problems 4.3

4.3.1. For advanced students. Design a heterodyning and recording system that enables you to recover the signal with its correct phase at its carrier frequency. For the example on p. 129, estimate the cost reduction (i.e. reduction of number of samples) for the heterodyning technique compared to brute force sampling. Hints: Heterodyning is a linear transformation in frequency space. Consider using the same frequency for both the carrier and local oscillator for heterodyning the signal. Do you need to record two heterodyned signals, i.e. heterodyne with cos $2\pi f_c t$ and sin $2\pi f_c t$? Can you heterodyne to transform the recorded signal back to its original carrier?

Problems 4.4

4.4.1. For advanced students. You have the basic theory needed to compute a synthetic seismogram for the reflection of a sonar signal at the (layered) sea floor. Compute $p(t)$ at 0.1 km ranges intervals between 0.1 and 0.5 km for the reflection from a botton at a depth of 200 m. (a) Let $c_o = 1500$ m/s, $\rho_o = 1$ Mg, $c_1 = 1.2c_o$, $\rho_1 = 2\rho_o$. Let the explosive source (section A 4.3) have $T_e = 10^{-2}$s. Look for change of wave form for $p(t)$ for beyond critical reflections. (b) Assume the bottom is layered and $\rho_1 = 1.3\,\rho_o$, $c_1 = c_o$, $h_1 = 15$ m; $\rho_o = 2\rho_o$, $c_2 = 1.2c_o$ and $h_2 = \infty$. Consider (2.10.7) to compute   for the bottom.

Hints: In (A 4.4.2), identify X_m as being the source spectrum and F_m as being the spectrum of the transmission function. Use (A 4.4.4) and (A 4.4.5) to construct X_m and F_m for $m > N/2$.

SUGGESTED READING

R. B. Blackman and J. W. Tukey, *The Measurement of Power Spectra*, Dover, New York, 1959. The use of correlation measurements for the estimation of power spectra. This is a pre-FFT "classic."

E. Oran Brigham, *The Fast Fourier Transform*, Prentice-Hall, Englewood Cliffs, N.J., 1974. A development of finite Fourier transformation theory and the Cooley–Tukey algorithms.

A. V. Oppenheim and R. W. Schafer, *Digital Signal Processing*, Prentice-Hall, Englewood Cliffs, N.J., 1975. The modern computation technology of the fast Fourier transformation (Cooley-Tukey algorithm), to develop all aspects of signal processing and spectral analysis.

E. A. Robinson, *Multichannel Time Series Analysis with Digital Computer Programs*, Holden-Day, San Francisco, 1967. Basic theory and useful programs, such as the fast Fourier transform.

V. V. Solodovnikov, *Introduction to Statistic Dynamics of Automatic Control Systems*, Dover, New York, 1960. An extended discussion of minimum phase and its relation to the Hilbert transformation and the amplitude response of a causal filter.

Y. W. Lee, *Statistical Theory of Communication*. Wiley, New York, 1960. An introduction to signals, probability, and random functions. A careful development of Fourier theory and correlation functions.

FIVE

OPERATING CHARACTERISTICS OF SONAR SYSTEMS

Sensing the ocean beyond visual range requires a sonar system. The question is, what kind of sonar system will do the job? Unless the problem is the same as those others have had, no system specially designed to solve it exists. The alternatives are to design special equipment or to use existing equipment in a different mode of operation. For either alternative the performance specifications of the system must be compared with the requirements. This involves first determining the required signal level, resolution, range, and information rate. These data are translated into signal frequency, source level (relative to the noise level), beam width, ping length, and display system specifications.

5.1. SONAR SYSTEMS

5.1.1. Passive Sonars

The "object" makes its own noise, and the passive sonar is simply a hydrophone that determines the presence and acoustical characteristics of the noise maker. A directional receiver can determine the direction to the source. The signals from several receivers or two parts of one receiver can be combined to determine the location of the noise maker (Fig. 5.1.1).

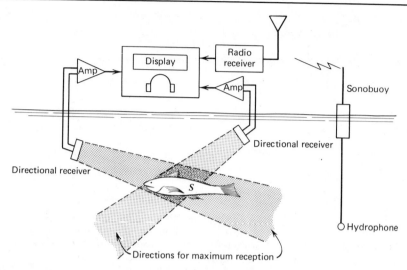

Figure 5.1.1.   Passive sonar. The source S makes a sound, and the signal may be received from directional receivers or hydrophones. It may be displayed versus the direction to the source. Cetaceans (whales, porpoises, and dolphins) make sounds that have an extremely wide frequency range. Just listening to the sound usually helps to identify the source.

Examples of noisy objects are submarines, earthquakes, natural fish, a fish carrying a sound source, and a transponder implanted in the sea floor to aid in position location. People are often adept at recognizing and identifying noises; therefore it helps to listen to an audio output.

5.1.2.   Active, Continuously Transmitting (CW) Sonars

The active system transmits a beam of sound much like a searchlight (Fig. 5.1.2). If something is in the beam, some of the sound is backscattered. The backscattered sound is received by a hydrophone or by the same "transducer," which has been switched to receive. This system may be most useful for the detection of moving objects. Motion of the object causes a Doppler frequency shift of the scattered signal relative to the transmitted frequency. A spectrum analyzer, a frequency counter, or even aural presentation can permit detection of the presence of a moving object in the sound field.

If the source ship is moving, sound scattered by the bottom or by fixed or moving bodies in the water shows a Doppler shift at the receiver.

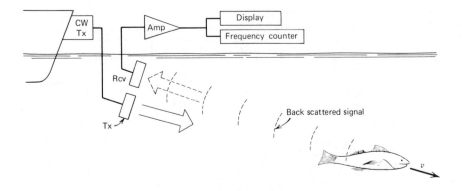

Figure 5.1.2. Continuously transmitting sonar. The transmitter Tx sends a CW-signal. The backscattered signal is picked up by the receiving transducer Rcv. The frequency shift of the scattered signal is proportional to the speed of the object in the direction of the sound path.

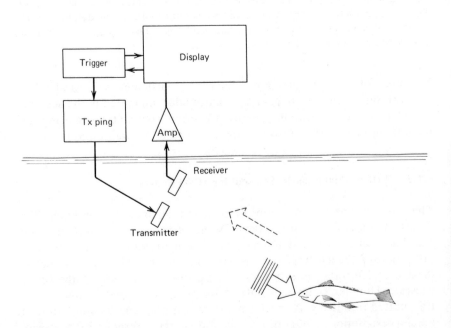

Figure 5.1.3. "Pinging," or active pulse, sonar. The trigger causes the source to transmit a tone burst or ping. The backscattered signal is received, amplified, and displayed. The trigger either starts the display or is provided by the display unit.

140

Doppler navigation equipment uses the frequency shift of backscattered sound from the bottom to measure the ship speed.

5.1.3. Active Pulse Sonars

Pulse sonars emit a short burst of sound energy and listen for the echo (Fig. 5.1.3). The signal is usually displayed as a function of travel time and direction. The common depth sounder is an example of this kind of instrument. At a much more sophisticated level, we can include coded types of transmission and frequency modulation in this class because the signal processor yields a display of travel time and range, much like that for the short ping. Since the output of the matched receiver is a short pulse, we regard the system as having the same operational characteristics as a pulse sonar.

All the systems depend on transducers to change sound signals in the water into electrical signals, or vice versa. Therefore we need to consider how transducers can be selected to give the user the directivity, sensitivity, and frequency response to suit his needs.

5.2. TRANSDUCERS AND THEIR DIRECTIVITIES

Today's most common transducers are blocks of piezoelectric materials that expand, contract, or change shape when electrical voltages are applied, and vice versa. Since 1950, ceramic materials such as barium titanate, lead zirconate, and more exotic mixtures have been used because of their high output, mechanical strength, and ability to be shaped. A ceramic transducer disk commonly has a thin conducting electrode covering each face. A voltage applied across the electrodes causes the thickness of the disk to increase; reversal of the voltage causes the thickness to decrease. When the element is in water, these expansions and contractions cause the condensations and rarefactions that are sound signals. Electromechanical transducers are reciprocal; that is, an alternating pressure on the element creates an alternating voltage across the electrodes.

A scanning sonar system uses very directional transducers. In transmission, it can beam or focus the sound power into a given direction. In reception, it receives predominantly from the desired direction. Perhaps the most important use of directionality is to separate the backscattered signals from closely spaced objects. Directional transducers also reduce the noise coming from other directions. The discussion is easier if we talk solely about transmission and remember that the directional responses of the transducer are the same for its use as a source and as a receiver.

Shaped sound reflectors can be used, as mirrors are used in optics, to focus the sound signals from a small source (Fig. 5.2.1a). The subject is taken up in Section 5.3.3. The more common approach has been to construct an array of small transducer elements (Fig. 5.2.1b). If the individual transducers are driven with the same signal, the face of the array can move as a piston and the main radiation is perpendicular to the plane containing the transducers.

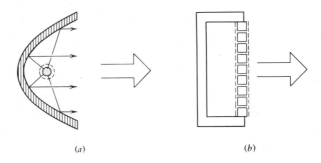

(a)   (b)

Figure 5.2.1. Transducers in cross section. (a) Line source and parabolic cylindrical mirror. (b) Baffled array. The large open arrow indicates the main radiation. The transducer elements are shown as expanding to the dashed lines. Contraction is not shown.

Time delays can be inserted in each transducer line to ensure that the main radiation is in a chosen direction. Generally the back of the array is baffled to eliminate radiation in the back direction. If unbaffled, back and front radiation will occur and the interference of the two alters the performance (Section A5.1).

The physical means of transduction need not be the piezoelectric effect. Magnetostrictive, electrodynamic, and thermoacoustic phenomena have also been used. The applications of these transduction phenomena have been as varied as the mind of man is fertile, and the records of the patent office attest to this fertility.

5.2.1. Radiation Pattern

To determine the radiation pattern of a rectangular transducer, we start with the pressure radiated by a line of small elements (Fig. 2.4.2). Each element is imagined to be a small sphere (relative to λ), which radiates as a point source. There are N sources.

For the nth source, we use (3.1.9)

$$p_n = \frac{P_{0n}R_0 \exp[i(\omega t - kR_n)]}{R_n}$$

The nth source position is found from (2.5.12)

$$y_n = \frac{nW}{N-1}$$

where W is the width of the source.

Its range, from (2.4.7), is

$$R_n \simeq R - \frac{nW(\sin \phi)}{N-1}$$

where ϕ is defined in Fig. 5.2.2.

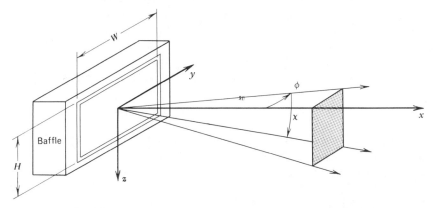

Figure 5.2.2. Geometry for radiation from rectangular transducer.

The sum over N sources is

$$p = \sum p_n \qquad\qquad (5.2.1)$$

$$= \frac{P_{0n}R_0}{R} \exp[i(\omega t - kR)] \sum_{n=0}^{N-1} \exp\left[\frac{iknW(\sin \phi)}{N-1}\right]$$

where $1/R_n \simeq 1/R$ has been factored out of the summation.

The summation was evaluated in (2.5.14) through (2.5.21) where a was the amplitude instead of $P_{0n}R_0/R$. The result is

$$p = D\frac{P_0 R_0}{R} \exp\left[i(\omega t - kR)\right] \tag{5.2.2}$$

where the directional response is

$$D = \frac{\sin\left(k\dfrac{W}{2}\sin\phi\right)}{k\dfrac{W}{2}\sin\phi} \qquad \text{line source} \tag{5.2.3a}$$

and $P_0 = NP_{0n}$ is the axial rms pressure of the line source at the reference distance R_0.

The generalization to a rectangular source in which all elements move in phase and with the same amplitude (Fig. 5.2.2) gives the radiation pattern of a piston

$$D = \frac{\sin\left(k\dfrac{W}{2}\sin\phi\right)}{k\dfrac{W}{2}\sin\phi}\frac{\sin\left(k\dfrac{H}{2}\sin\chi\right)}{k\dfrac{H}{2}\sin\chi} \qquad \text{rectangular piston}$$

$$\tag{5.2.3b}$$

where W, H are the width and height of the piston.

The radiation pattern for a circular piston is derived in Section A5.2,

$$D = \frac{2J_1(ka\sin\phi)}{ka\sin\phi} \qquad \text{circular piston} \tag{5.2.3c}$$

where a is the radius of the circular piston and ϕ is the angle with the axis; J_1 is the Bessel function of the first kind (Table A5.2.1).

Sometimes we specify the axial pressure in terms of the motion of the piston. The connection is derived in Section A5.1, where it is shown that for a pulsating sphere (A5.1.5)

$$p \simeq \frac{ik(\rho c)}{4\pi R} U_a S_a \exp\left[i(\omega t - kR)\right] \qquad \text{for} \quad ka \ll 1 \tag{A5.1.5}$$

where ρ = ambient density (subscript A has been dropped for simplicity)
$\quad U_a$ = rms radial velocity of sphere
$\quad S_a$ = surface area of sphere
Comparing the amplitude of (A5.1.5) with (5.2.2), where D is set equal to unity, we find

$$P_0 R_0 = \frac{k\rho c}{4\pi} U_a S_a \tag{5.2.4}$$

Therefore the radiated pressure is proportional to volume per second and frequency of the source and the ρc of the medium.

The acoustic intensity, being proportional to P^2, is proportional to D^2. The dependence of D^2 on $kW \sin \phi$ is shown in Fig. 5.2.3 for $\chi = 0$. The

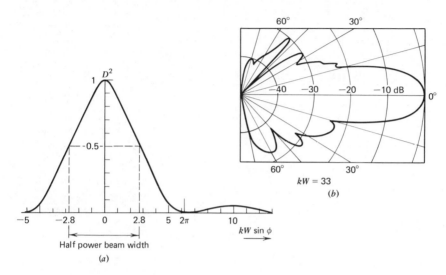

Figure 5.2.3. Directional response of a rectangular piston source. (a) D^2 as a function of $kW \sin \phi$, linear scale. Given by (5.2.36) with $\chi = 0$; D^2 is listed in Table A5.3.1. (b) Polar plot of experimental measurements of a transducer. The measurements were made on a 26 cm, square transducer at 30 kHz. The response is plotted in decibels relative to the axial value, dB $= 10 \log D^2$. (Eckart, 1968, p. 145.)

graph of D^2 also represents the diffraction pattern of a rectangular aperture in the plane wave or Fraunhofer approximation. The beam-width, defined in the trade as the angle from "half power" to "half power" of the main lobe, depends on the piston dimension and the radiated frequency. More properly, these are "half intensity" points.

The directional response of a circular piston is shown in Fig. 5.2.4. The response of a circular piston of diameter $2a$ is close to that of a square piston of width $W \approx 2a$ in the plane $\chi = 0$.

The resolution of a directional receiver is measured by its ability to separate closely spaced objects. Borrowing the Rayleigh criterion from optics, the equal strength signals from two distant objects are resolvable when the maximum of the beam pattern from one is on the first minimum of the other. Figure 5.2.3 reveals that for the rectangular piston this

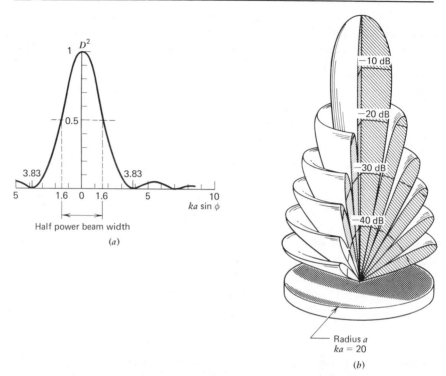

Figure 5.2.4. Directional response of a circular piston transducer, radius a. (a) Linear plot of D^2 versus $ka \sin \phi$, from (5.2.3c); D is listed in A5.3.1. (b) Theoretical three-dimensional polar response at $ka = 20$. (Eckart, 1968, p. 143.)

corresponds to $\sin \phi \geq 2\pi/(kW)$. The same criterion gives $\sin \phi \geq 3.83/(ka)$ for the circular piston (Fig. 5.2.4). The directional responses are used to estimate the resolution of the sonar system. When the sources are unequal, the magnitudes of the side lobes become important. The same criterion holds in the insonification of two adjacent objects.

5.2.2. Transmitting Directivity Factor

The transmitting *directivity factor* of a transducer Q_t is a single number measure of the directional properties of a source, including side lobes; it is defined as the ratio of the sound intensity in the axial beam direction $(\phi = 0, \chi = 0)$ to the intensity at the same position that would be caused by a point source radiating the same total power. The intensity in the axial direction

$$I_{ax} = \frac{P_{ax}^2}{\rho c}$$

is related to the intensity in other directions by

$$P^2 = P_{ax}^2 D^2 \tag{5.2.5}$$

where P_{ax} is measured at R in the axial direction, in the far field.

The power radiated in an isotropic medium Π is obtained by integration at fixed range R over all directions.

$$\Pi = \int \frac{P^2}{\rho c} \, dS$$

Use (5.2.5) and multiply and divide by R^2

$$\Pi = \left[\frac{P_{ax}^2 R^2}{\rho c} \right] \int_{surface} \frac{D^2}{R^2} \, dS$$

For a spherical surface at constant radius R, dS/R^2 is the differential solid angle $d\Omega$ and

$$\Pi = \frac{P_{ax}^2 R^2}{\rho c} \int_{4\pi} D^2 \, d\Omega \tag{5.2.6}$$

The intensity due to the point source radiating the same power Π is $\Pi/4\pi R^2$. Therefore using (5.2.5) and (5.2.6), we find the transmitting directivity factor to be

$$Q_t = \frac{I_{ax}}{\Pi/(4\pi R^2)} \tag{5.2.7}$$

$$Q_t = 4\pi \left[\int_{4\pi} D^2 \, d\Omega \right]^{-1} \tag{5.2.8}$$

The analytical integration is often difficult to perform. The answers may be obtained by numerical integration. The Q_t's for circular and rectangular piston transducers of dimensions $\gg \lambda$ are

$$\text{rectangle of sides } H, W \qquad Q_t \approx \frac{k^2 HW}{\pi} \tag{5.2.9a}$$

$$\text{circle of radius } a \qquad Q_t \approx (ka)^2 \tag{5.2.9b}$$

The directivity index for transmission DI_t is used in the sonar equation. It is defined by

$$DI_t \equiv 10 \log_{10} Q_t \tag{5.2.10}$$

The values for the large (re λ) rectangular piston and circular piston are

$$\text{rectangle}\qquad DI_t = 10 \log_{10} \frac{k^2 HW}{\pi} \qquad (5.2.11)$$

$$\text{circle}\qquad DI_t = 20 \log_{10} ka \qquad (5.2.12)$$

These expressions are very useful because they enable us to estimate the acoustic pressure in the illuminated area from the power radiated and the dimensions of the transducer. In other words, Q_t and DI_t are measures of the effect of funneling the power into the beam instead of letting it spread out omnidirectionally. The directivity index describes the increase of source level SL compared to the ominidirectional case. Hence for a directional source the SL, measured on axis, is changed in (3.4.4) to

$$SL = SL \text{ (omnidirectional)} + DI_t \qquad \textbf{(5.2.13)}$$

$$= 10 \log_{10} \left[\frac{P_0}{P_r}\right]^2 + DI_t$$

and in terms of the source power in watts for average water it is

$$SL = 10 \log_{10} \Pi + DI_t + 50.8 \text{ dB re 1 Pa} \qquad \textbf{(5.2.14)}$$

Two other reference pressures are commonly used: 1 dyne/cm^2 = 0.1 Pa = 1 μbar, for which the constant in (5.2.14) becomes +70.8 dB re 1 dyne/cm^2, and 1 micropascal (1 μPa) = 10^{-6} Pa, for which the constant is +170.8 dB re 1 μPa.

Illustrative Problems. To compare the pressure levels due to an omnidirectional source and a directional source in the ocean, we first compute the rms pressure for an omnidirectional sound source operating in seawater. The values are source power = 1 kW, frequency = 100 kHz, and distance = 1 km. The sound speed is assumed to be constant, $c = 1473$ m/s.

Solution. We use the sonar equation (3.4.2). The signal level is

$$SPL = SL - TL$$

(a) The source level for the omnidirectional source is conveniently written by letting $DI_t = 0$ in (5.2.14).

$$SL = 10 \log_{10} \Pi + 50.8$$

$$SL = 10 \log_{10} (1000) + 50.8$$

$$SL = 80.8 \text{ dB re 1 Pa,} = 200.8 \text{ dB re 1 } \mu\text{Pa (at 1 m)}$$

(b) The transmission loss is given by (3.3.8) for an isotropic medium $TL \approx 20 \log_{10} (R/R_0) + \alpha(R - R_0)$. From Fig. 3.3.1 the approximate absorption for seawater is found to be

$$\alpha = 2 \times 10^{-2} \text{ dB/m}$$
$$R = 10^3 \text{ m}$$
$$TL = 60 + 20 = 80 \text{ dB}$$

(c) On combining (a) and (b), we obtain the SPL at the field position

$$SPL \approx 80.8 - 80 \text{ dB}$$
$$\approx 0.8 \text{ dB re 1 Pa}, = 120.8 \text{ dB re 1 } \mu\text{Pa}$$

or

$$P \approx 1.1 \text{ Pa} \qquad \text{at} \quad R = 1 \text{ km}$$

Assume that the directional transducer has dimensions 0.12×0.06 m. For the same parameters as in the preceding example, how much would the SPL be increased along the axis of the transducer?

Use (5.2.11) with

$$k = 2\pi f/c$$
$$= 2\pi(10^5/1473) \text{ m}^{-1} = 427 \text{ m}^{-1}$$
$$H = 0.06 \text{ m}, \quad W = 0.12 \text{ m}$$
$$DI_t = 26.2 \text{ dB}$$
$$SL = SL(\text{omni}) + DI_t$$
$$\approx 80.8 + 26.2 = 107.0 \text{ dB re 1 Pa}, = 227.0 \text{ dB re 1 } \mu\text{Pa (at 1 m)}$$
$$SPL = 107.0 - 80 = 27 \text{ dB re 1 Pa}, = 147 \text{ dB re 1 } \mu\text{Pa at 1000 m}$$
$$= 20 \log (P/P_r)$$
$$P = 22 \text{ Pa} \quad \text{at} \quad R = 1 \text{ km}.$$

The effect of the directionality of the transducer is to increase the axial signal by 26.2 dB, that is a pressure about 20 times greater than for the omnidirectional source.

5.2.3. Peak Power and Average Power

Pinging sonar systems, such as that in Fig. 5.1.3, transmit short bursts of signal and wait until the next transmit cycle. This mode of operation leads to the use of the term "peak power." Referring to Fig. 5.2.5, suppose the transducer radiates the power Π during the time Δt and zero power during the remainder of the cycle time $T - \Delta t$. The peak power is Π. The

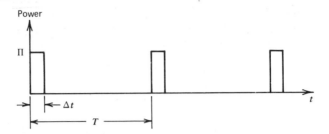

Figure 5.2.5. Signal transmissions for pinging sonar. The pulse is on for duration Δt and repeats after time T. The duty cycle is $\Delta t/T$.

average power is

$$\text{average power} = \Pi\frac{\Delta t}{T} \tag{5.2.15}$$

The ratio $\Delta t/T$, the fraction of time during which power is being radiated, is called the "duty cycle."

The effective duty cycle can be modified by transmitting a long coded signal of duration T_s and compressing it in the correlation receiver to Δt. This can simplify sonar design for two reasons: the maximum useful pressure level at the transducer is limited by the cavitation effect (Section 5.4.2), and by increasing the duty cycle, the amplifiers can operate for a longer time at lower peak power levels. In this way cavitation is avoided and electronic power amplifier requirements are lowered. On reception the correlation receiver compresses the long coded transmission into a short pulse having a peak value $T_s/\Delta t$ times the input to the filter.

5.2.4. Receiving Directivity Factor

The output of a directional receiver in a noise field depends on the directional response of the array and the directionality of the noise field. We begin with the description of the noise field. For a simple model we assume the noise field (1) is due to many uncorrelated (random phases) sources of noise, (2) is at very large distances from the transducer, and (3) comes in equally from all directions (isotropic noise field). The noise signal is passed through a narrow pass band filter having the bandwidth Δf centered at f. The ambient noise spectral density $2S_{nn}(f)$ (e.g. in Fig. 4.2.5) is the noise power per hertz received by an omnidirectional transducer. Then the noise power in the bandwidth Δf is

$$\sigma_{no}^2 = 2\Delta f S_{nn}(f) \tag{5.2.16}$$

when $S_{nn}(f)$ is nearly constant over Δf.

The noise level measured by an omnidirectional hydrophone is

$$NL_0 = 10 \log_{10}\left(\frac{\sigma_{no}^2}{P_r^2}\right) \text{ dB re } P_r^2$$

To calculate the response of a directional transducer in an isotropic noise field, initially we assume that one source is located in the direction of the polar coordinates ϕ and χ. We calculate the mean square signal for that direction. The calculation is repeated for each direction, and we assume that all these contributions add as the sum of squares (addition of powers). This uses our assumption that the sources are uncorrelated. By assuming n sources per unit area, the product of the $1/R^2$ amplitude dependence and the area of the large sphere cancel. The noise power output of the directional transducer is

$$\sigma_{nd}^2 = \Delta f \int_{4\pi} \frac{2S_{nn}(f)}{4\pi} D^2 \, d\Omega \tag{5.2.17}$$

where $2S_{nn}(f)/4\pi$ is the spectral density of the noise field per steradian of solid angle. Assuming that $S_{nn}(f)$ is isotropic, we write

$$\sigma_{nd}^2 = \frac{\Delta f S_{nn}(f)}{2\pi} \int_{4\pi} D^2 \, d\Omega \tag{5.2.18}$$

The *receiver directivity factor* Q_r is defined as the ratio of the noise power received by an omnidirectional transducer to that received by a directional transducer in the same noise field.

$$Q_r = \frac{\sigma_{no}^2}{\sigma_{nd}^2} \tag{5.2.19}$$

When the field is isotropic, we use (5.2.16) and (5.2.18) for the narrow band Δf and get

$$Q_r = 4\pi \Big/ \int_{4\pi} D^2 \, d\Omega \tag{5.2.20}$$

where Q_r is the same as Q_t (5.2.8) because we have assumed that the noise field is isotropic. The *directivity index* of a receiver is

$$DI_r \equiv 10 \log_{10} Q_r \tag{5.2.21}$$

The noise level of a directional transducer in an isotropic noise field is written in terms of the reference pressure P_r

$$NL_d \equiv 10 \log_{10} \frac{\sigma_{nd}^2}{P_r^2} \qquad (5.2.22)$$

Using (5.2.18), (5.2.20), and (5.2.21), we write

$$NL_d = 10 \log_{10} \frac{2\Delta f S_{nm}(f)}{P_r^2} - DI_r = NL_0 - DI_r \qquad (5.2.23)$$

These formulas are useful for making gross estimates of equipment performance. Actually the noise field can be very. directional as well as frequency dependent.

The ambient noise is the collection of background noises. There are a multiplicity of highly variable sources in the ocean. Let us briefly consider the effect of wind. Wind makes waves, whitecaps, bubbles, and droplets, which all are sources of ambient noise. Sources near the surface seem to radiate omnidirectionally at low wind speeds, whereas at higher wind speeds the sources radiate as $\cos \chi$ and $\cos^2 \chi$, where χ is measured from the vertical. The result is that the noise, as heard by a deep transducer, is predominantly from the horizontal at low sea states and appears to come more directly from the surface at high wind speeds.

Illustrative Problem. The circular piston transducer is used as a receiver. The radiation pattern for $ka = 20$ is in Fig. 5.2.4. The bandwidth is 1600 Hz centered on 100 kHz. For 10 m/s wind speed, compare the noise for this transducer with that of an omnidirectional transducer. The noise field is assumed to be omnidirectional.

Solution. With the aid of (5.2.23), write

$$NL_d = 10 \log_{10} \frac{2\Delta f S_{nn}(f)}{P_r^2} - DI_r \text{ dB}$$

From Fig. 4.2.4 at 100 kHz and 10 m/s wind speed, we have

$$10 \log_{10} \frac{2 S_{nn}(f)}{P_r^2} = 30 \text{ dB re } 1 \ \mu\text{Pa}^2/\text{Hz}$$

Over a 1.6 kHz bandwidth we have

$$NL_0 = 10 \log_{10} \frac{2\Delta f S_{nn}(f)}{P_r^2} = 30 + 10 \log_{10}(1600)$$

$$= 62 \text{ dB re } 1 \ \mu\text{Pa}^2$$

From the transducer calculations and for omnidirectional noise, $(ka = 20)$,

$$DI_r = DI_t = 20 \log_{10} ka$$
$$= 26 \text{ dB}$$

The noise level for the directional receiver is

$$NL_d = 62 - 26 \text{ dB}$$
$$= 36 \text{ dB re } 1 \ \mu\text{Pa}^2$$

5.3. NEAR AND FAR FIELDS OF TRANSDUCER ARRAYS

5.3.1. Pistons

Because of the simplicity of its design, the piston transducer has been the most completely studied source of acoustic waves in a fluid. Although its radiated pressure field at great distances $(kR \gg 1)$ has been very well known for many years (see Section 5.2.1), only the advent of high speed computers permitted the calculation of the very complicated acoustic field closer to the source. Figure 5.3.1 is a three-dimensional plot of the pressure field near a circular piston of size $a/\lambda = 2.5$. There is an inner circle of high pressure amplitude at the face of the transducer, and two somewhat lesser pressure amplitude rings out toward the periphery of the radiating disk. Along the axis of the source the pressure amplitude

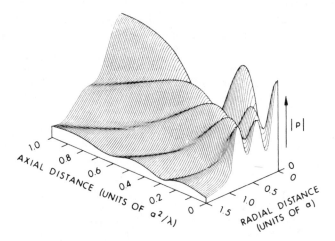

Figure 5.3.1. Three-dimensional plot of pressure amplitude in the near field of a circular piston $(a/\lambda = 2.5)$. (Lockwood and Willette, 1973.)

oscillates as the range increases, reaching a final, slowly changing maximum at $R = a^2/\lambda$ before it decreases into the far field. Other studies show that an increase in the ratio a/λ produces an increase in the number of pressure maxima and minima in this region.

The situation shown in Fig. 5.3.1 is identified as the "near field" of the source. We explain the gross features by using the Huygens wavelets that issue from a source of uniform particle velocity at the face. The near field is caused by constructive and destructive interferences of the radiation from different subareas of the transducer face (Fig. 5.3.2). At distances

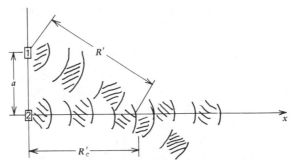

Figure 5.3.2.   Interfering radiation from a piston source. Only radiation from the edge and the center of the piston source are shown. The ratio a/λ is approximately 1.5. Segments of the radiating condensations are shaded.

"close" to the source, these radiating wavelets may differ in travel path by $\lambda/2$. When this happens there can be complete cancellation. At every point the wavelets add as described in Section 2.4.

We seek a criterion for distinguishing between a range that is in the complicated near field and one that can be identified as far field with radiating pressure simply decreasing as $1/R$, due to uniform spherical divergence. An approximate answer is rather easy to derive. The critical range R_c will occur at a position at which it is no longer possible for wavelets traveling the longest path (from the rim of the source) to interfere destructively with those traveling the shortest path (from the center of the source). As Fig. 5.3.2 indicates, a pressure minimum cannot occur when $R' - R_c' < \lambda/2$. From the geometry, and using the binomial expansion, we have

$$R' = (R_c'^2 + a^2)^{1/2} \simeq R_c'\left(1 + \frac{a^2}{2R_c'^2}\right)$$

$$R' - R_c' \simeq \frac{a^2}{2R_c'} < \frac{\lambda}{2} \qquad (5.3.1)$$

Therefore the range beyond which there can be no minimum is $R_c' \approx a^2/\lambda$.

The last axial maximum occurs at approximately $R' = a^2/\lambda$, and at ranges greater than this the sound pressure starts to decrease monotonically. It remains only to decide at what range R_c the monotonic decrease is close enough to $1/R$ to satisfy our needs for the far field approximation. The U.S. Institute (now American National Standards Institute) *arbitrarily* decided that the criterion for a circular piston will be that the far field begins at the critical range

$$R_c = \frac{\pi a^2}{\lambda} \qquad (5.3.2)$$

For a square piston of side W, the far field begins at

$$R_c = \frac{W^2}{\lambda} \qquad (5.3.3)$$

and for a piston of area A in any shape the relation is

$$R_c = \frac{A}{\lambda} \qquad \textbf{(5.3.4)}$$

Beam pattern measurements must be made at ranges greater than R_c.

5.3.2. Array Steering

When the transducer is an array of smaller elements, each one separately controlled, it is possible to adjust not only the amplitude but also the time delays or phases of the signals from or to the individual elements. One of the most common phase adjustments is used to electronically steer the array.

Consider an array of elements in a line perpendicular to the direction $\phi = 0°$ (Fig. 5.3.3). Let the array be used for receiving, rather than for sending, for present purposes. If the time delays τ_n are zero, each element will sense the acoustic variations from a distant source in the direction $\phi = 0°$. To listen to a source in some other direction $\phi \neq 0°$ without mechanically turning the transducer, we use electronic steering. That is, we adjust the time delays of the nearest receiving elements of the array by the amounts required to give the signals from the farthest elements time to catch up. Electronic delay lines are commonly used to make the time delays. Let the signal at the zeroth hydrophone be $A_0 F(t)$ where A_0

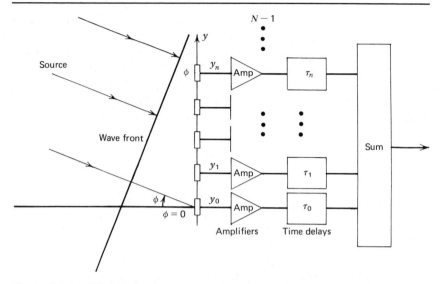

Figure 5.3.3. Electronic time delay steering. The array elements are at y_0, $y_1, \ldots$. The time delays are $\tau_0, \tau_1, \ldots$. The amplifiers can also have individual gain settings for shading the array response.

is the amplitude. From Fig. 5.3.3, the wave front arrives at the nth hydrophone Δt_n before reaching the zeroth hydrophone.

$$\Delta t_n = \frac{y_n \sin \phi}{c} \tag{5.3.5}$$

where c is sound speed. The sum of the signals for N channels is

$$F_N(t) = \sum_{n=0}^{N-1} A_n F(t - \Delta t_n) \tag{5.3.6}$$

If the time delays τ_n are inserted into each line, the output is

$$F_N(t) = \sum_{n=0}^{N-1} A_n F(t - \Delta t_n + \tau_n) \tag{5.3.7}$$

If τ_n is chosen to equal Δt_n, then

$$F_N(t) = F(t) \sum_{n=0}^{N-1} A_n \qquad \text{for} \quad \Delta t_n = \tau_n \tag{5.3.8}$$

Except for the increased amplitude, $F_N(t)$ is identical to $F(t)$ when the time delays are chosen as in (5.3.8). If the condition $\Delta t_n = \tau_n$ is not satisfied, $F_N(t)$ is not equal to $F(t)$. This is easy to demonstrate for an impulsive wave front. This method of steering an array of receivers is known as a *delay and sum system*. The only assumption is that the signals in each channel are the same except for their time delays.

If the transducer is to be used for sending in direction ϕ instead of receiving, the time delay adjustments are all the same as described in the previous paragraph.

Arrays are built in many configurations: cylinders, spheres, and so on. By using time delays, almost any shape can be steered to receive signals from any direction. When the arrays are built around a structure, however, diffraction effects can cause the performance to be peculiar.

We have assumed that the incident waves in a receiving array are plane waves. This is equivalent to saying that the curvature of the wave front is small over the dimensions of the array (e.g., less than $\lambda/8$). The plane wave assumption is useful for small arrays and distant sources. However the time delays can be chosen to match a wave front of any known curvature and to focus the array to receive signals from any position.

5.3.3. Shaped Transducers, Mirrors, and Lenses

In the mid-1940s, with the advent of preformed ceramic piezoelectric materials such as barium titanate, it became possible to make transducer elements into shapes such as rods, hollow cylinders, hollow spheres, sections of spheres, and paraboloids. Each of these forms emits a distinctive acoustic field when used as a source and has its own directionality when used as a receiver. The versatility of the single elements can be vastly expanded if they are arrayed in mosaics of special shapes. This has been done both to increase the size of the transducer and to focus the sound.

A disadvantage of a mosaic must be mentioned: when the transducers are large and closely spaced, they interact with one another. This can affect both the radiation pattern and the power amplifiers that drive the elements of a mosaic source. The interaction of transducer elements is a specialized problem beyond the level of our discussion.

A larger directional system can also be made by mounting a transducer element in front of a reflector. If the sound wavelength is small compared to the dimensions of the reflector, all the ideas of geometrical optics are applicable. The transducer in Fig. 5.3.4 was made for use with side-scanning sonar systems. It was used on the French bathyscaph *Archimède*

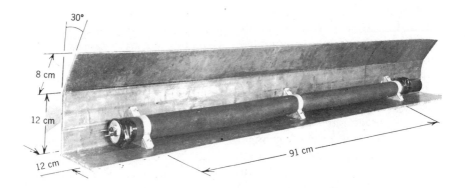

Figure 5.3.4. Reflector and transducer for side-scanning sonar. The operating frequency is 30 kHz. The axis of the transducer is 5 cm, one wavelength, out from the corner.

during her 1964 dives in the Puerto Rico Trench. The long cylindrical transducer was assembled from short barium titanate cylinders mounted in an oil-filled hose. The mirror was basically a corner reflector that in theory supplements the source by three images in the reflecting planes. The axis of the transducer was one sound wavelength out from the corner. Side-scanning sonars use very narrow horizontal beams ($\sim 1°$) and wide vertical beams ($\sim 30°$). In this example, the vertical beam was depressed about 15° below the horizontal plane. The bend of the upper reflecting plane improved the insonification of the ocean bottom at close range.

Recall that the impedance ratio for the reflector material and water

$$m \equiv \frac{\rho c_{\text{material}}}{\rho c_{\text{water}}} \tag{5.3.9}$$

is the significant quantity in determining reflection (2.9.5). A mirror reflects well when it has a ρc very different from that of water. Although heavy metals such as lead and steel have a large ρc compared with water, they may not make good mirrors because the compressional waves in water can change into flexural and compressional waves in the mirror. These waves might cause the mirror to be frequency sensitive and to have a poor radiation pattern. The reflector in Fig. 5.3.4 was designed to be used to depths below 8000 m. It was constructed of lead epoxied on aluminum to provide a large value of m, to damp waves traveling in the mirror, and to impart strength.

Another way to obtain a very large difference of ρc is to use a pressure release surface as a mirror. Reflective material that has found favor because it is inexpensive, light, and easy to fabricate is Corprene, a combination of cork and neoprene with $m = 0.08$. This kind of reflector is usually limited to shallow depths because it compresses readily.

Other pressure release techniques that have been used include thin, air-filled, hexagonal cell structures with the air chambers covered by a thin diaphragm facing the water; an epoxy matrix filled with hollow plastic beads; and a screen of parallel, air-filled, thin-walled aluminum tubes separated by a fraction of wavelength.

Acoustical lenses are often used to focus the sound field onto a receiving transducer. Three conditions must be satisfied if the lens is to be effective: the lens aperture must be large compared to the wavelength of the sound if it is to have a significant gathering effect; the lens material must have a speed of sound different from that of water,

$$h = \frac{c_{\text{material}}}{c_{\text{water}}} \neq 1 \tag{5.3.10}$$

in order to refract; the lens impedance must be close to that of the water, $m \simeq 1$, if it is to permit the sound to pass through without large reflections at the interfaces.

Two examples of acoustical converging lenses appear in Fig. 5.3.5. We assume that the aperture is large compared to the wavelength; thus a ray treatment can be used. Snell's law is then assumed to apply to each ray at each interface. In Fig. 5.3.5a, $h < 1$ and the lens is convex. For example, the lens material could be carbon tetrachloride $\rho_l = 1.59 \times 10^3 \text{ kg/m}^3$, $c_l = 0.938 \times 10^3 \text{ m/s}$, so that $h = 0.63$, $m = 0.99$. The lens shape is convex, ensuring that the rays along the shorter axial lens path take the same time to reach the focus as the outer rays traveling the longer, higher speed, water path. The arrangement closely resembles that of an optical lens in air because the speed of light in a glass lens is less than it is in the medium. The carbon tetrachloride would have to be in a thin, noncorrosive, convex envelope. It is essential to eliminate all bubbles in a liquid lens, to minimize absorption and scatter.

The second example of a converging lens (Fig. 5.3.5b) is concave, rather than convex, and uses a lens material in which the speed of sound is greater than that in water. Such a material is polystyrene, for which $\rho_l = 1.1 \times 10^3 \text{ kg/m}^3$, $c_l = 2.4 \times 10^3 \text{ m/s}$, and $h = 1.6$. Since $m = 1.7$, there

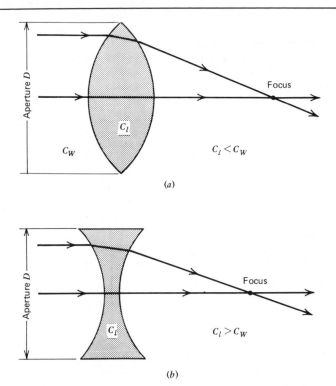

Figure 5.3.5. Two examples of converging lenses. (a) The speed of sound of the material of the lens is lower than that of the water. (b) The speed is greater in the lens than in the water. The aperture D is assumed to be the full, "unstopped" diameter of the lens.

would be some reflection at the interfaces partially canceling the convenience of using a solid lens instead of a liquid one. The degrading effects of shear waves, internal dissipation, limited aperture, and external reflection at nonnormal incidence are extremely difficult to calculate theoretically, in lens design.

For practical lens diameters acting on common underwater sound frequencies, the ratio of the aperture diameter to wavelength, D/λ, which is a measure of the energy-gathering ability of a lens, is *very* small compared to the value for optical lenses. For example, it is not uncommon to find an optical lens diameter of about 5 cm, operating in the middle range of optical frequencies where the wavelength is about 5×10^{-5} cm. The acoustic lens designer will always envy that ratio $D/\lambda \approx 10^5$. Because the ratio D/λ is never very large for acoustic lenses, the

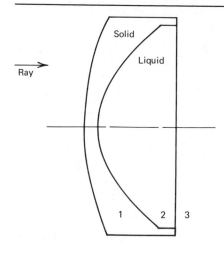

Figure 5.3.6. A compound lens design to minimize spherical aberration. (Tannaka and Koshikawa, 1973.)

popular easy-to-use ray diagrams, such as in Fig. 5.3.5, are crude guides to the behavior of underwater lenses.

The compound acoustic lens design in Fig. 5.3.6 uses a solid concave material (1) (polymethylmethacrylate) of speed ratio $h_1 = \dfrac{c_l}{c_w} = 1.85$ in combination with a liquid convex lens (2) (Dow-Corning 500 silicone) of ratio $h_2 = 0.61$. The liquid is enclosed by a 2 mm thin plane sheet of polyethylene (3), which is "transparent" to sound. The interface between regions (1) and (2) is a spheroidal surface, of major axis 21.99 cm and minor axis 20.83 cm. These radii, as well as placing liquid with $h_2 < 1$ next to the solid with $h_1 > 1$, helps to concentrate rays at a unique focus, thereby minimizing spherical aberration.

The lens has been used at 800 kHz ($\lambda = 1.88$ mm), where it produced the diffraction patterns in Fig. 5.3.7 for angles of incidence 0°, 5°, and 10°. The effective lens diameter was 19.9 cm compared to the actual diameter of 26 cm, thus the resolving power (Section 5.2.1) was

$$\phi_R = \frac{3.83}{ka} = 0.66°$$

With its focal distance of 19.9 cm, the photographer's so-called f number $\equiv f/D$ was $f/1$, a very fine lens. The lens has also been used at higher frequencies at which the central sound pressure peaks are even narrower and the resolving power even greater than shown. To achieve the same lens action at lower frequencies would require proportionately larger lens apertures.

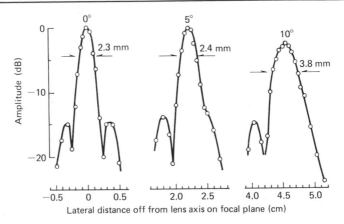

Figure 5.3.7. Radiation patterns for three angles of incidence (800 kHz) with lens of Fig. 5.3.6. (Tannaka and Koshikawa, 1973.)

5.4. HIGH POWER TRANSDUCERS

5.4.1. The Virtual End-Fire Array*

The human ear is a nonlinear device. In the presence of intense sounds, the listener is convinced that the original sounds have frequencies that are not detected by objective measurements with a linear receiver. In the early days of loudspeaker development, equipment too small to effectively radiate bass sounds could be driven hard and both the loudspeaker and the human ear would generate the missing tones as "difference" frequencies.

Sound propagation in water is also nonlinear. In 1963 Westervelt pointed out that if two intense sound beams were coaxial, the nonlinearity of the medium would create new propagating frequencies that would be the sum and difference of the primary frequencies. For example, when two primary beams of frequencies 500 and 600 kHz are superimposed, a secondary beam of 100 kHz and another of 1100 kHz will be generated with sources distributed all along the intense part of the primary beams. The difference frequency is particularly attractive for technical applications because it will have an extremely narrow beam at relatively low frequency; in the example given, the difference frequency of 100 kHz will have the narrow beam width of a 550 kHz source. In addition, the bandwidth of the difference frequency will be greatly increased.

---

* The term "parametric array" is also used.

The phenomenon can be understood by considering the pressure at a point x along a section of a primary plane wave.

$$p_1(x_0, t) = \sqrt{2} P_1 \cos(\omega_1 t - k_1 x_0) \qquad (5.4.1)$$

For simplicity, let the distance from the source be such that $k_1 x_0 = n\pi$ so that

$$p_1(x_0, t) = \sqrt{2} P_1 \cos \omega_1 t \qquad (5.4.2)$$

Another primary wave of different frequency would give

$$p_2(x_0, t) = \sqrt{2} P_2 \cos \omega_2 t \qquad (5.4.3)$$

Now assume that the two primary waves are coaxial and that both are very intense so that they interact nonlinearly. Then instead of the amplitude at a point being simply the sum of the two contributing amplitudes, the amplitude of the p_1 wave will be modulated by the presence of p_2 and will become $P_1(1 + m \cos \omega_2 t)$ where m is the modulation amplitude. When this is inserted into (5.4.2) and is added to (5.4.3), we obtain the sum

$$p = \sqrt{2}(P_1 \cos \omega_1 t + P_2 \cos \omega_2 t + P_1 m \cos \omega_1 t \cos \omega_2 t) \qquad (5.4.4)$$

The third term in parentheses is the nonlinear interaction and $m = m(P_1, P_2)$ is a measure of its strength. The product $(\cos \omega_1 t)(\cos \omega_2 t)$ can be rewritten by using the trigonometric relation

$$2 \cos x \cos y = \cos(x + y) + \cos(x - y) \qquad (5.4.5)$$

and p becomes

$$p = \sqrt{2}\left[P_1 \cos \omega_1 t + P_2 \cos \omega_2 t + \frac{P_1 m}{2}(\cos \omega_\Sigma t + \cos \omega_d t) \right] \qquad (5.4.6)$$

The third term in the brackets is a new pressure given in terms of the two new frequencies,

$$\begin{aligned} \omega_\Sigma &\equiv \omega_1 + \omega_2 \\ \omega_d &\equiv \omega_1 - \omega_2 \end{aligned} \qquad (5.4.7)$$

The nonlinear interaction has created sum and difference frequencies. Furthermore, these new frequency components are generated at all points of intense interaction along the beam. They therefore constitute a volume distribution of equivalent (or virtual) secondary sources, in what is called an end-fire array. Previously real end-fire arrays were built by assembling a properly phased line of simple real sources. These arrays have highly directional beam patterns.

To evaluate the interaction, we move to Westervelt's wave equation for the nonlinear secondary tones, which we write in one dimension:

$$\frac{\partial^2 p_s}{\partial x^2} - \frac{1}{c^2}\frac{\partial^2 p_s}{\partial t^2} = -\frac{\beta}{\rho c^2}\left[\frac{1}{c^2}\frac{\partial^2 (p_1+p_2)^2}{\partial t^2}\right] \qquad (5.4.8)$$

where β is discussed in Section A2.5 and p_s is the secondary or "scattered" acoustic pressure. This is a wave equation for the secondary pressure p_s with a source term on the right-hand side which accounts for the generation of the nonlinear wave components. Inserting the expressions for attenuating primary plane waves,

$$p_1 + p_2 = \sqrt{2}\,P_1 \cos\,(\omega_1 t - k_1 x)\,\exp\,(-\alpha_1 x)$$
$$+ \sqrt{2}\,P_2 \cos\,(\omega_2 t - k_2 x)\,\exp\,(-\alpha_2 x) \qquad (5.4.9)$$

into the right-hand side of (5.4.8) we get

$$\frac{\partial^2}{\partial t^2}(p_1+p_2)^2 = -4\omega_1^2 P_1^2 \cos\,[2(\omega_1 t - k_1 x)]\,\exp\,(-2\alpha_1 x)$$
$$-4\omega_2^2 P_2^2 \cos\,[2(\omega_2 t - k_2 x)]\,\exp\,(-2\alpha_2 x)$$
$$-2\omega_\Sigma^2 P_1 P_2 \exp\,[-(\alpha_1 + \alpha_2)x]\cos\,[\omega_\Sigma t - k_\Sigma x]$$
$$-2\omega_d^2 P_1 P_2 \exp\,[-(\alpha_1 + \alpha_2)x]\cos\,[\omega_d t - k_d x] \qquad (5.4.10)$$

The new second harmonic terms (first two terms) verify our graphical proof of the development of harmonics in shock waves (Section A2.5.1). The higher harmonics of shock waves (not shown) are the result of further nonlinear interactions.

The third and fourth terms are the sum and difference tones. Since the difference frequency can be exploited, we consider it alone. Next, we use Huygens' principle to calculate the directional dependence of the low frequency component of p_s having the frequency ω_d. These components are sources of Huygens' wavelets in the volume shown in Fig. 5.4.1. We

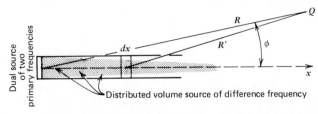

Figure 5.4.1. Geometry for generation of difference frequency by nonlinear effects of two high frequency coaxial primary beams. The difference frequency is generated throughout the primary beam volume, wherever the pressure amplitude of the primaries is large. The signal is observed at Q.

assume that the diameter of the volume is small compared to the wave length of the low frequency component and element of volume is $S_0 \, dx$ where S_0 is the area of the high frequency sound beams. The wavelets diverge spherically and have the low frequency attenuation coefficient α_d. Ignoring obliquity factor and constants that depend on the complete solution, the sound pressure at Q is the integral over all sources

$$p_d(R, \phi) \sim \frac{\omega_d^2 \beta P_1 P_2 S_0}{\rho c^4}$$

$$\times \int_0^\infty \frac{\exp\left[-\alpha_d R' - (\alpha_1 + \alpha_2)x + i(\omega_d t - k_d x - k_d R')\right]}{R'} \, dx + \text{C.C.}$$

$$R' \equiv (R^2 - 2xR \cos \phi + x^2)^{1/2} \simeq R - x \cos \phi$$

where C.C. means the complex conjugate and the sum gives the cosine; we give the limit as infinite for convenience; in practice, the length of the volume is small compared to R because $\exp\left[-(\alpha_1 + \alpha_2)x\right]$ tends to zero rapidly. Substitution of the approximation for R' and $1 - \cos \phi = 2 \sin^2 (\phi/2)$ gives a simple integral. The approximate value of $p_d(R, \phi)$ is

$$p_d(R, \phi) \sim \frac{\omega_d^2 \beta P_1 P_2 S_0}{2\rho c^4} \frac{\exp\left[-\alpha_d R + i(\omega_d t - k_d R)\right]}{R} \frac{1}{\dfrac{\alpha_1 + \alpha_2}{2} + ik_d \sin^2 \dfrac{\phi}{2}} + \text{C.C.}$$

Determination of the amplitude factor requires solution of three dimensional form of (5.4.8) and we refer the reader to Westerfelt (1963) and Beyer (p. 311–316, 1974) for solutions;

$$p_d(R, \phi) = \frac{\beta \omega_d^2 P_1 P_2 S_0}{4\pi R \rho c^4} \exp(-\alpha_d R) \cos(\omega_d t - k_d R - \varepsilon)$$

$$\times \left(\alpha_e^2 + k_d^2 \sin^4 \frac{\phi}{2}\right)^{-1/2} \tag{5.4.11}$$

where R, ϕ = polar coordinates of the field position (Fig. 5.4.1)

S_0 = cross-sectional area of primary beams

α_d = pressure attenuation coefficient of difference frequency (Np/m)

$\alpha_e = (\alpha_{e1} + \alpha_{e2})/2$ = mean pressure attenuation coefficient of primary frequencies (Np/m)

ε = phase angle = $\arctan\left[(k_d/\alpha_e) \sin^2 (\phi/2)\right]$

$\beta \approx 3.4$ (see Section A2.5.1)

The geometry is in Fig. 5.4.1. The difference frequency is generated wherever the product P_1P_2 is large. Therefore the contributions from parts of the beam near the source are the greatest. Once generated, the difference frequency is "launched" and is on its own as it propagates with the attenuation α_d appropriate to that frequency. Since the primary frequencies are relatively high compared to the difference frequency, they die off, and the difference frequency remains and radiates with relatively small attenuation. The secondary pressure is proportional to $1/R$ and has a very narrow beam pattern defined by its directional response

$$D_d = \left[1 + \left(\frac{k_d}{\alpha_e}\right)^2 \sin^4\frac{\phi}{2}\right]^{-1/2} \tag{5.4.12}$$

The sound pressure level will be down 3 dB at the angle ϕ_d given by

$$\sin\frac{\phi_d}{2} = \left(\frac{\alpha_e}{k_d}\right)^{1/2} \tag{5.4.13}$$

For the usual conditions of difference tone generation, α_e/k_d is small, thus

$$\text{beam width} = 2\phi_d \approx 4\left(\frac{\alpha_e}{k_d}\right)^{1/2} \tag{5.4.14}$$

Not only is the difference frequency beam width much narrower than would be expected from a piston of the size used for the primaries, but (5.4.12) predicts that there will be no side lobes in the radiation pattern of the difference frequency.

Figure 5.4.2 shows the beam patterns of two primary waves, one at 418 kHz, the other at 482 kHz, and the difference tone they produced at 64 kHz. Notice the striking absence of side lobes down to about -40 dB in the beam pattern of the difference tone.

Illustrative Problem

(a) Given a difference tone generation due to primary frequencies $f_1 = 418$ kHz and $f_2 = 482$ kHz, calculate the 3 dB beam width.

(b) Calculate the size of piston that would have been required to produce the same half power beam width.

(c) Calculate the radius of the transducers used. (Assume piston radiation for the primaries.)

(d) Calculate the reduction in source radius in using the end-fire array for the same beam width.

Solution

(a) We write

$$\text{avg } f = \frac{418 + 482}{2} = 450 \text{ kHz}$$

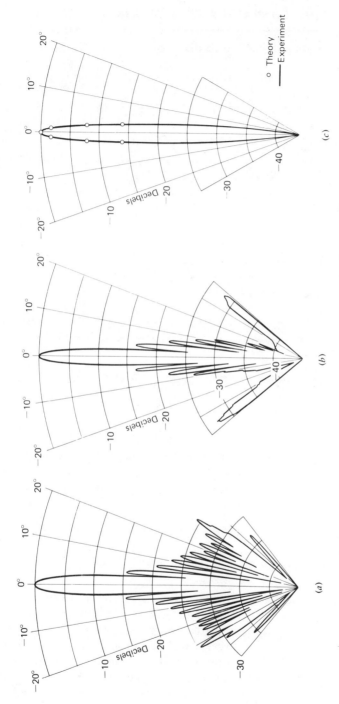

Figure 5.4.2. Parametric transmitting array directivity patterns. (*a*) Beam pattern of 482 kHz primary. (*b*) Beam pattern of 418 kHz primary. (*c*) Difference frequency beam pattern. Theory, open points; experiment, solid lines. (Muir 1974).

167

At 450 kHz, the attenuation is 0.15 dB/m or 1.7×10^{-2} Np/m.

$$f_d = 482 - 418 = 64 \text{ kHz}$$

$$k_d = \frac{\omega_d}{c} \approx 268 \text{ m}^{-1}$$

$$2\phi_d \approx 4 \left(\frac{\alpha_e}{k_d} \right)^{1/2}$$

$$\approx 0.032 \text{ rad} \qquad \text{or} \qquad 1.8°$$

(b) From Table A5.2.1 for a circular piston the -3 dB angle is given by

$$k_d a \sin \phi_d = 1.6$$

Therefore we require

$$a = 1.6 (k_d \sin \phi_d)^{-1}$$
$$a = 37 \text{ cm}$$

(c) The -3 dB angle in Fig. 5.4.2 is approximately 2°. The -3 dB angle for a circular piston is at $ka \sin \phi = 1.6$. Therefore for $f = 450$ kHz,

$$a \text{(primary)} = \frac{1.6}{(18.8)(0.035)} = 2.6 \text{ cm}$$

(d) The radius has been reduced to $2.6/37 = 7\%$.

Another advantage of the virtual end-fire array is the increase in bandwidth. The bandwidth of a primary source normally is some percentage of the central frequency, say $\pm 5\%$. If the primaries can be operated within this percentage bandwidth with essentially no depreciation of output, the difference frequency can range over this same absolute bandwidth *at its lower frequency*. For example, if the primary frequencies are $500 \pm 5\%$ (i.e., 475 to 525 kHz) and $600 \pm 5\%$ (i.e., 570 to 630 kHz), the difference frequency will be 100 kHz ± 25 kHz. Such a wide bandwidth at low frequencies is generally difficult to achieve for primary radiation.

A further advantage of the virtual end-fire array is that its beam width is relatively insensitive to changes of the difference frequency.

The great virtues of the parametric source are bought at the cost of very low power efficiency. To calculate the efficiency of the nonlinear source we assume that the two primary sources radiate the same power $W_1 = W_2$, that the rms field pressures are equal $P_1 = P_2$, and that the attenuation of the difference tone is negligible. The powers radiated through the primary beam area S_0 are

$$\Pi_1 = \Pi_2 = \frac{S_0 P_1^2}{\rho c}$$

Rewriting (5.4.11) in terms of the total primary power $\Pi_0 = \Pi_1 + \Pi_2$, the rms pressure of the difference tone, on axis, is

$$P_d = \frac{\beta \omega_d^2 \Pi_0}{\sqrt{28} \pi R c^3 \alpha_e}$$

or, using (5.4.14)

$$P_d = \frac{\sqrt{2} \beta f_d \Pi_0}{2 R c^2 \phi_d^2} \qquad \text{(5.4.15)}$$

where

$$f_d = \frac{\omega_d}{2\pi}$$

The power radiated by the difference tone Π_d is the product of the average intensity by the beam area $\pi(R\phi_d)^2$. Approximating the average intensity by the axial intensity $P_d^2/\rho c$, we get

$$\Pi_d = \frac{\pi \beta^2 f_d^2 \Pi_0^2}{(2\rho c^5 \phi_d^2)} \qquad \text{(5.4.16)}$$

and the percentage efficiency is

$$\% \text{ effic} = \frac{100\Pi_d}{\Pi_0} = \frac{\pi \beta^2 f_d^2 \Pi_0}{2\rho c^5 \phi_d^2} \times 10^2 \qquad \text{(5.4.17)}$$

The efficiency is generally a fraction of 1% (see Problem 5.4.2). There are three evident parameters for increasing the efficiency; increased difference frequency, increased primary power, or decreased beam width (lower primary frequency with constant primary beam area).

Only increased power offers direct promise for increased efficiency without sacrificing the advantage of the nonlinear source. That avenue is restricted by saturation effects and beam broadening (Section A2.5.1) and cavitation (Section 5.4.2) at high intensities.

When gas bubbles are present in an intense sound beam, they may be driven into large amplitude, nonlinear oscillations at or near their resonance frequencies. This effect can substantially increase the level of the difference frequency, though with some loss in the radiation directionality.

Even though the present efficiencies are very low, nonlinear transducers may have a bright future. They are one solution to the design of transducers having a uniform response over a wide range at low frequencies. A seismic system having response from 50 to 1000 Hz would be very desirable for profiling the deep ocean bottom. The nonlinear transducer system could be combined with coded transmission and matched filters to gain both bandwidth and usable signal levels.

5.4.2. The Cavitation Limit

As technology permits the generation of higher and higher sound intensities, the swings of pressure finally reach such great values that the medium ruptures or "cavitates" at the rarefaction phase. At and near sea level, minute bubbles, more or less of micron size, are always present in the ocean. For sound sources near the sea surface, these cavitation nuclei permit rupture to occur at pressure swings of the order of 1 atm (10^5 Pa or 220 dB re 1 μPa) depending on the frequency, duration, and repetition rate of the sound pulse.

There are many physical phenomena associated with a cavitating bubble. When violently oscillating bubbles are close to a solid surface, the stresses associated with the emitted shock waves and acoustic streams result in rapid erosion of the toughest of metals or plastics. The high pressures and high temperatures occurring during transient bubble collapse cause luminescence of the bubble gas. Chemical reactions can be initiated or increased in activity, and living cells and macromolecules are ruptured by cavitation. Most important to the user of acoustics in the ocean, cavitation oscillations near a sound source produce harmonic distortion and streams of jetting, exploding bubbles that reduce the transmission of sound from the transducer, while they radiate highly audible noise.

Even at relatively low levels of acoustic intensities, the output wave form of a transducer is affected by the ambient bubbles that are particularly common near the sea surface. As the CW sound pressure amplitude increases, ambient bubbles begin to oscillate nonlinearly and a second harmonic of the applied frequency is generated. At sea level the amplitude of the second harmonic is less than 1% of the fundamental as long as the pressure amplitude of the fundamental is less than about 180 dB re 1 μPa rms (Rusby, 1970). When the signal is up 20 dB to 0.1 atm there is about 5% distortion.

When the peak pressure amplitude is somewhat greater than 1 atm, the absolute pressure for a sound source at sea level will be less than zero during the rarefaction part of the cycle. This negative pressure, or tension, is the trigger for a sharply increased level of distortion and for the issuance of broad band noise, if the CW sound frequency is below 10 kHz. Any attempt to increase the sound pressure amplitude appreciably beyond the ambient pressure will cause total distortion and the generation of a cloud of bubbles so large that the far field radiated output of the transducer will actually *decrease* with increasing application of voltage.

The foregoing description for a low frequency, CW source at sea level identifies the external effects of cavitation rather than the detailed dynamics of the phenomenon. Laboratory measurements have identified two types of transient cavitation that take place in regions of very large sound pressure amplitudes—gaseous and vaporous cavitation. Gaseous cavitation is recognized by streamers of relatively stable, hissing bubbles that jet away from regions of high alternating sound pressure; it is assumed that the relative amounts of gas and water vapor in the bubbles are constant. Vaporous cavitation is characterized by violently intermittently collapsing single bubbles, which thereby radiate shock waves and broad band noise; it is believed that water vapor passes through the bubble walls from the liquid during the expansion phase and condenses at the inner surface during the contraction phase of a cycle of the bubble motion.

The nuclei for cavitation are minute, stable bubbles assumed to be caught in crevices of solid particles. Under the influence of an alternating acoustic radius, if the pressure swing is great enough (Hsieh-Plesset threshold), bubbles grow during pulsation by a process called "rectified diffusion." During a cycle of this action, more gas diffuses inward from the liquid to the bubble than is squeezed out. This is because the surface area of the bubble in the expanded phase of the cycle is greater than the area during the contracted phase causing the net gas diffusion to be inward for a complete cycle. Growth of very small bubbles ($\lesssim 1\ \mu$) by rectified diffusion takes place in a sound field until a critical radius is reached. If the sound pressure swing is constant at that point (Blake threshold), the bubble will expand explosively.

A large number of laboratory experiments have been conducted to determine the threshold acoustic pressure amplitude required for cavitation, for CW sounds. The dependence on frequency is shown in Fig. 5.4.3 from the excellent review article by Flynn (1964). The threshold, in fact, depends on the criterion chosen to define the existence of cavitation. Sometimes visual observation is used. More commonly, the broad band noise of the cavitating bubbles is an early and reasonably consistent clue to the onset. Finally, the generation of subharmonics of half-fundamental frequency are a valuable criterion often used to define the onset of cavitation. The limiting curves of Fig. 5.4.3 reveal this uncertainty of

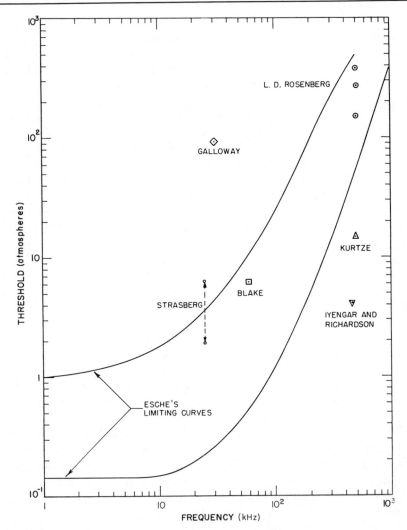

Figure 5.4.3. CW cavitation threshold measurements; all data for sea level. From Flynn (1969).

definition. Most importantly, they show the very large increase in CW sound pressure amplitude that is required for cavitation if the sound frequency is above 10 kHz.

It seems clear that at sea there are ample cavitation nuclei available at all radii, ensuring that cavitation can readily take place if the acoustic pressure amplitude is great enough. Figure 5.4.4 summarizes the numbers

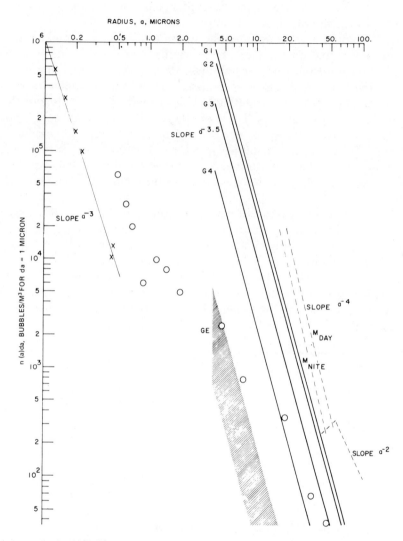

Figure 5.4.4. Bubble populations in various media: degassed water, ($\times$, $\bigcirc$) (Messino et al., 1963) tap water, G (Gavrilov, 1969); coastal ocean at depth of 3.3 m, M, dashed lines (Medwin, 1970). $n(a)\,da$ is the number of bubbles per cubic meter, with radius between a and $a + da$, where $da = 1\,\mu$. $G1$, $G2$, $G3$, $G4$, and GE are Gavrilov's values after stilling for 25 minutes, 1 hour, 3 hours, 5 hours, and the estimated terminal value.

of ambient bubbles determined by experiments on seawater, tap water, and degassed, filtered, distilled water.

The sound field of an ocean-going transducer consists of regions close to the source and on the face of the source, where the sound pressure level is much greater than, or less than, the source level referenced to 1 meter. Furthermore, acoustic radiation pressures and streaming (Section A2.5) rapidly move the cavitating bubbles to new positions, where they suffer different growth rates. Therefore there exist "hot" and "cold" spots where cavitation activity is significantly different from the prediction average. To watch the cavitating near field of a transducer is to experience an unforgettable performance of randomly driven and exploding streamers, resembling the conclusion of an Independence Day pyrotechnics display.

The dependence of onset of cavitation (solid line) and of 10% distortion (dashed line) on pulse duration for 7 kHz sound at two depths in the ocean is illustrated in Fig. 5.4.5. For pulse durations of less than 100 ms, the acoustic intensity required for 10% distortion or cavitation is constant and about twice the value for CW sound. Higher intensities can be radiated at greater depths before initiating cavitation or distortion because there are fewer cavitation nuclei, the radii of these bubbles are much smaller, and the ambient pressure is higher.

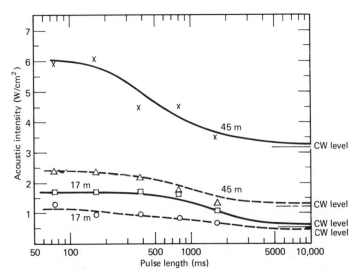

Figure 5.4.5. Onset of cavitation (solid lines) and 10% amplitude distortion (dashed lines) in seawater as a function of pulse length, 7 kHz low duty cycle. (Rusby 1970.)

When a change of duty cycle is studied, it is found that starting with CW sound, higher acoustic intensities are permitted as the on time decreases until an asymptote is reached (Fig. 5.4.6). It is not surprising to see that the asymptotic levels of intensity that can be reached at very small duty cycles are greater for the shorter pulse durations.

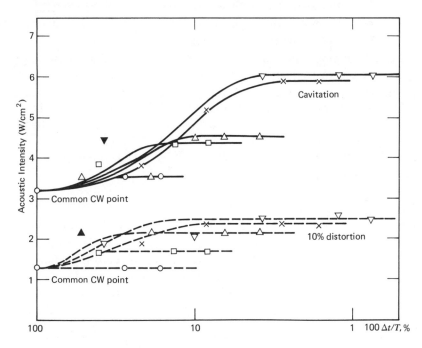

Figure 5.4.6. Onset of cavitation and 10% amplitude distortion at 45 m in seawater as a function of duty cycle for different pulse lengths, 7 kHz. Pulse durations; 74 ms, ∇; 165 ms, ×; 380 ms, △; 800 ms, ☐; 1650 ms, ○. (Rusby, 1970.)

PROBLEMS

Problems 5.2

5.2.1. Use (5.2.3b) with $\chi = 0$ and (5.2.3c) to compare graphically the radiation patterns of a circular piston of radius a and a rectangular piston of width $W = 2a$.

5.2.2. Plot the polar radiation pattern of a piston transducer of radius 10 cm when radiating frequency 15 kHz in the ocean.

5.2.3. Plot the polar radiation pattern of a square transducer of side W, at the "corner" angle $\phi = \chi$.

5.2.4. Perform the following operations.

    (*a*) Calculate the directional response of a 26 cm, square piston transducer in the plane $\chi = 0$. The radiated frequency is 30.0 kHz. Plot in decibels, using polar coordinates.

    (*b*) Compare these theoretical results with the experimental directional response in Fig. 5.2.3*b*.

5.2.5. Assuming that the circular transducer of Fig. 5.2.4 is radiating sound of frequency 20 kHz, what is the piston radius?

5.2.6. The approximation is sometimes made that the directional response of a square piston of width W, in a plane parallel to one edge, is about the same as for a circular piston of diameter $2a = W$. Use Table A5.3.1 to calculate the directional response of a circular piston of $k(2a) = kW = 33$ and compare with Fig. 5.2.3*b*.

Problems 5.3

5.3.1. Given a circular array (radius a) of line transducers, arranged as the staves of a barrel. The positions of the elements are defined by the angle θ with reference to the zeroth element. Find an expression for the delay of the nth hydrophone at angle θ_n when the array is electronically steered toward angle ϕ.

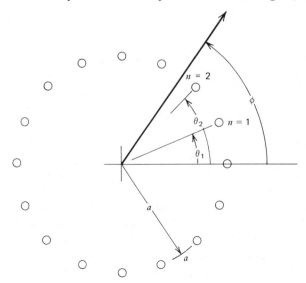

5.3.2. Calculate the diameter of a lens similar to the one in Fig. 5.3.6, that could be used to give a resolution of 0.66° for sound of frequency 15 kHz. Recall the Rayleigh criterion in Section 5.2.1.

5.3.3. A calibration is needed for the output of a circular piston source of radius 20 cm over the frequency range 5 to 50 kHz. Where should the hydrophone be placed to ensure that it is safely out of the near field.

Problems 5.4

5.4.1. Do the following.
(a) Plot the intensity beam pattern of the difference tone of the situation described in the illustrative problem in Section 5.4.1 where $\alpha_e = 0.017$ Np/m, $k_d = 268$ m^{-1}.
(b) Compare with the beam pattern produced by a circular piston large enough to yield the same beam width.

5.4.2. An end-fire array is advertised as having a mean primary frequency of 200 kHz, a difference frequency of 12 kHz, and source input powers $W_1 = W_2 = 200$ W. Using the value $\beta = 3.5$ for water and $\alpha_e = 0.08$ dB/m $= 0.01$ Np/m for the 200 kHz primaries, do the following.
(a) Calculate the efficiency of this commercial device.
(b) Calculate its beam width.
(c) Consider a circular piston source that gives the same beam width at the same frequency as this nonlinear source. Calculate the ka value required for the piston source.
(d) What is the diameter of this simple piston relative to that of the piston used for the primary beams of the nonlinear source? (Assume that the primary beams come from piston radiation and that the beam width of the primary equals the beam width of the difference frequency.)

5.4.3. A virtual end-fire array is to be designed to operate with primary frequencies 95 and 105 kHz and primary power outputs of $W_1 = W_2 = 5$ kW. Assume that the source transducer is 50×50 cm.
(a) Calculate the beam width of the difference tone $2\phi_d$.
(b) Calculate the source level ($R_0 = 1$ m) of the difference tone.
(c) Calculate the efficiency of the nonlinear source.

SIX

SCATTERING AND ABSORPTION BY BODIES AND BUBBLES

6.1 PERCEPTION OF BODIES OR BUBBLES BY SCATTERING PHENOMENA

We begin by considering how the presence of bodies and bubbles is perceived when we look for them in the water. In "clear" water we often see numerous bright and dark objects. We sense the presence of these objects by receiving the light they scatter toward us (bright objects) or by the shadows they cast when they appear as dark spots relative to the background.

Essentially the same process occurs when sound is used to sense the presence of objects. Since audible sound waves and visible light waves are different in character and are greatly different in wavelength, they give different information about what is in the water. For example, certain plankton have almost the same optical properties as the water that surrounds them and are almost invisible in water. When the density or compressibility of such plankton differ from the values for water, these bodies can be readily found by the use of sound waves.

What we call "muddy" or "turbid" water has many very fine particles floating in it that so scatter and absorb light that it is nearly impossible to use optical or visual systems. And yet these very fine particles hardly affect typical sound signals in the ocean; thus muddy water is beautifully transparent to sound.

178

Gas bubbles in water are another matter. Tiny, almost invisible, gas bubbles can have a resonance frequency the same as the sound signal; when this happens, the bubble is a very strong scatterer and absorber of sound waves. Bubbly water for sound resembles muddy water for light. Gas bubbles are not all bad. Many fish control their buoyancy by means of an air-filled swim bladder (i.e., a gas bubble). The scattered signal from a swim bladder is often so large that individual fish can be observed, and a density of one small fish in $1000\,m^3$ can be detected at a great distance by the scattered sound. Because bubbles in water have such strong effects on sound waves, this chapter discusses the scattering and absorption of sound by bubbles in detail.

A procedure for studying the scattering of sound by objects or bubbles is represented in Fig. 6.1.1. The Huygens wavelets originating at the

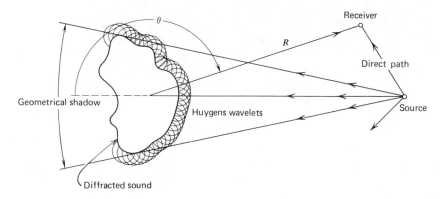

Figure 6.1.1. Scattering of sound by an object. The incident wave front is omitted for clarity. The scattered Huygens wavelets appear in the plane of the diagram. The scatter angle ϕ (out of the plane of the diagram) is not shown. Forward scatter would be in the direction $\theta = 0°$, backscatter is at $\theta = 180$.

object go in all directions. Wavelets that originate at the edge of the geometrical shadow propagate as diffracted waves into the shadow region to a degree that depends on the wavelength compared to the size of the object.

Measurements of the scattered sound are made by illuminating (insonifying, really) the object and determining the sound scattered to the receiver. For an arbitrary object the actual measurements would depend on the scattering direction defined by the coordinates θ, ϕ, as well as the incident direction, and would be incredibly tedious. Numerical techniques are the usual approach for calculating scatter from an arbitrary body such as that in Fig. 6.1.1.

Most analytical studies have been limited to relatively simple objects such as spheres, cylinders, disks, ellipsoids, and sectors of planes. We devote most of our energies to the sphere. Fortunately this is quite practical because gas bubbles are generally spherical and the scattering from many more complicated bodies can be understood by using the sphere as a prototype.

6.1.1. The Scattering Function

Let us consider the sound scattering characteristics of an object. Returning to Fig. 6.1.1, we assume that the sound speed is constant and that the absorption losses are negligible in the water. This permits us to use spherically spreading waves in calculating the amplitudes. Since in a real ocean the corrections can be inserted and the sonar equations modified, this is not a limitation. The source is assumed to be far enough from the object that *the object is well within the first Fresnel zone* (Fig. 2.6.2) and the incident sound waves are approximately plane. The ping length $c\Delta t$ is assumed to be much larger than the dimension of the object, ensuring that a steady state scattering situation exists. The receiver is assumed to be at very large range from the scattering object.

With the exception of forward scatter, we can separate the sound waves scattered by the object from the direct sound wave by using a pinging technique and observing the scattered signal when there is no signal from the source.

Let I_p be the intensity of the incident, effectively plane wave, sound signal at the scatterer. The intensity of the signal scattered to the receiver is I_s. Assuming spherically spreading scattered Huygens wavelets at the receiver distance R and no absorption loss, we have the proportionality

$$I_s \sim \frac{I_p}{R^2} \qquad (6.1.1)$$

Since at long ranges $I_p = P_p^2/(\rho_A c)$ and $I_s = P_s^2/(\rho_A c)$, (6.1.1) can equally well be written

$$P_s^2 \sim \frac{P_p^2}{R^2} \qquad (6.1.1a)$$

The constant of proportionality depends on the frequency of the sound, the size, the shape, and the orientation of the object with respect to the source and the receiver. For a given body this dependence can be written

$$I_s = \frac{I_p}{R^2} \mathscr{S}(\theta, \phi, \theta_p, \phi_p, f) A(\theta_p, \phi_p) \tag{6.1.2}$$

where θ, ϕ, and θ_p, ϕ_p are the spherical coordinates of the scattered sound and the incident sound, respectively, A is the cross section of the projected area of the scatterer viewed from the source, and f is the sound frequency. This defines the scattering function of a body $\mathscr{S}$ which is dimensionless.

Often $\mathscr{S}$ is a very complicated function of angle as well as of the sound frequency and the dimensions of the object. For simplicity, we assume that the angle at which the sound is incident on the body is fixed and that we need specify only the scattered angles θ (Fig. 6.1.1) and ϕ, out of the plane of incidence. When the source and receiver are at different positions, the geometry is called "bistatic" scatter.

For simplicity and practicality, we concentrate on two cases—the backscattered sound $\theta = 180°$ ($\phi = 0°$), sometimes called "monostatic" scatter, and the total scattered sound. Backscattering is, of course, the most common geometry for remote sensing. The total scattered energy along a sound path contributes to the attenuation of the sound, sometimes very significantly.

6.1.2. Backscatter and Target Strength

The backscattering cross section is a measure of the ability of a body to scatter sound back to a receiver that is at the same location as the transmitter. It has the dimensions of an area and is defined as

$$\sigma_{bs} \equiv \mathscr{S}_{bs} A = \frac{I_{bs} R^2}{I_p} = \left(\frac{P_{bs}}{P_p}\right)^2 R^2 \tag{6.1.3}$$

where $\mathscr{S}_{bs} \equiv \mathscr{S}(\theta, \phi, f)|_{\theta=\pi} \equiv$ backscattering function
$I_{bs} =$ backscattered intensity at receiver
$P_{bs} =$ backscattered pressure at receiver

In some of the literature σ_{bs} is called the differential scattering cross section for backscatter and is designated $d\sigma_s$.

Much of the sonar literature uses the concept "target strength" $\equiv TS$. The target strength is a decibel measure of the backscatter. As in Chapter 3, we divide both sides of (6.1.3) by length squared to obtain a dimensionless equation before taking the logarithm. On the left side we divide by the 1 square meter of area A_1 and on the right we divide by its equal

R_1^2, where $R_1 = 1$ m. Then target strength is defined by

$$TS \equiv 10 \log_{10} \frac{A\mathscr{S}_{bs}}{A_1} = 10 \log_{10} \frac{\sigma_{bs}}{A_1} \quad \text{dB} \quad \text{re } A_1 \tag{6.1.4}$$

Combining (6.1.3) with (6.1.4), we get the operational definition

$$TS = 10 \log_{10} \frac{I_{bs} R^2}{I_p R_1^2} = 10 \log_{10} \left(\frac{P_{bs} R}{P_p R_1}\right)^2 \text{dB} \tag{6.1.5}$$

With our assumption of isothermal water and no absorption loss, we get P_p in terms of the source pressure P_0 at reference range R_0,

$$P_p^2 = \frac{P_0^2 R_0^2}{R^2}$$

so that

$$TS = 10 \log_{10} \left[\left(\frac{P_{bs}}{P_0}\right)^2 \frac{R^4}{R_0^2 R_1^2}\right] \text{dB} \tag{6.1.6}$$

To put pressures in decibel notation, multiply and divide the argument in the brackets by P_r^2 before taking the log. Then

$$TS = SPL_{bs} - SL + 2TL \text{ dB} \tag{6.1.7}$$

where $SPL_{bs} = 20 \log_{10} (P_{bs}/P_r)$
     $SL = 20 \log_{10} (P_0/P_r)$
     $P_r = $ reference pressure $(=1 \, \mu bar$ or $1 \, \mu$Pa or 1 Pa$)$

In this development TL, the one-way transmission loss, is simply $20 \log_{10} (R/R_0) = 20 \log_{10} (R/R_1)$. In a homogeneous absorbing medium, TL is given by (3.3.8), and for a stratified absorbing medium (3.4.5) should be used. If the return path (2) is different from the send path (1), $2TL$ should be replaced by $TL_1 + TL_2$.

Illustrative Problem. A body, presumably a big fish, is detected at a range of 1 km by backscatter of 20 kHz sound that has an $SPL_{bs} = +80$ dB re $1 \, \mu$Pa. Assume that the sonar source level is 220 dB re $1 \, \mu$Pa. What is the target strength of the fish?

Solution. To calculate the target strength we need the total loss. From Fig. 3.3.1 we get $\alpha \simeq 3 \times 10^{-3}$ dB/m. Assuming isothermal water, (3.3.8) gives

$$TL = 20 \log \frac{R}{R_0} + \alpha R$$

$$= 20 \log \frac{10^3}{1} + (3 \times 10^{-3})(10^3)$$

$$= 63 \text{ dB}$$

Using (6.1.7), we find

$$TS = SPL_{bs} - SL + 2TL$$
$$= +80 - 220 + 2(63)$$
$$= -14 \text{ dB}$$

6.1.3. The Total Acoustical Cross Section

The concept of total acoustical scattering cross section is used to describe the cross section of a body that is implied by the total power it scatters over all angles. It may be larger or smaller than the geometrical cross section, depending on the frequency. The concept is useful in specifying the scatter from a distribution of omnidirectional scatterers and in calculating transmission loss due to scatter.

The total scattering cross section is defined by

$$\sigma_s \equiv \frac{\Pi_s}{I_p} = \frac{\displaystyle\int_0^{4\pi} I_s R^2 \, d\Omega}{I_p} = \int_0^{4\pi} \mathscr{S} A \, d\Omega \qquad \textbf{(6.1.8)}$$

where Π_s is the total power scattered over all angles and $d\Omega$ is the increment of solid angle.

When the scattered intensity I_s is independent of angle, the integration easily yields

$$\sigma_s = \frac{4\pi R^2 I_s}{I_p} = 4\pi\sigma_{bs} \qquad \text{omnidirectional} \qquad \textbf{(6.1.9)}$$

Some experimenters have reported values of σ_s obtained by multiplying the backscatter measurements of $\mathscr{S}_{bs} A = \sigma_{bs}$ by 4π. This operation is valid only when the scatter is known to be omnidirectional.

A body may absorb energy from the sound wave. Without going into the details of this conversion of acoustic energy to heat, we call the sound power absorbed by an object Π_a and, similar to (6.1.8), define the absorption cross section

$$\sigma_a \equiv \frac{\Pi_a}{I_p} \qquad \textbf{(6.1.10)}$$

The sum of the sound power scattered *and* absorbed by an object is the extinguished power.

$$\Pi_e = \Pi_s + \Pi_a \qquad \textbf{(6.1.11)}$$

The extinction cross section of a body is the sum of its absorption and scattering cross sections.

$$\sigma_e = \sigma_s + \sigma_a \qquad \textbf{(6.1.12)}$$

6.2. SCATTERING FROM SPHERES

The marine acoustician is interested in spheres not because there are spherical fish or plants in the sea but because scattering from spheres has been studied more thoroughly than any other scattering problem in acoustics and because the results of these studies are readily applied to marine animals. These studies provide clues about the change of scatter strength with sound frequency and the dependence of sound scatter on the size, compressibility, and material density of marine bodies. From these studies we learn that an acoustically small, nonspherical body whose dimension is less than the wavelength scatters in the same way as a sphere of the same volume and same average physical characteristics. These ideas are based on the physics of the problem. They will not change with time nor with the technology of the season.

6.2.1. Small Nonresonant Spheres ($ka \ll 1$), Rayleigh Scatter

When the sound wavelength is much greater than the sphere's circumference, there are two effects that can cause scatter. (a) If the sphere elasticity E_1 (=compressibility^{-1}) is less than that of the water E_0, incident condensations and rarefactions compress and expand the sphere, and a spherical wave is reradiated. This monopole radiation occurs also if $E_1 > E_0$ but the phase is different. (b)* If the sphere density ρ_1 is much greater than that of the medium ρ_0, the sphere's inertia will cause it to lag behind as the plane wave in the fluid swishes back and forth. The motion is equivalent to the medium being at rest and the sphere being in oscillation. This action generates a dipole type of reradiation (Section A5.1). When $\rho_1 < \rho_0$, the effect is the same but the phase is different. In either event when $\rho_1 \neq \rho_0$ the scattered pressure is proportional to $\cos \theta$ where θ is the angle between the scattered direction and the incident direction.

The scattering function for a small nonresonant sphere was first derived

---

* The subscript A for ambient density is omitted for convenience.

by Rayleigh (Section 335, 1896) and is

$$\mathscr{S} = \frac{(ka)^4}{\pi} \left[\frac{e-1}{3e} - \left(\frac{g-1}{2g+1} \right) \cos \theta \right]^2 \qquad ka \ll 1 \qquad \textbf{(6.2.1)}$$

where a = radius of sphere

$k = 2\pi/\lambda$ = wave number in the medium

$e = E_1/E_0$ = ratio of elasticity of sphere to medium; [E is usually calculated from $c^2 = E/\rho_A$ (2.8.8)]

$g = \rho_1/\rho_0$ = ratio of density of sphere to medium

and $\mathscr{S}_{bs}$ follows by setting $\theta = 180°$ *in* (6.2.1).

The total scattering cross section, obtained by integrating using (6.1.8) and $A = \pi a^2$, is

$$\sigma_s = \pi a^2 \int_0^{4\pi} \mathscr{S} \, d\Omega = 4\pi a^2 (ka)^4 \left[\left(\frac{e-1}{3e} \right)^2 + \frac{1}{3} \left(\frac{g-1}{2g+1} \right)^2 \right] \quad \textbf{(6.2.2)}$$

We notice that $\mathscr{S}$ has zero isotropic scatter (first term in squared brackets in (6.2.1)), when the marine body has the same elasticity as water $e = 1$. It has zero dipole scatter (the $\cos \theta$ term) when the density of the sphere is the same as that of water $g = 1$. The terms may add or cancel depending on the relative magnitudes of e and g.

The other important factor that describes scatter from any small body $(ka)^4$ characterizes what is called "Rayleigh" scatter. When $ka \ll 1$ the backscatter function, being proportional to k^4, is proportional to frequency to the fourth power.

When a sound search frequency is in the region of ka much less than unity and if $e \geq 1$ and $g \geq 1$, the backscatter is very small. Figure 6.2.1 shows this for a rigid sphere, in which case $e \gg 1$ and $g \gg 1$. The full range of ka values is given. It is evident that the scattering function leaves the Rayleigh scattering region at approximately $ka = 1$. This is the criterion on which the biomass pyramid (Fig. 1.6.1) was based. Physically, the acoustical cross section in Rayleigh scatter is very much smaller than the geometrical cross section because the sound waves bend around, and are hardly affected by, acoustically small nonresonant bodies. It is interesting to observe that ocean waves scatter from rocks and islands in very similar ways.

When a multifrequency sound impinges on "small" nonresonant scatterers, the backscattered energy is very much stronger for the higher frequency components. Quoting Lord Rayleigh's words in 1896 (Vol. II, Section 296): "If the primary (incident) sound be a compound musical note, the various component tones are scattered in unlike proportions.

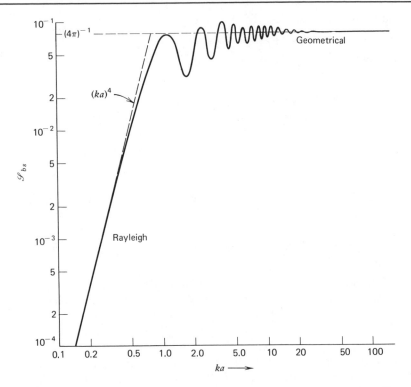

Figure 6.2.1. Backscattering function for a rigid, fixed, sphere of radius a. Derived from Stenzel (1938).

The octave, for example, is sixteen times stronger relatively to the fundamental tone in the secondary (scattered) than it is in the primary sound. There is thus no difficulty in understanding how it may happen that echoes returned from such $(ka \ll 1)$ reflecting bodies as groups of trees may be raised an octave." The Rayleigh scatter effect for broad band signals exists also in optics, where it causes the predominantly blue appearance of the sky. The effect occurs because the wave number k of the blue end of the sunlight spectrum is approximately double that of the red end, and the scattered intensity is therefore 16 times greater for blue light.

For a gas bubble at sea level, $e \ll 1$ and $g \ll 1$, the scatter is very large and is omnidirectional because the elasticity term dominates. Highly compressible bodies are also capable of resonating in the $ka \ll 1$ region. The scatter near the resonance frequency can be very much greater than the value given by (6.2.2). The subject of resonant bubbles is treated separately in Section 6.3.

6.2.2. Larger Spheres, Geometrical Scatter, and Interior Modes

For $ka \gg 1$, the behavior is described as "geometrical" scatter. The situation can be understood by using the high frequency ray approximation as illustrated in Fig. 6.2.2.

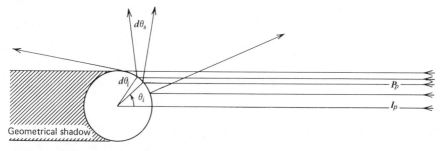

Figure 6.2.2. Geometry for ray acoustics solution for geometrical scatter from a rigid, fixed sphere, $ka \gg 1$. For convenience, θ_i is measured as shown.

For simplicity, assume a rigid, perfectly reflecting sphere; no energy penetrates into the sphere. Each ray is reflected with its angle of reflection equal to its angle of incidence. The description is sometimes called "ray acoustics." The incoming power in the angle increment $d\theta_i$ at incident angle θ_i is calculated in polar coordinates at the sphere surface. The surface area for the ring of width $a\,d\theta_i$, and ring radius $a \sin \theta_i$ is $dS = 2\pi(a \sin \theta_i)a\,d\theta_i$, and its component perpendicular to the incoming rays is

$$dS_\perp = (2\pi a^2 \sin \theta_i\,d\theta_i) \cos \theta_i$$

The incoming power through the ring at angle θ_i is

$$d\Pi_i = I_p 2\pi a^2 \sin \theta_i \cos \theta_i\,d\theta_i$$

The scattered power from this ring is emitted at angle $\theta_s = 2\theta_i$ within increment $d\theta_s = 2d\theta_i$. Using polar coordinates at the range R, the scattered power is found to be

$$
\begin{aligned}
d\Pi_s &= I_{gs} 2\pi R^2 \sin \theta_s\,d\theta_s \\
&= I_{gs} 2\pi R^2 (\sin 2\theta_i) 2\,d\theta_i \\
&= I_{gs} 8\pi R^2 \sin \theta_i \cos \theta_i\,d\theta_i
\end{aligned}
\tag{6.2.3}
$$

where I_{gs} is the intensity for geometrical scatter.

Since there is no power loss, we equate $d\Pi_i = d\Pi_s$, and for geometrical scatter, independent of the angle of scatter, we obtain

$$I_{gs} = I_p \frac{a^2}{4R^2} \qquad (6.2.4)$$

The total and backscattering cross sections, backscattering function, and target strength for a rigid sphere in geometrical scatter are as follows:

$$\sigma_s = \frac{\Pi_s}{I_p} = \frac{(4\pi R^2)I_{gs}}{I_p} = \pi a^2 \qquad (6.2.5)$$

$$\sigma_{bs} = \mathcal{S}_{bs} A = \frac{a^2}{4} \qquad (6.2.6)$$

$$\mathcal{S}_{bs} = \frac{1}{4\pi} \qquad (6.2.7)$$

$$TS = 10\log_{10}\frac{a^2}{4A_1}\ \text{dB} \qquad (6.2.8)$$

This value of $\mathcal{S}_{bs}$ is the asymptote being approached at the far right of Fig. 6.2.1.

This asymptote indicates that the backscattering function of the rigid sphere for $ka > 1$ is crudely constant. The backscattered intensity is therefore proportional to the cross-sectional area of the scatterer. This is called the region of geometrical scatter. In this region both $\mathcal{S}_{bs}$ and σ are very much larger than for $ka < 1$. The combination of Rayleigh scatter and geometrical scatter operate as if the rigid sphere is a high pass backscattering filter with cutoff at approximately $ka \approx 1$.

Light follows essentially the same backscattering laws as sound. One big difference between the two is that the wavelength of visible light is of order 5×10^{-5} cm. Therefore almost all bodies in the sea have optical cross sections equal to their geometrical cross sections. However the same particles are very much smaller than the wavelength of sounds in the sea (generally 1 cm or more). Therefore they scatter sound in the Rayleigh region. This is why ocean water is turbid for light but transparent to sound.

In the case of a rigid sphere there is no interior motion. For any fluid sphere, when the interior wavelength λ_i is comparable to or less than the radius, interior wave propagation becomes important. Then standing waves are set up, and these determine the *modal resonance frequencies* of the body.

When there is motion within the body, the scattering functions and scattering cross sections have peaks and troughs at frequencies that correspond to the modes stimulated in the interior of the sphere. The magnitude of the response depends on the damping effect of the sphere material.

For acoustical studies of marine life we must consider spheres of relative densities g and relative elasticities e close to unity. The data are in Chapter 7.

The backscattering function due to interior resonances is presented in Fig. 6.2.3. The case of both small density and elasticity, curve C, shows strong modal resonance responses.

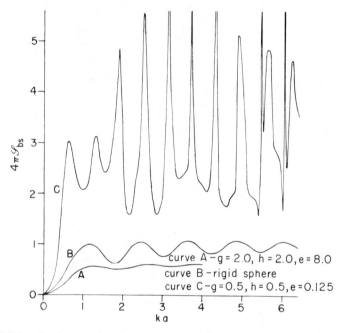

curve A — g = 2.0, h = 2.0, e = 8.0
curve B — rigid sphere
curve C — g = 0.5, h = 0.5, e = 0.125

Figure 6.2.3. Backscattering function for various values of relative density g, relative speed h, and relative elasticity, $e = gh^2$; k is the wave number in water. (Anderson, 1950.)

Figure 6.2.4a shows the dependence of the backscattering function on e for a body of the same density as water, whereas such dependence appears as a function of $e = g$, for a sphere of the same speed of sound as the medium, in Fig. 6.2.4b. Except for small values of e, the backscattering function for fluid spheres in the ocean is generally less than for geometrical scatter from a rigid sphere. The effect of modal resonances is

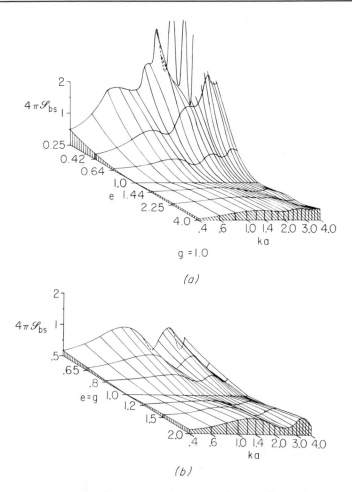

Figure 6.2.4. Backscattering function for fluid spheres of dimensions comparable for a wavelength. The $4\pi\mathscr{S}_{bs}$ appears (a) as a function of ka and relative elasticity, e for relative density $g = 1$ and (b) in terms of ka and $e = g$ for $h = 1$. k is the wave number in water. (Anderson, 1950.)

evident and is particularly exaggerated for e less than unity (compressible bodies).

Some marine life consists of a fluid body surrounded by an elastic shell (a carapace). A fluid-filled shell has the possibility of longitudinal waves in the fluid interior plus longitudinal and transverse waves in the shell. The backscatter is very sensitive to the relative thickness of the shell.

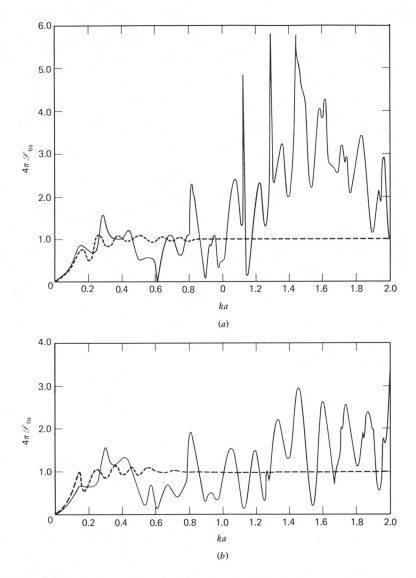

Figure 6.2.5. Backscattering function for a hollow aluminum sphere filled with water to a distant source of continuous waves: a = outer radius, b = inner radius; theory for the rigid sphere indicated by dashed line. (*a*) $b/a = 0.925$. (*b*) $b/a = 0.95$. (Hickling, 1964.)

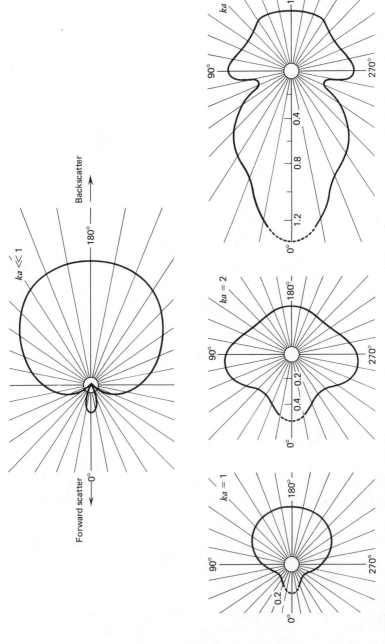

Figure 6.2.6. Polar scattered radiation patterns in the far field ($kR \gg 1$) for a fixed, rigid sphere as a function of different values of ka. The sound is incident from the right. The length of the radius vector from the center gives the relative pressure. The scattered amplitudes for $ka \ll 1$ have been greatly enlarged. Cases $1 \le ka \le 10$ are from Stenzel (1938).

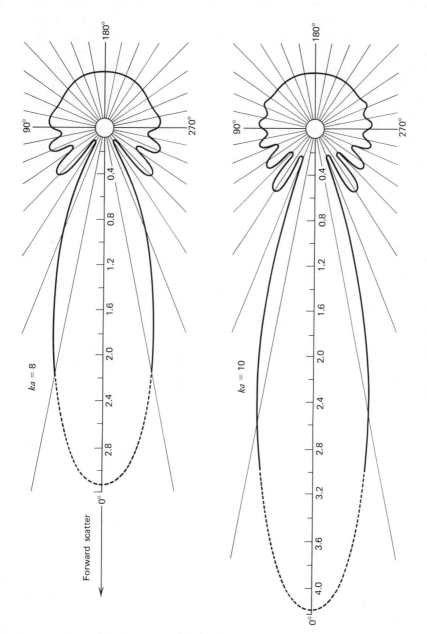

Figure 6.2.6. (*Continued*)

Wide excursions in the scatter take place compared with the constant value for the rigid sphere at $ka \gg 1$. The problem has been solved for a water-filled spherical shell of aluminum (Fig. 6.2.5). The scattering behavior is qualitatively the same for a marine animal with an elastic shell.

The scattered radiation patterns for the rigid sphere, $e \gg 1$ and $g \gg 1$, has been calculated for $ka \ll 1$, from (6.2.1) and is shown in Fig. 6.2.6. When $ka \approx 1$, $\mathcal{S}$ becomes a very complicated function of frequency; notice $\mathcal{S}_{bs}(f)$ for a rigid body in Fig. 6.2.1, particularly for $1 < ka < 10$; $\mathcal{S}(\theta)$ also goes wild as shown for a rigid body at $ka = 1, 2, 4, 8, 10$ in Fig. 6.2.6. We notice that as ka increases, an omnidirectional pattern develops supplemented by a growing forward-scattered lobe. In practice the scatter near 0° is difficult to isolate from the incident wave that travels about the same path in about the same time. The inner 10° cones of Fig. 6.2.6 appear as dotted lines to suggest this difficulty. The complexity of the radiation patterns and imperfections in real solid spheres also limit the precise experimental verification of this theory.

6.3. THE BUBBLE

A bubble differs from a rigid sphere in two major ways. First, the ρc of the bubble gas is much less than that of the surrounding medium (e.g., at sea level, $\rho c_{water}/\rho c_{air} = 5 \times 10^3$). Second, a bubble can resonate. When the bubble senses a frequency at or near its natural frequency it very effectively absorbs and scatters that sound. At resonance, the scattering and absorption cross sections of a typical bubble at sea are of the order 10^3 times its geometrical cross section. In addition to being a source of scatter and absorption, bubbles change the compressibility of the water and cause the phase speed of sound to be a function of frequency. The medium is then said to be dispersive. During stormy periods the water within meters of the sea surface contains enough bubbles to alter the speed of sound by several meters per second (depending on the frequency), compared to the values calculated from the temperature, salinity, and depth (3.2.20).

6.3.1. Resonance

We now calculate the natural pulsation frequency of a bubble, assuming for simplicity that the motion is completely accounted for by two factors: (a) the compressibility of the enclosed gas, and (b) the liquid mass moved by the bubble as it pulsates. These two components are all that is needed to produce a harmonic oscillation at a natural frequency. For the present we assume that the damping is negligible and that there are no effects due to surface tension or thermal conductivity. These corrections will be applied later.

BUBBLE STIFFNESS

We start with the adiabatic gas relation, $pV^\gamma = \text{constant}$, where p is the instantaneous total pressure within the bubble volume V. The constant γ is the ratio of specific heats for the enclosed gas. Differentiating yields

$$\frac{dp}{dV} = -\frac{\gamma p}{V}$$

Write $dp \equiv p_i = \text{instantaneous incremental pressure} = p - P_A$, where $P_A \gg |p_i|$ is the static ambient pressure. Use $V = \frac{4}{3}\pi a^3$ and $dV = 4\pi a^2 \xi$, where $\xi = \text{radial displacement during oscillation}$. Therefore

$$p_i = -\frac{3\gamma P_A \xi}{a}$$

Assuming that there is no surface tension, the restoring force F_r acting over the entire spherical surface is

$$F_r = 4\pi a^2 p_i = -(12\pi\gamma P_A a)\xi \qquad (6.3.1)$$

Equation (6.3.1) is in the form of Hooke's law; the bubble is an elastic system in which stress is proportional to strain. The quantity in parentheses is the stiffness constant s', where the prime refers to the previous assumptions.

$$s' = 12\pi\gamma P_A a \qquad \textbf{(6.3.2)}$$

BUBBLE MASS

The major part of the oscillating mass is due to the liquid adjacent to the bubble rather than the mass of the gas. The inertial force acting over the bubble surface is determined by calculating the pressure of the re-radiated sound. Later we verify that, at resonance, the acoustic wavelength in the water is much greater than the radius of the bubble. For now, we assume that $ka \ll 1$ and that the scattered pressure is isotropic and is given by

$$p_s = \frac{P_s R_1}{R} \exp[i(\omega t - kR)] \qquad \textbf{(6.3.3)}$$

The acoustic momentum equation (2.7.3) gives the particle acceleration in term of the acoustic pressure in the medium. In spherical coordinates, the radial component is

$$\rho_A \ddot{\xi} = -\frac{\partial p_s}{\partial R} \qquad (6.3.4)$$

where a dot indicates differentiation with respect to time and ρ_A is the ambient water density. Use (6.3.3) in (6.3.4) for the condition at the bubble surface to obtain

$$\rho_A \ddot{\xi}]_{R=a} = \frac{P_s R_1}{R^2} (1 + ikR) \exp\left[i(\omega t - kR)\right]\Bigg]_{R=a}$$

Since $ka \ll 1$, this simplifies to

$$p_s]_{R=a} = \rho_A a \ddot{\xi}]_{R=a} \qquad (6.3.5)$$

The equivalent mass is found by calculating the inertial force F_m at the surface

$$F_m]_{R=a} = -4\pi a^2 p_s]_{R=a} = -4\pi a^3 \rho_A \ddot{\xi}]_{R=a} \qquad (6.3.6)$$

Equation (6.3.6) allows us to identify an effective mass

$$m = 4\pi a^3 \rho_A \qquad \textbf{(6.3.7)}$$

equivalent to a volume $4\pi a^3$ of water that rides with the bubble as it pulsates.

From (6.3.1) and (6.3.7), Newton's second law for a radial displacement ξ is then

$$F_r = ma$$

$$-12\pi\gamma P_A a\xi = 4\pi a^3 \rho_A \ddot{\xi} \qquad (6.3.8)$$

This equation describes simple harmonic motion caused by a combination of localized inertia and restoring force. The solution is $\xi = \xi_0 \cos(2\pi f'_R t - \phi)$ where the amplitude ξ_0 and the phase ϕ depend on the initial conditions of the motion. The natural frequency f'_R, obtained by substitution of ξ into (6.3.8), depends only on the constants of the system.

$$f'_R = \frac{1}{2\pi}\left(\frac{s'}{m}\right)^{1/2} = \frac{1}{2\pi a}\left(\frac{3\gamma P_A}{\rho_A}\right)^{1/2} \qquad \textbf{(6.3.9)}$$

The subscript R refers to resonance, and the prime refers to the previous assumptions. For an air bubble in water at depth Z meters, we can use $\rho_A = 1.03 \times 10^3 \text{ kg/m}^3$, $P_A \approx 10^5 (1 + 0.1Z) \text{ N/m}^2$, $\gamma = 1.40$, to obtain a simplified expression for the resonance frequency in hertz.

$$f'_R \cong \frac{3.25(1+0.1Z)^{1/2}}{a \text{ (meters)}} = \frac{3.25 \times 10^6}{a \ (\mu)}(1+0.1Z)^{1/2} \qquad \textbf{(6.3.10)}$$

Defining k_R, the wave number in water at the resonance frequency, we note that the sea level value of $k_R a_R$ is 0.0136 for an air bubble. Therefore our previous assumption that $ka \ll 1$ near resonance is verified.

The assumption that volume stiffness is the only restoring force must be modified for bubbles of small radii because surface tension is then a significant additional force. Also, the assumption that the gas vibrates adiabatically is no longer valid if the bubble radius is very small. Then the temperature at all points of the gas volume stays close to the water temperature, and the oscillation is more nearly isothermal.

The introduction of these two effects is a lengthy process that is discussed in Appendix A6.1. Then (6.3.9) becomes

$$f_R = \frac{1}{2\pi a}\left(\frac{3\gamma P_A b\beta}{\rho_A}\right)^{1/2} \qquad \textbf{(6.3.11)}$$

where b and β are functions, defined in (A6.1.26) and (A6.1.27), that correct for the adiabatic assumption and surface tension. The two corrections are to some extent mutually counteracting; thus the simpler equations (6.3.9) and (6.3.10) are in error by less than 8% for air bubbles of radius greater than 2 μ, at sea level.

We have been describing a clean, free bubble. Bubbles at sea are of other gases, as well as air, and they may have organic skins or other detritus on their surfaces or, indeed, be parts of plankton or fish bodies! The resonance frequency is of course altered by these changed conditions. The value γ ranges from 1.0 to 1.67 depending on the gas. Other surfaces of free bubbles have been considered in the literature (e.g., Fox and Herzfeld, 1954; Harvey et al., 1944; Turner, 1961).

6.3.2. Damping

A sound beam propagating through bubbly water attenuates and backscatters. The physical causes of the attenuation are thermal conductivity and shear viscosity absorption principally at the bubble wall, as well as reradiation (scatter) of sound out of the beam. For a single bubble the motion is described by adding a damping force to the stiffness and inertial forces of (6.3.8). Without damping there would be no limit to the amplitude of the bubble pulsation at resonance; with it, the amplitude reaches a predictable maximum at the resonance frequency and has predictable reduced motion at nearby frequencies.

The peakedness, or sharpness of the change of amplitude with frequency near resonance, is defined in terms of the Q of the system

$$Q \equiv \frac{f_R}{\Delta f} \qquad (6.3.12)$$

where $\Delta f = f_U - f_L$, and $f_U = f_R + \Delta f/2$ and $f_L = f_R - \Delta f/2$ are the upper and lower frequencies at which the power which is absorbed and scattered by the bubble drops to one-half the value at the resonance peak.

Another useful descriptor of the width and peak value of the resonance curve is the damping constant *at resonance* δ_R defined as

$$\delta_R = \frac{1}{Q} \qquad (6.3.13)$$

The dependence of the damping constant on the physical characteristics of the bubble gas is discussed in Section A6.1; δ_R is made up of the damping constants due to reradiation or scatter δ_{Rr}, plus shear viscosity δ_{Rv}, and thermal conductivity δ_{Rt}. The subscript R indicates the resonance value.

$$\delta_R = \delta_{Rr} + \delta_{Rt} + \delta_{Rv} \qquad (6.3.14)$$

The resonance damping constant due to reradiation is

$$\delta_{Rr} = k_R a_R \qquad (6.3.15)$$

which may be calculated from the resonance condition (6.3.11). For an air bubble at sea level $\delta_{Rr} = 0.0136$.

The dependence of the resonance damping constants on the resonance frequency is plotted in Fig. 6.3.1 for an air bubble at sea level. Single bubble experiments performed in the frequency range 1 to 300 kHz verify these curves, which have been derived from the physical constants of the gas and water.

Illustrative Problem. A laboratory experiment in which bubbles were generated by breaking waves (Glotov, 1962) showed that the largest number per unit volume were of radius 60 μ. Calculate the resonance frequency and the half power frequencies. Assume sea level pressure.

The approximate resonance frequency at sea level can be calculated from (6.3.10)

$$f'_R = \frac{3.25 \times 10^6}{60} = 54.2 \times 10^3 \, \text{Hz} = 54.2 \, \text{kHz}$$

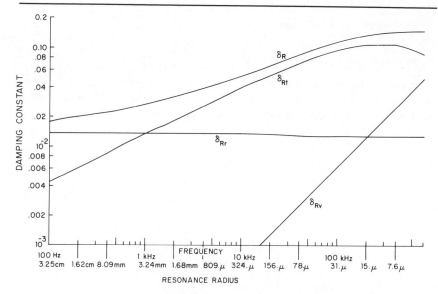

Figure 6.3.1. Resonance damping constants for an air bubble at sea level.

The resonance frequency corrected for thermal conductivity and surface tension can be calculated from Fig. A6.1.1 where $f_R/f'_R = 0.96$ for a 60 μ bubble. Therefore $f_R = 52.0$ kHz.

To calculate the half power points, use $\delta_R = 0.095$ from Fig. 6.3.1. Then (6.3.12) and (6.3.13) give

$$\Delta f = f_R \, \delta_R = (52.0)(0.095) = 4.94 \text{ kHz}$$

$$f_U = f_R + \frac{\Delta f}{2} = 54.5 \text{ kHz}; \qquad f_L = f_R - \frac{\Delta f}{2} = 49.5 \text{ kHz}$$

6.3.3. Scattering Function and Cross Sections

When a bubble is insonified by a plane wave

$$p_p = P_p e^{i\omega t} \tag{6.3.16}$$

of wavelength much greater than the bubble radius, $ka \ll 1$, the bubble is driven into oscillation and reradiates or "scatters" a spherically symmetrical pressure wave given by (6.3.3). The ratio of the scattered to incident intensity for a damped bubble, $|P_s|^2/|P_p|^2$ is derived in Section A6.1.

The scattering cross section is

$$\sigma_s = 4\pi \frac{|P_s|^2}{|P_p|^2} = \frac{4\pi a^2}{[(f_R/f)^2 - 1]^2 + \delta^2} \qquad (6.3.17)$$

The proof is given in Section A6.1, which defines the general damping constant δ, a function of frequency. The assumption is sometimes made that δ is slowly varying so that $\delta \approx \delta_R$, the damping at resonance.

Since bubble scatter is omnidirectional, $\mathscr{S}_{bs} = \sigma_s/4\pi^2 a^2$ and $TS = 10\log_{10}\sigma_s/4\pi R_1^2$ as given by (6.1.3) and (6.1.4); $\sigma_s/\pi a^2$ is plotted in Fig. 6.3.2.

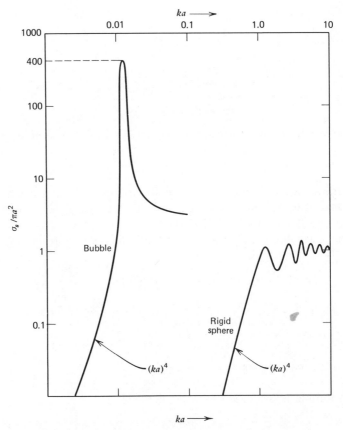

Figure 6.3.2. Variation of ratio of total scattering to geometrical cross section $\sigma_s/\pi a^2$ with ka for a fixed rigid sphere and a bubble. The height of the bubble resonance peak was calculated for the sea level resonance 52 kHz. The low frequency tail of the bubble resonance curve may also be calculated from (6.2.2) for $ka \ll 1$ and $e \ll 1$, $g \ll 1$.

We are particularly interested in the exaggerated effects that occur at the resonance frequency where (6.3.17) gives

$$\sigma_s = \frac{4\pi a^2}{\delta_R^2} \qquad \text{resonance} \qquad (6.3.18)$$

The acoustical importance of bubbles can be appreciated when their values of σ_s are compared with those of the rigid sphere (e.g., Fig. 6.3.2).

To understand the amplified acoustical cross sections it helps to realize that the bubble at resonance is effectively a "hole" of very low acoustic impedance compared to that of the water medium. This hole distorts the incoming acoustic field over a very large volume surrounding the bubble. The distorted field represents a power flow toward the bubble from a section of the incoming wave that is very much greater than the bubble cross section. As viewed from a great distance, the absorption and scatter caused by the bubble thereby identify acoustical cross sections that can be far larger than the geometrical cross section.

6.3.4. Absorption and Extinction Cross Sections

The extinction cross section of a bubble can be calculated from

$$\sigma_e = \frac{\Pi_e}{I_p} \qquad (6.3.19)$$

The extinguished power (scattered and absorbed) is obtained from the rate at which work is done on the bubble by the incident pressure. The calculation is in Section A6.1.3. The result is

$$\sigma_e = \frac{\Pi_e}{P_p^2/\rho_A c} = \frac{4\pi a^2 (\delta/\delta_r)}{[(f_R/f)^2 - 1]^2 + \delta^2} \qquad (6.3.20)$$

where $\delta_r = ka$. Figure 6.3.3 gives σ_s and σ_e as a function of radius for four frequencies. The resonance peaks are clearly seen. For radii smaller than the resonance size, the cross sections are approximately proportional to a^6; a^4 of this rapid variation is the Rayleigh scattering effect that comes from the denominator of (6.3.20) because f_R is proportional to $1/a$ (6.3.9). For radii much larger than the resonance size, the denominator of (6.3.20) is approximately constant; therefore the scattering and extinction cross sections are proportional to the geometrical cross section. Each "approximately" in these statements is needed when we assume that the damping is constant.

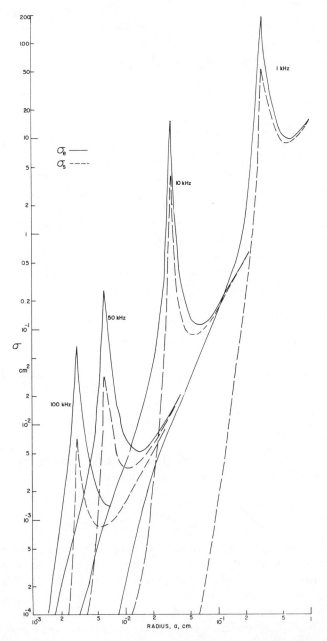

Figure 6.3.3. Extinction and scattering cross sections of air bubbles of different radii at sea level when bubbles are insonified by four different sound frequencies. The extinction cross section $\sigma_e = \sigma_a + \sigma_s$ is much greater than the scattering cross section except for $a \gg a_R$, where σ_a is negligible and $\sigma_e \simeq \sigma_s$. (Medwin, 1976.)

Comparison of (6.3.20) with (6.3.17) shows that at resonance where $\delta \to \delta_R$ and $\delta_r \to \delta_{Rr}$, and so on, the acoustical cross sections are proportional to the damping constants that cause them, yielding

$$\frac{\sigma_e}{\sigma_s} = \frac{\delta_R}{\delta_{Rr}} \quad \text{and} \quad \frac{\sigma_a}{\sigma_s} = \frac{\delta_{Rt} + \delta_{Rv}}{\delta_{Rr}} \qquad (6.3.21)$$

6.4. MULTIPLE SCATTERING AND ATTENUATION

The ocean is full of unidentified scatterers that clutter and sometimes obscure the backscatter from our targets. Some of them not only scatter our signal, they absorb it as well. The net result of the scatter and absorption by these bodies and bubbles is sometimes great enough to cause significant attenuation. When the absorption is large, it is accompanied by dispersion; that is, the phase velocity of propagation becomes a function of sound frequency (Section A6.2).

6.4.1. Multiple Scattering by Bubbles

When scatterers are widely and randomly spaced, the scattering cross sections of the individuals simply add, and the backscattering cross section per unit volume s_v becomes a useful concept.

$$s_v = \sum_i N_i \sigma_{\mathrm{bsi}}$$

where $N_i = $ (number/volume) of scatterers of backscattering cross section σ_{bsi}. This subject is considered in the next chapter.

When scatterers are packed together closely enough for their scattered fields to interact, the picture is completely changed. Interaction occurs approximately at separations less than the wavelength. The result of the interaction is that the resonance curve of a single bubble is broadened and the scattering cross section of a group of bubbles is less than the sum of the individual cross sections.

A study of the scatter from a group of bubbles is summarized by Fig. 6.4.1. The target strength for the single spherical bubble is drawn with the reference resonance peak arbitrarily set at 0 dB. For simplicity, the curve is drawn under the assumption that $\delta_R = 0.0136$. When a line of closely packed bubbles is considered, the target strength is broadened about the resonance position and the resonance frequency is lowered. For comparison the two cases are drawn with the same reference peak.

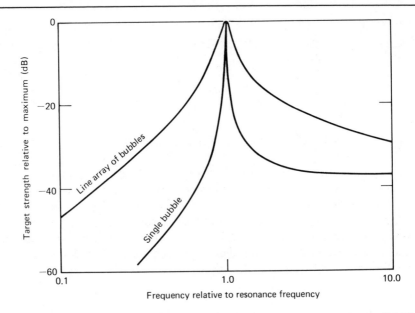

Figure 6.4.1.   Response curve shapes for bubbles. Redrawn from Weston (1966).

Although the bubble here has a $Q = 1/\delta_R \approx 75$ at sea level, the Q of the *line* of bubbles is only 7. Whereas the sea level resonance frequency of the spherical bubbles is given by $f_R \approx 325/a$, the resonance frequency of the line of bubbles is $f_R = 114/a$, where a is in centimeters.

   If bubbles are packed randomly in a three-dimensional array, as in the wake of a ship, the face of the bubble cluster acts as a pressure release surface, and reflection takes place at this interface.

   A closely packed bubble cluster in a *regular* array would produce additional resonance effects dependent on the array spacing.

6.4.2.   Attenuation by Bubbles

BUBBLES OF ONE SIZE ONLY

Assume that we have water containing bubbles of radius a and that the bubbles are far enough apart to prevent interaction effects. Effectively this will be true when the separation is greater than $\sqrt{\sigma_e}$. (Sometimes the stronger criterion, separation $> \lambda$, is used.)

   If the incident plane wave intensity is I_p, the power extinguished by each bubble is $I_p\sigma_e$, where σ_e is calculated from (6.3.20). Assuming N resonant bubbles per unit volume, the change of intensity over a distance dx is

$$d\bar{\imath} = -I_p\sigma_e N\, dx$$

Integrating

$$I_x = I_p \exp\left(-\sigma_e Nx\right) \qquad \text{or} \qquad P_x^2 = P_p^2 \exp\left(-\sigma_e Nx\right)$$

After traversing distance x, the change in sound pressure level becomes

$$\Delta SPL = 20 \log_{10} \frac{P(x)}{P_p} = -10\sigma_e Nx \log_{10} e$$

The excess attenuation per unit distance due to bubbles is

$$\alpha_b = -\frac{\Delta SPL}{x} = 4.34\sigma_e N \text{ dB/distance} \tag{6.4.1}$$

where $\sigma_e N =$ scattering cross section per unit volume. The length units must be consistent, for example,

$$[\sigma_e] = \text{m}^2, \qquad [N] = \text{m}^{-3}, \qquad [\alpha_b] = \text{dB/m}$$

BUBBLES OF MANY SIZES

When there are bubbles of several sizes, the number per unit volume must be defined in terms of the range of radii of interest. For example, an investigator might be interested in the number of bubbles per unit volume of radius between 59 and 60 μ, or the number per unit volume over radius range 59 to 61 μ. Since the number depends on the radius increment da, we define

$$n(a)\, da = \frac{\text{number of bubbles of radius between } a \text{ and } a + da}{\text{volume}}$$

It is common to set da equal to one micron.

The extinction cross section per unit volume S_e for sound traversing a random mixture of noninteracting bubbles is calculated by using (6.3.20) in the integration.

$$S_e = \int_0^\infty \sigma_e n(a)\, da = \int_0^\infty \frac{4\pi a^2 (\delta/\delta_r) n(a)\, da}{[(f_R/f)^2 - 1]^2 + \delta^2} \tag{6.4.2}$$

The units of S_e are generally given as $\text{m}^2/\text{m}^3 = \text{m}^{-1}$. The attenuation due to a mixture of bubbles is obtained by repeating the derivation that led to (6.4.1), where S_e replaces $\sigma_e N$ for the scattering cross section per unit volume. The result is

$$\alpha_b = 4.34 S_e \text{ dB/m} \qquad (6.4.3)$$

Evaluation of the integral in (6.4.2) can be done by approximation, using the knowledge that the major contribution is near the resonance frequency, or by graphical or numerical methods. The approximation technique (Wildt, 1946) assumes that only bubbles of radius close to the resonance radius contribute to S_e and that the number of bubbles per unit volume and the damping constant are constant over the integration. The approximate attenuation is (loc. cit., p. 470)

$$\alpha_b = \frac{85.7 a_R^2 n(a_R)}{k_R} \qquad (6.4.4)$$

Sometimes this is written in terms of the fraction of the water that is in the form of bubbles $u(a_R)\, da_R$, which is the product of the number of bubbles per unit volume by the volume of each bubble at resonance

$$u(a_R)\, da_R = \frac{4 \pi a_R^3}{3} [n(a_R)\, da_R] \qquad (6.4.5)$$

In terms of $u(a_R)$ the approximate solution is written

$$\alpha_b = \frac{20.5 u(a_R)}{k_R a_R} \text{ dB/m} \qquad (6.4.6)$$

When the number of bubbles per unit volume is known, the graphical solution or numerical integration can be used to find S_e. In Fig. 5.4.4 the curves labeled M show the results of an acoustical determination of bubble population near the sea surface. The method is described in Section A6.3. The fourth power dependence of number on bubble radius is redrawn in Fig. 6.4.2, together with the extinction cross section per bubble from Fig. 6.3.3. The area under the product curve represents S_e. Then α_b is calculated from (6.4.3).

6.4.3. Absorption and Attenuation by Multiple Bodies

Consider the attenuation caused by viscosity and thermal conductivity at the boundaries of marine bodies. In the case of shear viscous losses, Epstein and Carhart (1953) have shown that the attenuation is proportional to $(1-g)$, where g is the ratio of the body density to the medium density. Since g for marine bodies is close to unity (Table 7.1.1), the loss caused by shear viscosity is very small.

$f = 50$ kHz

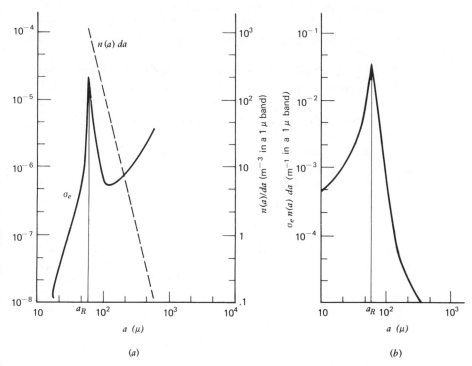

(a)   (b)

Figure 6.4.2. Extinction cross section per unit volume $S_e = \int_0^\infty \sigma_e n(a)\, da$ can be calculated from the curves σ_e for single bubbles and $n(a)\, da$, which represents a typical a^{-4} dependence of number of bubbles per unit volume (cubic meter) in a $1\,\mu$ band near the sea surface (see Fig. 5.4.4). The σ_e curve (a) is from Fig. 6.3.3 for 50 kHz. (b) Product of the two curves, representing the extinction cross section per unit volume in a $1\,\mu$ radius increment at the radius indicated. The area under that curve represents S_e for the bubble mixture. The major part of S_e comes from bubbles of radius near resonance.

Thermal conductivity losses are proportional to $(\gamma - 1)$, where γ is the ratio of specific heats of the body. Since marine bodies are liquidlike, and since γ is close to unity for liquids, there is very little loss of energy by thermal conductivity.

The physical reasons for small losses are easily stated. The typical small marine body, with $g \approx 1$, is almost completely entrained by the sound wave, thereby has small shear viscous losses. Also, since $e \approx 1$, it experiences small changes of volume in response to the acoustic pressures around it, thus exhibiting small temperature changes and small thermal conductivity losses.

The added attenuation in water containing a large number of non-resonating marine bodies is proportional to the frequency squared and can be described as an increase in the shear and bulk viscosity compared to clean water.

For example, in a laboratory study of suspended algae (*Scenedesmus*) consisting of individual ellipsoidal cells clustered into planar groups of diameter approximately 50 μ, Meister and St. Laurent (1960) found attenuations proportional to concentration in distilled water and proportional to sound frequency squared from 15 to 27 MHz ($ka \ll 1$). There was no measurable attenuation due to scatter, and the total attenuation could be expressed by the same equation as for clean water but with different viscosities [recalling (3.3.5)].

$$\alpha_F = \frac{4.34\omega^2}{\rho_F c_F^3}\left(\frac{4\mu}{3} + \mu'\right) dB/m$$

The shear viscosity μ increased linearly with the algae concentration.

The attenuation experiment gave the bulk viscosity $\mu' = 44\mu$ independent of concentration or temperature from 5 to 20°C. Since $\mu'_F \approx 2.8\mu_F$ for clean freshwater, the experiment showed that the algae increased the compressibility losses even more than the shear losses. Figure 6.4.3 is a

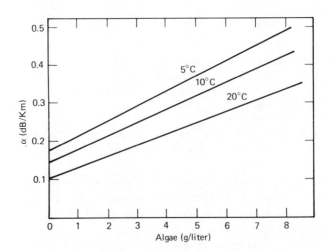

Figure 6.4.3. Attenuation in suspension of *Scenedesmus* algae in distilled water as a function of concentration, at frequency 21 kHz. Extrapolation from Meister and St. Laurent (1960).

graph of attenuation as a function of algae concentration, extrapolated down to frequency 21 kHz on the basis of the ω^2 dependence given by (3.3.5).

Although the experiment was for freshwater algae in distilled water, it can be assumed that sound absorption by suspensions in seawater would follow a similar behavior. It should be noted, however, that although the body attenuations of Fig. 6.4.3 are larger than the attenuation in clean water, they are smaller than the attenuation in seawater (Fig. 3.3.1).

6.4.4. Resonating Marine Bodies

Body resonance of a spherical marine body takes place at a frequency that can be calculated from the stiffness and mass as in the case of the bubble (Section 6.3.1).

$$f_R = \frac{1}{2\pi}\left(\frac{s}{m}\right)^{1/2} \tag{6.4.7}$$

Assume that the body is a fluid of elasticity E_i surrounded by a skin or elastic membrane. Using Hooke's law for the bulk elasticity of the interior fluid, we write

$$E_i = \frac{F/A}{\Delta V/V} \tag{6.4.8}$$

where $A = 4\pi a^2$
$\quad V = \frac{4}{3}\pi a^3$
$\quad \Delta V = 4\pi a^2 \xi$
$\quad \xi = $ radial displacement

The restoring force is

$$F_S = -F = -12\pi E_i a\xi \tag{6.4.9}$$
$$= -s_b\xi \tag{6.4.10}$$

where the body stiffness is $s_b = 12\pi E_i a$. Comparison with (6.3.2) shows that E_i replaces γP_A for a gas bubble.

There is also restoring force due to the elasticity of the body wall tissue. Assuming that it is rubberlike, the tissue stiffness (McCartney and Stubbs, 1970) is

$$s_t = 48\pi\mu_1 t \tag{6.4.11}$$

where $t = $ tissue thickness
$\quad \mu_1 = $ real component of shear modulus of elasticity of tissue

Therefore the total stiffness for the fluid sphere is $s = s_b + s_t$

$$s = 4\pi a\left(3E_i + 4\mu_1\frac{3t}{a}\right) \qquad (6.4.12)$$

The effective oscillating mass of the sphere fluid can be calculated from its kinetic energy. This will be much less than the radiation mass if the interior density $\rho_i \lesssim \rho_A$ and $ka \leq 1$; therefore we use the radiation mass due to the oscillating water mass alone, which we calculated for the bubble (6.3.7).

$$m = 4\pi a^3\rho_A$$

Using (6.3.7) and (6.4.12) in (6.4.7), the body resonance frequency is found to be

$$f_R = \frac{1}{2\pi a}\left[\frac{3E_i + 4\mu_1(3t/a)}{\rho_A}\right]^{1/2} \qquad (6.4.13)$$

For marine bodies $E_i = \rho_i c^2 \simeq 2 \times 10^9$ Pa (see Table 7.1.1) compared to $\gamma P_A \simeq 1.5 \times 10^5$ Pa for bubbles at sea level. Therefore the stiffness of a marine body would be of the order 10^4 greater than that of a bubble. Adding the shear elasticity term would further increase the restoring force, which would result in a body resonance frequency at least 10^2 greater than for a bubble of the same radius.

For example, shrimp eggs of radius $115\ \mu$ would be expected to have a body resonance at least 2.8 MHz depending on μ_1 and t. In fact Gruber and Meister (1961) experimentally found a strong resonance at approximately 7 MHz for shrimp eggs (*Artemia*). The resonance was accompanied by dispersion of the speed of sound (Watson and Meister, 1963). The damping constant at resonance was found to be $\delta_R = 0.7$.

The attenuation at the resonance frequency can be calculated from (6.4.1) by using the extinction cross section derived for bubbles (6.3.20), with the proper value of δ_R and δ_{R_r} from (6.3.15).

Higher frequency modal resonances are also present, as described in Section 6.2.2.

6.5. OTHER SCATTERING EFFECTS

This chapter has been concerned to this point with continuous waves that completely surround the scattering body. Also, all the solutions have been "steady state" solutions during continuous rather than transient irradiation. This section considers what happens when these restrictions are removed.

6.5.1. Transients

The transient response of a scatterer when it begins to be insonified depends on (a) the temporal growth of the sound wave at the front of the wave, which in turn depends on the transducer that originally generated the wave; (b) the spatial extent of the wave as it partially or completely blankets the scatterer; and (c) the damping constants of each of the modes of the scatterer.

When the scatterer is large enough ($ka \geq 1$) for more than one mode to be excited, each mode has its own damping constant, therefore its own growth rate, steady state amplitude and decay rate. Consequently, the displacement at any point of the scatterer and the external scattering from it are involved functions of time and coordinates.

6.5.2. Creeping Waves

The steady state analysis of CW scatter into modes of reradiation that come from specific standing waves in the interior of the body is less valuable in describing transients. Furthermore, in the case of large ka, the shadow behind the scatterer is adequately defined only when a very large number of modes have been considered.

An alternative description of this problem is based on Sommerfeld's explanation of the bending of radio waves as they propagate around the earth. This technique was applied by Franz to explain the diffraction of electromagnetic waves around electrically conducting cylinders and spheres. Franz called the diffraction "creeping waves." Later, acousticians studied diffraction around bodies by reexpressing the scattered radiation in terms of circumferential waves. One class of circumferential waves that evolves from this description propagates around the outside of the body; such waves have a speed slightly slower than the speed of waves in the free liquid and a large attenuation that increases with increasing angle, and depend somewhat on the elastic properties of the scatterer. The Franz wave, or creeping wave, depends principally on the geometry of the body. They are compressional waves and are really the same diffraction phenomenon in the geometrical shadow region considered qualitatively in Chapter 2 from the point of view of Huygens wavelets (Fig. 6.5.1).

6.5.3. Mode Conversion

When sound is incident on a *fluid* scatterer, the longitudinal (compressional) wave in the medium generates longitudinal waves in the scatterer.

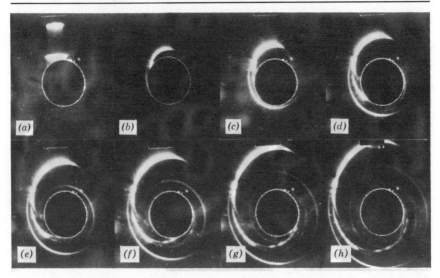

Figure 6.5.1. A time sequence of Schlieren photographs of a 3 μs pulse of 6.6 MHz, incident on type 6061 solid aluminum cylinder. The source is at the top radiating downward, $ka = 353$. (Neubauer and Dragonette, 1970.)

If the scatterer is a solid or has a solid shell, however, an incident compressional wave will generate waves of *both* the compressional and shear form.

This "mode conversion" can be made dramatically clear by Schlieren photographs. The Schlieren technique, which shows density differences in transparent media, has been used to identify both the incident pulse and the various scattered waves from an aluminum cylinder in water. In Fig. 6.5.1 the cylinder has a diameter of 2.54 cm and can support longitudinal waves with a speed of 6370 m/s and shear waves of speed 3136 m/s. The sound pulse has a duration of 3 μs, (pulse length 4.5 mm in water). The frequency is 6.6 MHz, and the speed is 1485 m/s. The pulse is directed toward the upper left quadrant of the cylinder. The figure gives eight time sequences. In Fig. 6.5.1a there is a double exposure showing the advancing pulse at two different times separated by about 10 μs. The cylinder appears as a dark circle with an illuminated periphery.

The sound frequency and radius yield $ka = 353$, which is high enough for us to use elementary ray concepts to follow the sequence of the waves developed. That is, for reflection, incident angle = reflection angle; for refraction, Snell's law applies. In Fig. 6.5.1b the specular reflection develops as an unwinding crescent. This is because the ray that struck the top of the cylinder has traveled farthest and succeeding rays, reflecting at points counterclockwise from the top of the cylinder, lag with respect to the topmost one. A Huygens construction can be used to show the evolution of the crescent.

Meanwhile some acoustic energy has entered the cylinder. Entry occurs at incidence angles between 0° (normal incidence) and a critical angle dependent on the speed of the wave in the solid. In aluminum the critical angle, calculated from (3.2.2) is 13.5° for longitudinal waves and 28.3° for the slower, shear waves. The efficiency with which the incident longitudinal pulse is converted to an interior shear wave as well as to a compressional wave in the cylinder depends on the angle of incidence, and is greatest for the range 13.5° to 28.3°. Because the cylinder is opaque, the internally generated shear waves and the interior longitudinal waves are not revealed by the Schlieren technique. However their reradiated, scattered waves are clearly seen.

PROBLEMS

Problems 6.2

6.2.1. Starting with (6.2.1) show that the total scattering cross section for a sphere is given by (6.2.2).

6.2.2. Perform the following.

(a) Calculate $\mathcal{S}_{bs}$ and TS for a grain of sand of radius 10 μ in water. Assume $E_1 = 5 \times 10^{10}$ N/m^2, $\rho_1 = 2500.$ kg/m^3, and $ka \ll 1$.

(b) Calculate $\mathcal{S}_{bs}$ and TS for a nonresonant gas bubble of radius 10 μ in water. Assume $c_1 = 330.$ m/s, $\rho_1 = 1.0$ kg/m^3, and $ka \ll 1$.

(c) Calculate $\mathcal{S}_{bs}$ and TS for a small rigid body $(g \gg 1,\ e \gg 1)$, assuming $ka \ll 1$.

(d) Compare the three answers and comment on the effects of the elasticity and density ratios e and g in determining the scatter in the three cases.

6.2.3. Plot the scatter $\mathcal{S}/(ka)^4$ as a function of angle θ for the grain of sand, the bubble, and the rigid body in Problem 6.2.2.

6.2.4. Show that the target strength of a fixed, rigid sphere is given by

(a) $20 \log_{10} \left(\dfrac{a}{2R_1} \right)$ for $ka \gg 1$

(b) $20 \log_{10} \left(\dfrac{5}{6} \dfrac{k^2 a^3}{R_1} \right)$ for $ka \ll 1$

6.2.5. A distribution of rigid spheres, one each, of radii 0.1, 0.2, 0.5, 1, 2, and 5 cm is known to exist in a limited region of the ocean.

(a)   The volume is insonified at frequency 10 kHz. Calculate the target strength of the 10 spheres.

(b)   The frequency is changed to 5 kHz to decrease the signal attenuation. How much have the target strengths changed?

(c)   The frequency is changed to 20 kHz. Calculate the increase in the target strength.

(d)   Assuming incoherent scatter, what is the increase in the backscatter cross section for each sphere, and for the assemblage, as the frequency is increased from 5 to 10 to 20 kHz.

6.2.6. Assume that the distribution of spheres of the previous problem are to be detected by sonar, that the available source level is 220 dB re 1 μPa at 5, 10, or 20 kHz, and that the background noise spectrum level is negligible at the three frequencies. Taking into account the *TL*, do the following.

(a)   Calculate the optimum frequency (of the three) if detection is to be at 1 km range.

(b)   Recalculate the optimum frequency for detection at 10 km range.

(c)   Recalculate the optimum frequency for detection at 500 m range.

(d)   How are the conclusions changed if there is background noise described by the ambient noise curves for wind speed 4.5 m/s? Assume that the maximum signal to noise ratio determines optimum detection.

Problems 6.3

6.3.1. A small grain of sand (radius 100. μ) drops into the ocean and traps an air bubble of 10 μ in a crevice. Calculate and compare the target strengths of the dust particle and its bubble, assuming that each is spherical and independent of the other, at frequencies 10, 50, 100, and 500 kHz.

6.3.2. Plot the ratio of the resonance scattering cross section to the resonance absorption cross section for an air bubble in water at sea level for the frequency range 100 Hz to 1 MHz.

6.3.3. Plot the resonance curve of σ_e versus f for one octave on each side of the resonance frequency for a 60 μ radius bubble. Calculate and indicate the half power points. Assume that the damping constants at resonance do not change over the two octaves.

6.3.4. Derive (6.3.21).

6.3.5. A bubble of radius 20 μ leaves the sediment at a water depth of 100 m, in isothermal water. Assuming that the bubble expands isothermally so that $P_A V = $ constant, calculate and plot a graph of the resonance frequency as a function of depth.

Problems 6.4

6.4.1. Show that for air bubbles near the sea surface in water of temperature 15°C, the attenuation due to bubbles is

$$\alpha_b = 1.5 \times 10^9 \, u(a_R) \, \text{dB/m}$$

where $u(a_R)$ has units μ^{-1} and the total volume fraction of bubbles of resonance radii between a_{R1} and a_{R2} is

$$u(a_{R2} - a_{R1}) = \int_{a_{R2}}^{a_{R1}} \frac{\alpha_b(f)}{1.5 \times 10^9} \, da$$

where $[\alpha_b]$ is in decibels per meter and $[a]$ is in microns.

6.4.2. Calculate the following.

(a) An attenuation $\alpha = 0.1$ dB/m is measured near the sea surface at a frequency of 15 kHz. Taking account of the attenuation in clean sea water, and assuming that the excess attenuation is due to bubbles, calculate the radius, the volume fraction of resonant bubbles, and the number per unit volume.

(b) The same attenuation is measured at 100 kHz. Calculate the volume fraction of bubbles and the number per unit volume.

SEVEN

MEASUREMENTS OF FISH AND OTHER LIFE IN THE SEA

The application of sonar to measurements of marine life in lakes and oceans requires interdisciplinary research by biologists and acousticians. The spatial sampling problems are difficult because many forms of marine life actively respond to the presence of food, predators, and light. Even plankton move with ocean currents and change depth with light. During a sonar fish survey, the distribution constantly changes and the separation of spatial and temporal changes requires many simultaneous observations.

The location of the body that causes an echo may depend on environmental factors. For example, some fish and zooplankton move toward the surface at night and go down during the day. Perhaps because of temperature preference or the concentration of prey, some species of fish concentrate within a small layer region of a thermal gradient.

We approach the problem of remote acoustical sensing in two ways: by estimating how the backscatter of the individual animal depends on its physical structure (Section 7.1), and by devising signal processing procedures that permit us to assign estimates of the biomass (Sections 7.2–7.5).

7.1. SCATTERING CHARACTERISTICS OF MARINE LIFE

We are interested in what is called the "inverse" problem. We want to use sonar measurements to identify the scatterers in the ocean. But to do

216

so we first examine how typical marine bodies interact with a sound wave.

One approach is to study individual marine bodies experimentally. The scattering of sound from these isolated bodies, either in the laboratory or under captive conditions at sea, is then the model for the real world. We give the results of two of these studies.

A physical model constructed by a theoretician to represent a myctophid might consist of a gas-filled prolate ellipsoidal bubble (the swim bladder) situated within a larger prolate ellipsoidal fluid which has a density and an elasticity close to those of the seawater (the fish body). To solve for the sound scatter from such a theoretical body as a function of sound frequency would be a formidable problem, even with the assistance of a high speed computer.

Our immediate purpose in this section is much simpler—namely, to examine the dependence of the backscattering function on the physical description of the marine body, by extending to fish and plankton the results of our acoustical studies of such simple objects as the fluid sphere and the gas bubble.

7.1.1. Nonresonant Bodies

To apply the results of Chapter 6 to nonresonant marine animals, we need values of the relative density and elasticity of these bodies compared to seawater. Table 7.1.1 lists some of these values. This section assumes that there are no resonant cavities in the body.

When the sound frequency is so low that λ is very much greater than any linear dimension of the body, two assumptions are reasonable: (a) it is sufficient to use the average values of the constants of the body; (b) the scattering function is about the same as from a sphere of equivalent volume.

The backscattering cross section of the small sphere is found from (6.1.3) and (6.2.1) with $\theta = 180°$.

$$\sigma_{bs} = A\mathscr{S}_{bs} = \left[\frac{(ka')^4}{\pi}\right]\left[\frac{(e-1)}{3e} + \frac{(g-1)}{(2g+1)}\right]^2 \pi a'^2 \qquad (7.1.1)$$

where $ka' \ll 1$ and a' is the equivalent radius calculated by equating the volume of the scatterer to the volume of the equivalent sphere.

Numerical Example. By means of simultaneous trawling, the Euphausiidae, shrimplike creatures up to several centimeters long, have been frequently identified as colocated with scattering layers. Euphausiids belong to the phylum Arthropoda, class Crustacea, and are generally

TABLE 7.1.1. RELATIVE PHYSICAL CONSTANTS OF MARINE ANIMAL TISSUE COMPARED WITH SEAWATER*

Material	Relative Density, $g = \rho/\rho_0$	Relative Bulk Modulus, $e = E/E_0$	Relative Speed of Sound, $h = c/c_0$
Seawater, 13°C	1.0	1.0	1.0
Prawns, Sergestidae, and Oplophoridae	—	0.93 ± 0.03 (1)	—
Euphausia pacifica	1.03 (3)	1.15 (2)	—
preserved (19.5°C)	1.016 (7)	—	1.033 (7)
fresh (7.5°C)	1.038 (7)	—	—
Cyclothone, Chauliodus, Gobidae, *Sternoptyx, Argyropelecus, Melamphaes, Gonostoma,* Myctophidae (*Diaphus, Lampanyctus*)	—	1.04 ± 0.03 (1)	—
Fish flesh	1.03–1.06 (6)	—	1.03–1.08 (4)
Fish bone	2.04 (4)	—	3.75 (4)
Brine shrimp Unhatched	1.19 (5)	—	—
Hatched	1.12 (5)	—	—
Copepod (*Callanus marshallae*) fresh (9.5°C) (references, $\rho_o = 1.026$ g/cm^3 and $c_0 = 1485$ m/s) (7)	1.020 (7)	—	1.007 (7)

*The subscript 0 refers to seawater. Numbers in parentheses indicate the following references:

1. L. P. Lebedeva, "Measurement of the bulk modulus of elasticity of animal tissues," *Sov. Phys. Acoust.* **10**, 410–411 (1965).
2. J. T. Enright, "Estimates of the compressibility of some marine crustaceans," *Limnol. Oceanogr.* **8**, 382–387 (1963).
3. Peter Beamish, "Quantitative measurements of acoustic scattering from zooplanktonic organisms," *Deep-Sea Res.* **18**, 811–822 (1971).
4. R. W. G. Haslett, "Backscattering of acoustic waves in water by an obstacle, II: Determination of the reflectivities of solids using small specimens," *J. Phys. Soc.* **79**, 559–571 (1962).
5. John D. Watson and Robert Meister, "Ultrasonic absorption in water containing plankton in suspension," *J. Acoust. Soc. Am.* **35**, 1584–1589 (1963).
6. R. McM. Alexander, "Physical aspects of swim bladder function," *Camb. Phil. Soc., Biol. Rev.* **41**, 141–176 (1966).
7. C. F. Greenlaw III, *Acoustic backscattering from marine zooplankton*, M.S. Thesis, Oregon State University, June, 1976.

considered to be planktonic organisms. It is implied that their distribution is determined by water motion and that they are incapable of horizontal movement to other regions. Adult *Euphausia pacifica* reach lengths of approximately 50 mm in Monterey Bay, California. If their average diameter is estimated at 5 mm, and the average length of an equivalent cylinder for the trunk is 15 mm, the approximate volume of such a euphausiid is 300 mm^3 and the equivalent spherical radius is $a' = 4.15$ mm. (See Fig. 7.4.3 for sketches.)

What are the backscattering functions and target strengths of a euphausiid at 12 and 30 kHz?

From Table 7.1.1 we take the relative elastic constant $e = 1.15$ and the relative density $g = 1.03$. Then (7.1.1) becomes

$$\sigma_{bs} = A \mathcal{S}_{bs} = (k^4 a'^6)(2.84 \times 10^{-3})$$

In this case the effect of the elasticity goes in the same direction as the effect of the higher density. In some animals however the elasticity may be lower than that of seawater because of the presence of stable gaseous formations associated with the life activity or because there is a body fluid that is simply more compressible than water (recall that compressibility = 1/elasticity).

At 12 kHz we have $ka' = 0.209$ and, using (6.1.4), we get

$$\sigma_{bs} = A \mathcal{S}_{bs} = 9.26 \times 10^{-11} \text{ m}^2$$

$$TS = 10 \log_{10} \frac{A \mathcal{S}_{bs}}{A_1} = -100.3 \text{ dB re } 1 \text{ m}^2$$

Another common sonar frequency has been 30 kHz, where ka' is 0.522. The length of the trunk of the euphausiid is still a fraction of a wavelength. Again we use (7.1.1) and (6.1.4) to estimate

$$\sigma_{bs} = \mathcal{S}_{bs} A = 3.63 \times 10^{-9} \text{ m}^2$$

$$TS = -84.4 \text{ dB re } 1 \text{ m}^2$$

If the equivalent radius of the organisms had been twice as great, because of the sixth power dependence on radius, the backscatter cross section would have been $2^6 = 64$ times as great. Then the target strength of the euphausiid would be greater by $10 \log_{10} 64 = 18$ dB.

Because of the a^6 dependence of $\mathcal{S}_{bs} A$, small changes in dimension can have a profound effect on the backscatter cross section and target strength. Since there will be a distribution of sizes in any deep sea population, the a^6 factor means that the total backscatter will be determined almost completely by the largest members of the group. In

Rayleigh scatter ($ka \ll 1$) the sound frequency determines a cutoff size of scatterer. Small members of the population are not sensed acoustically unless their numbers are vastly greater than those of the large scatterers. An increase in sound frequency does not significantly affect the backscatter from those scatterers of size $ka' = 1$, but it does bring a fourth power (k^4) increase in the backscatter from the smaller members of the population. By measuring the volume scattering coefficient over a wide range of frequencies, it is possible to obtain data concerning the distribution of sizes of organisms.

When the frequency becomes high enough, the wavelength approaches the size of the scatterer and the $(ka)^4$ dependence no longer holds. For example, at 100 kHz, the length of the trunk of the euphausiid is the same as the wavelength, 15 mm. Then there are several sources of Huygens wavelets along the body. For N sources radiating in phase, the pressures add and we write

$$\text{in phase: } (\sigma_{bs})_{IP} \simeq \left[\sum_{i=0}^{N-1} (\sigma_{bsi})^{1/2} \right]^2 \tag{7.1.2}$$

If the sources radiate in random phase on the average, the intensities add and we write

$$\text{random phase: } (\sigma_{bs})_{RP} \simeq \sum_{i=0}^{N-1} \sigma_{bsi} \tag{7.1.3}$$

Numerical Example. Calculate the backscattering function and target strength of a euphausiid at high frequencies.

For a gross estimate of the source strength along the euphausiid at high frequencies, we divide the trunk into three squat cylinders, each of diameter 5 mm and length 5 mm, each segment having a volume 100 mm^3. The three "near spheres" have the equivalent radius $a'' = 2.9$ mm.

At 100 kHz, $ka'' = 1.2$ for each "sphere." Figure 6.2.4a includes the backscattering function for $g \simeq 1$, $e \simeq 1.15$, $ka'' = 1.2$, but it is not possible to read the graph accurately. Figure 6.2.1 shows that when the $(ka)^4$ behavior is extrapolated to roughly $ka = 0.8$, $\mathcal{G}_{bs}$ reaches its asymptotic value, and that it is roughly constant if we ignore diffraction peaks and interior resonances. For $a'' = 2.9$ mm, $ka'' = 0.8$ at $f = 66$ kHz.

Therefore, for the individual spheres the maximum values are roughly $\sigma_{bs} = A\mathcal{G}_{bs} = k^4 a''^6 (2.84 \times 10^{-3}) = 9.8 \times 10^{-9}$ m^2 at 66 kHz and higher.

There are two extreme ways in which the scatter from the elementary "spheres" add.

If the three spheres scatter in phase, (7.1.2) gives

$$\sigma_{bs} \approx [3(9.8 \times 10^{-9})^{1/2}]^2$$
$$\approx 8.8 \times 10^{-8} \, m^2$$
$$TS = -71 \, dB$$

For random phase addition of the three equal sources, (7.1.3) gives

$$\sigma_{bs} \approx 3(9.8 \times 10^{-9})$$
$$\approx 2.9 \times 10^{-8} \, m^2$$
$$TS = -75 \, dB$$

In the region above $ka' = 1$, the backscatter is sensitive to the shape and orientation of the scatterer as well as the physical constants and ka'. Scattering from different sections of the body can interfere constructively or destructively, internal resonances may exist, and large swings of σ_{bs} can occur. Section A7.1 examines the statistics of variations.

7.1.2. Resonant Swim Bladders of Fish

Many fishes have gas-filled swim bladders that assist in buoyancy control. The volume of the gas bubble depends on the lipid (fat) content of the fish. Crudely, the swim bladder bubble volume represents about 5% of the animal volume for ocean fish and 7% for freshwater fish.

The discussion of the scattering of sound by fish having swim bladders is based on the resonance phenomenon of a gas-filled bubble. The swim bladders are nearly prolate ellipsoids. Since the sound wavelength at resonance is much larger than the dimensions of the bladder, we replace it by a sphere having the same volume. When the ratio of the major axis to the minor axis is greatly different from unity, corrections shown in Fig. 7.1.1 are used.

A simplified formula for the swim bladder resonance frequency is adapted from the equation for a free bubble (6.3.9), which becomes

$$f_{RF} \approx \frac{1}{2\pi a'} \left(\frac{3\gamma P_A}{\rho_A} \right)^{1/2} \tag{7.1.4}$$

where f_{RF} is the resonance frequency of the fish bladder (Hz), γ is the ratio of specific heats of the swim bladder gas ($\gamma = 1.4$ for air), P_A is the ambient pressure $= 10^5(1 + 0.1z)$ Pa, where z is depth (m) and ρ_A is the density ($\approx 1035 \, kg/m^3$); $a' = (3 \text{ bladder volume}/4\pi)^{1/3}$ is the radius of a spherical bubble of the same swim bladder volume. At shallow depth, the

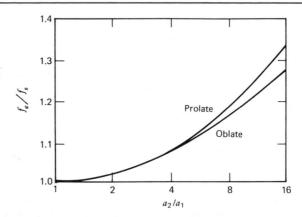

Figure 7.1.1. Dependence of bubble resonance frequency on ratio of ellipsoid axes a_2/a_1; f_e is the ellipsoid frequency and f_s the sphere. (Weston, 1967.)

shear elasticity of the tissue increases the resonance frequency. A detailed description of swim bladder resonance is given in Section A7.2.

The backscatter due to resonance is limited by the damping of the bubble oscillation. The backscattering cross section can be calculated from the (6.3.17), for a free bubble which becomes

$$\sigma_{bs} = \frac{\sigma_s}{4\pi} = \frac{a'^2}{[(f_{RF}/f)^2 - 1]^2 + \delta_{RF}^2} \tag{7.1.5}$$

A damping factor $\delta_{RF} = 0.2$ is often used. Details are in Section A7.2.

The importance of the fish bladder in the scattering of sound has been demonstrated conclusively by three separate experiments on a live, captured bubble-carrying fish, on the body of the same fish without the swim bladder, and on the excised bladder (Batzler and Pickwell, 1970). At the bubble resonance frequency ($ka' \ll 1$) the scatter from the body without the bubble is virtually undetectable; the scatter from the excised bladder shows a resonance behavior with a very low damping constant; and the scatter from the live fish with bladder has a high damping constant, presumably due to the effects of the fish tissue. For example, an experiment with an anchovy that was resonant at frequency 1275 Hz showed values $\delta_{RF} = 0.22$ for the live fish and $\delta_R = 0.048$ for the separate, intact bladder.

The combination of scattering from the swim bladder and scattering from the body of the fish is summarized schematically by Fig. 7.1.2, which shows the backscatter from fish bodies of three lengths and from the swim bladders if they have them. At frequencies below bubble resonance, the

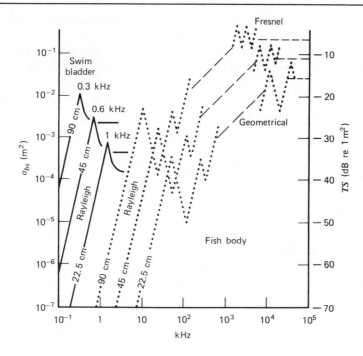

Figure 7.1.2. The relation between σ_{bs} and target strength and sound frequency for fish of 22.5, 45, and 90 cm length at 50 fathom depth. The scattered contributions from the swim bladder and fish body are shown separately. The Fresnel interference plateaus, where the fish is longer than the first Fresnel zone, depend on R, λ, and L. Adapted from Forbes and Nakken, Eds. (1972).

backscattering cross section is proportional to f^6, is omnidirectional, and is dominated by the low frequency tail of the swim bladder resonance. The backscattering cross section has a maximum at the bubble resonance frequency f_{RF}. These curves were drawn for a damping constant approximately 0.4 with a resonance peak up about 8 dB compared to $f \gg f_{RF}$. At high frequencies $kL > 1$, where L is the length of the fish, the scatter from the fish body is important.

Earlier we made the assumption that the dimensions of the scatterer are small compared to the first Fresnel zone. When the length of the fish is greater than the first Fresnel zone and the length of the fish is perpendicular to the sound beam (dorsal aspect or side aspect), the phases of the Huygens wavelets correspond to the alternating phases of the outer Fresnel zones. These wavelets then tend to interfere and the scattering function tends to a constant as a function of frequency. We call this the Fresnel interference region. See section 2.6.

Experimental measurements of the scattering function for single fish have been made under captive conditions, and the data have been plotted and fitted by empirical equations.

McCartney and Stubbs (1970) give an equation for average target strength of the dorsal aspect of fish for $L > \lambda$.

$$TS = 24.5 \log_{10} L - 4.5 \log_{10} \lambda - 26.4 \text{ dB} \qquad \textbf{(7.1.6)}$$

where L and λ are in meters. Love (1969) gives

$$TS = 24.1 \log_{10} L - 4.1 \log_{10} \lambda - 23.5 \text{ dB} \qquad \textbf{(7.1.7)}$$

for the side aspect.

7.1.3. Resonant Plankton

Any gas-bubble-carrying plankter can have a profound sound scattering effect regardless of its small size. One outstanding example is *Nanomia bijuga*, a colonial hydrozoan jellyfish, order Siphonophora, suborder Physonectae, which has been identified as the probable primary cause of certain sound scattering layers off San Diego, California. Siphonophores consist of many specialized individuals, aligned along an axis that in the case of *Nanomia bijuga* can be as long as 75 cm. One of the major groups of siphonophores, the Physonectae, is identifiable by pneumatophores (Fig. 7.1.3), bubble-carrying individuals that operate as flotation elements for the colony.

Positive identification of siphonophores was accomplished by simultaneous viewing from deep-sea research submersibles and recording of sound scattering by an echo sounder. The pneumatophore of the almost-transparent plankter approximates a prolate spheroid with axes of the order of 3 mm $\times$ 1 mm and volumes of contained gas as large as 12.6 mm^3 and as small as 0.25 mm^3. The enclosed gas is principally carbon monoxide ($\gamma = 1.40$), resulting in resonance frequencies ranging from 7 to 27 kHz at 100 m depth, or 13 to 50 kHz at 400 m.

The siphonophore releases a bubble as it moves in the water column. Ejections of these free bubbles have been observed only in a laboratory, sea-level environment, but the range of radii and the ability to recharge the bubble in a few hours suggests that this secondary source of free bubbles may also be significant. Assuming that the same ejected volumes occur at depth, the observed voluntarily expelled bubbles would resonate over the range 12 to 58 kHz at 100 m, and 24 to 111 kHz at 440 m depth.

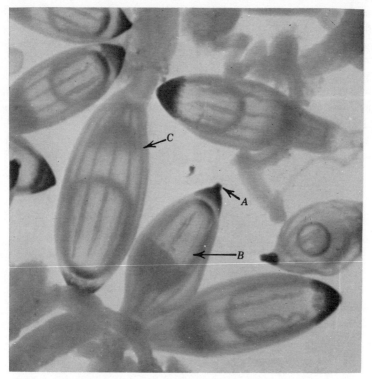

Figure 7.1.3. Pneumatophores of *Nanomia bijuga* (ca × 12) with contained gas bubbles: *A*, pore; *B*, gas gland; *C*, longitudinal muscle band. (Barham, 1963.)

The simple resonance bubble equation (6.3.9) can be used to calculate these frequencies.

To evaluate the effect of these siphonophores, it is necessary to consider how many scatterers are involved and the statistics of the size distribution. It is not yet clear how many of these populations would be close to the resonance size needed for a particular echo sounder frequency or what their geographical distribution is. However there can be no doubt that the acoustical effect of these pneumatophores is significant.

7.2. SIGNALS SCATTERED BY FISH OR OTHER BODIES

The following considerations apply to all marine bodies. For convenience we call the scattering body a "fish."

The sonar transmits a short ping which travels to the region of the fish.

When the distance is large the fish is completely within the central part of the first Fresnel zone and the incident wave front is approximately plane in the region. The scattered sound travels outward as if the fish were a source of sound.

The transmission loss equations apply in going from the source to the fish and from the fish to the receiver. For simplicity we assume a constant sound speed and spherically spreading wave fronts. The actual transmission losses are included later.

There are two principal sources of variability. First, the amplitude of the scattered signal depends on the position of the fish in the sonar beam (Fig. 7.2.1). It is tempting to replace the actual directional response of the

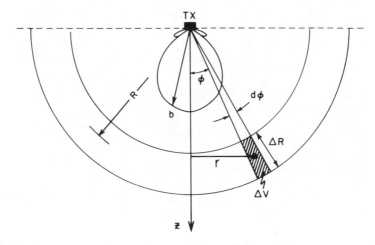

Figure 7.2.1 Geometry. The shell V is bounded by $R - \Delta R/2$ and $R + \Delta R/2$. The transducer face is circular and the sound pressure at the shell is symmetric about the z axis. b is the two way response or D^2 for the transducer.

sonar by an ideal beam pattern ($D = 1$ for $|\phi| < |\phi_1|$ and $D = 0$ for $|\phi| > \phi_1$). This is improper because it eliminates a systematic dependance of the amplitude of the signal on the location of the fish in the actual sonar beam.

Second, the scattering process has considerable variability because the fish is a complicated scattering object. We believe that the position of the fish in the sonar beam and the scattering at the fish are independent. The statistics of the scattering process are discussed in Section 7.5.

If the fish has a swim bladder, the resonance frequency varies from member to member. Our phenomenological description of the scattering process (at the fish) is used in the discussion of the signal processing procedures appropriate to bioacoustic studies.

7.2.1. Huygens Wavelets

The average fish is a long scattering object. For a qualitative description of the scattering process, we assume that the incident sound wave excites Huygens wavelet sources along the length and at its swim bladder if it has one (Fig. 7.2.2).

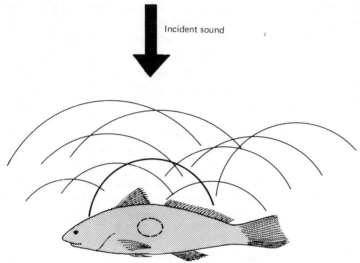

Figure 7.2.2. Fish as a source of backscattered Huygens wavelets. The swim bladder is dashed; its scatter is shown as stronger than that from elements of the body.

The scattering from the body of a fish has a source amplitude proportional to its length. The Huygens sources along the fish have the same directional properties as a transmitting array. For example, at low frequencies or large wavelengths relative to the fish's length, directional effects are negligible. At high frequencies the phases of the elementary wavelet sources are important and these depend on the attitude of the fish in the incident wave field. As the fish moves, turns, and swims, the scattered signal has many fluctuations.

If there is a swim bladder, it is the dominant scatterer at and below the resonance frequency. Even above its resonance frequency the bladder makes a substantial contribution.

The signal or "echo" at the receiver is the sum of all these contributions. Repeated pings at the same fish give a sequence of echoes that have different amplitudes because the fish or we have moved between each ping and the components add differently. When we receive the echoes from several fish, all the components from all the fish add in the receiver.

7.2.2. Backscattered Pressure

Most sonars use the same transducer for transmission and reception. The geometry is shown in Fig. 7.2.1. Since the transducer both transmits and receives, the directional response is D^2 where D is the pressure amplitude directional response. If different transducers are used for sending and receiving, D^2 is replaced by the product $D_s D_r$ for the two transducers. Since the transmitted pressure decreases as $1/R$ and the scattered pressure decreases as $1/R$, the echo is down by a factor $1/R^2$ compared to the source pressure.

For a long object, such as a fish, we write the backscattered pressure signal as proportional to an effective scattering length l, a complicated function that may be more or less than the length of the fish. In this simplified discussion we assume that the scattering process does not appreciably alter the time dependence of the signal. The signal or "echo" is received $2R/c$ seconds after it was radiated from the source and is

$$p_e(t) = lD^2 \frac{R_0}{R^2} p_0\left(t - \frac{2R}{c}\right) \qquad (7.2.1)$$

where l has the units of length. The sound pressure of the source at R_0 is $p_0(t - R_0/c)$; $p_e(t)$ is the result of the measurement of echo from the fish, and l is calculated from the measurement.

At the high operating frequencies of many fish-finding sonars (30 to 450 kHz), the lengths of the desired fish range from a few to many acoustic wavelengths. These frequencies are well above the resonance frequency of the swim bladder, and the entire fish body contributes to the scattered signal. The effective scattering length of the fish depends on its dimensions, relative density and elasticity compared to water, and orientation relative to the direction to the transducer. Therefore l depends on whether the sonar beam is horizontal or vertical. At low frequencies, at or near the resonance of swim bladders, the scattering function would be nearly omnidirectional.

As the fish swims, changing its direction and the shape of its body, repeated pings give a set of measurements of l. From the ith measurement

$$p_e(t)_i = l_i D_i^2 \frac{R_0}{R_i^2} p_0 \left(t - \frac{2R_i}{c} \right)$$ (7.2.2)

We expect the mean square of the set of l_i to be the same from one set of measurements to another. The integral over time of the squared ith echo is

$$\int p_e^2(t)_i \, dt = l_i^2 D_i^4 \frac{R_0^2}{R_i^4} \int p_0^2 \left(t - \frac{2R_i}{c} \right) dt$$ (7.2.3)

We can replace $p_0(t - 2R_i/c)$ by $p_0(t)$ on the right-hand side because the amplitude of the source pressure signal does not depend on travel time. For the present, we confine the fish to a small volume in which R_i and D_i are constant. The mean of (7.2.3) is

$$\left\langle \int p_e^2(t) \, dt \right\rangle_i = \langle l^2 \rangle D_i^4 \frac{R_0^2}{R_i^4} \int p_0^2(t) \, dt$$ (7.2.4)

or

$$\langle l^2 \rangle = \left\langle \int p_e^2(t) \, dt \right\rangle_i \frac{R_i^4}{R_0^2} D_i^{-4} \left[\int p_0^2(t) \, dt \right]^{-1}$$ **(7.2.5)**

We relate $\langle l^2 \rangle$ to the scattering function $\mathcal{S}_{bs}$ and the intercepted area A by defining a mean value of $\mathcal{S}_{bs}A$ and a mean backscattering cross section for the fish in the sense of the previous chapter.

$$\langle l^2 \rangle = \langle \mathcal{S}_{bs}A \rangle = \langle \sigma_{bs} \rangle$$ **(7.2.6)**

Since $\mathcal{S}_{bs}$, A, and σ_{bs} depend on the aspect of the fish, $\langle l^2 \rangle$ also depends on the aspect of the ensemble of fish. The use of the symbol $\langle l^2 \rangle$ is appropriate if bioacoustical information suggests that there is only one predominant fishlike species of scatterer. Otherwise we use the average backscattering cross section $\langle \sigma_{bs} \rangle = \langle \mathcal{S}_{bs}A \rangle$.

7.3. ECHO SQUARED INTEGRATION AND REVERBERATION LEVELS

Two common methods of processing acoustic measurements of marine life are echo squared integration and echo counting. They can be used in parallel (Fig. 7.3.1). Echo counting is described in Section 7.5.

Using the geometry in Fig. 7.2.1 and system in Fig. 7.3.1, a time gate opens and closes, limiting the echoes to the scatterers within the range

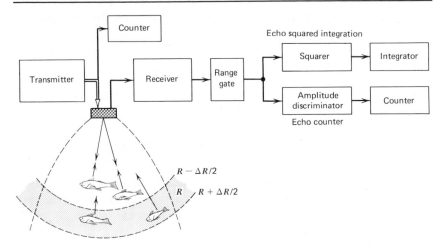

Figure 7.3.1. Signal processing systems. The range gate chooses the radial thickness of the volume (shaded areas) for measurement.

shell ΔR. The actual value of ΔR depends on the ping duration T_p, and T_G the duration of the gate opening. We develop expressions for ΔR later.

Again for simplicity we call the scatterers "fish." We regard the measurement of the echoes from many fish of the same species and size as being equivalent to repeated measurements of echoes from the same fish. The integral square of the echo from a single fish is given by (7.2.3). For many fish, the jth fish is at R_j, where the directional response is D_j; the sum over all fish within the gate opening time is

$$\int_{T_1}^{T_2} p_e^2(t)\, dt = \sum_{j=1}^{N} l_j^2 \frac{R_0^2}{R_j^4} D_j^4 \int_{T_1}^{T_2} p_0^2\left(t - \frac{2R_j}{c}\right) dt \qquad \textbf{(7.3.1)}$$

where $t = 0 = $ time of start of sound radiation
$\quad\quad\quad t = T_1 = $ time of gate opening
$\quad\quad\quad t = T_2 = $ time of gate closing
$\quad\quad\quad T_G = T_2 - T_1$
$p_0(t - 2R_j/c) = $ time dependence of the echo
The time dependence is the same as for the source signal delayed by traveling to the jth fish and returning to the transducer. We note that $p_e(t)$ represents all the echoes heard at time t and that for $\Delta R/R \ll 1$, the R_j are nearly equal and $R_j \approx R$.

The sum of integrals on the right-hand side of (7.3.1) is difficult to evaluate because $p_0(t - 2R_j/c)$ may be outside the gate, partially within the gate, or wholly within the gate. We can approximate the sum of integrals by choosing the operating conditions such that the contributions from $p_0(t - 2R_j/c)$ that are partially within the gate are negligible. Two examples of conditions are sketched in Fig. 7.3.2. The echoes are from a

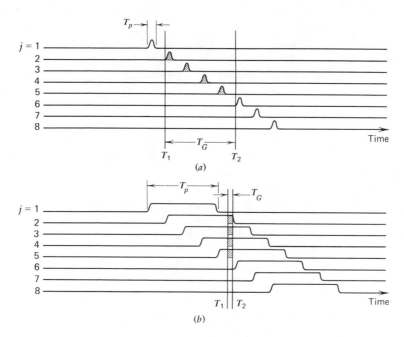

Figure 7.3.2. Use of gate for limit range region, giving the envelope of the echoes. The portions passed by the gate are shaded. For both conditions the echoes $j = 2$, 3, 4, and t contribute. The ΔR in range is $T_G c/2$ for a and $T_p c/2$ for b. (a) Short ping and long gate. (b) Long ping and short gate.

sequence of scatterers at increasing ranges. For a very short ping or transient signal (Fig. 7.3.2a), the range increment ΔR from which echoes pass through the gate is proportional to T_G and is

$$\Delta R \simeq \frac{T_G c}{2} \qquad \text{for} \quad T_p \ll T_G \qquad \qquad \textbf{(7.3.2)}$$

We use the approximate sign because we ignore the effect of being partially within the gate opening.

The long ping transmission is sketched in Fig. 7.3.2b. For a very short duration gate T_G relative to the length of the ping T_p, the number of contributions passed by the gate is proportional to the duration of the ping. Echoes of the more distant fish that are insonified by the head of the ping are heard at the same time as the echoes from closer fish, insonified by the tail of the ping. The range increment from which echoes are heard at the same time is proportional to the ping duration.

$$\Delta R \simeq \frac{T_p c}{2} \qquad \text{for} \quad T_p \gg T_G \tag{7.3.3}$$

We use different procedures to evaluate (7.3.1) for these two operating conditions. For both conditions, the fish are at random spacings and their echoes add as the sums of squares, or incoherently.

7.3.1. Short Ping and Long Gate

Within the gate, all the echoes have the same wave form even though their arrival times are different. Since each echo is zero before and after its duration T_p, we use the approximation

$$\int_{T_1}^{T_2} p_0^2\left(t - \frac{2R_j}{c}\right) dt \simeq \int_0^{T_p} p_0^2(t)\, dt \qquad \text{for} \quad T_p \ll T_G \tag{7.3.4}$$

Actually $p_0(t)$ can be any transient wave form that occurs within a maximum time duration T_p.

Next we replace l_j^2 by its average value $\langle l^2 \rangle$ or σ_{bs} and R_j by its average value R. With these changes and (7.3.4), (7.3.1) becomes

$$\int_{T_1}^{T_2} p_e^2(t)\, dt \simeq \langle l^2 \rangle \frac{R_0^2}{R^4} \int_0^{T_p} p_0^2(t)\, dt \sum_{j=1}^N D_j^4 \tag{7.3.5}$$

where both the number and direction of location of the fish are contained in the summation.

To evaluate the summation on D_j^4, we assume that the fish are uniformly distributed within the range shell and have a density N_f (the number of fish per cubic meter). We replace the summation by the product of N_f and the integral of D^4 over the volume

$$\sum D_j^4 = N_f\, \Delta V \tag{7.3.6}$$

where

$$\Delta V \equiv \frac{[(R + \Delta R/2)^3 - (R - \Delta R/2)^3]}{3}\, \Psi_D \tag{7.3.7}$$

$$\Delta V \simeq R^2\, \Psi_D\, \Delta R \qquad \text{for} \quad R \gg \Delta R$$

$$\Psi_D \equiv \int_{4\pi} D^4\, d\Omega \tag{7.3.8}$$

where $d\Omega$ is the incremental solid angle, Ψ_D is the integrated beam width factor, and D is the (amplitude) directional response of the transducer. When the directional responses of the source and receiving transducers are different, D^4 becomes $D^4 = D_s^2 D_r^2$. We give the results of numerical evaluations of (7.3.7) for common transducers combinations in Table 7.3.1.

TABLE 7.3.1. INTEGRATED BEAM WIDTH FACTOR Ψ_D FOR TRANS-DUCER COMBINATIONS*

Source	Receiver	Ψ_D	Approximation Condition
Piston, radius a	Same as source	$5.78/(ka)^2$	$ka \gg 1$
Rectangular, $L \times W$	Same as source	$17.4/(k^2 LW)$	$kL \gg 1,\ kW \gg 1$
Omnidirectional	Piston radius a	$12/(ka)^2$	$ka \gg 1$
Omnidirectional	Rectangular	$\pi^2/(k^2 LW)$	$kL \gg 1,\ kW \gg 1$
Omnidirectional	Omnidirectional	4π	

*The transducers are baffled to eliminate back radiation. Definitions are as follows:

$\Psi_D \equiv \int_{4\pi} D_r^2 D_s^2 \, d\Omega$

$D_s =$ directional response of the source

$D_r =$ directional response of the receiver

The substitution of (7.3.6) into (7.3.5) gives

$$\int_{T_1}^{T_2} p_e^2(t) \, dt \simeq N_f \langle l^2 \rangle \frac{R_0^2}{R^4} \Delta V \int_0^{T_p} p_0^2(t) \, dt \qquad \text{for} \quad T_p \ll T_G \quad (7.3.9)$$

Sometimes it is convenient to measure the average echo squared, and we define

$$\langle p_e^2 \rangle \equiv (T_G)^{-1} \int_{T_1}^{T_2} p_e^2(t) \, dt$$

We divide both sides of (7.3.9) by T_G and get

$$\langle p_e^2 \rangle \simeq N_f \langle l^2 \rangle \left(\frac{R_0^2}{R^4} \right) \left[\frac{\Delta V}{T_G} \right] \int_0^{T_p} p_0^2(t) \, dt \qquad \textbf{(7.3.10)}$$

Using $\Delta R = T_G c/2$, ΔV reduces to

$$\Delta V = \frac{T_G c R^2 \Psi_D}{2} \qquad \textbf{(7.3.11)}$$

Although T_G is factored out of the expression for the mean square echo, the condition that the duration of the ping is much shorter than the opening of the gate still holds.

7.3.2. Volume Backscattering Coefficients

The *volume backscattering coefficient* $N_f\langle l^2 \rangle$ is proportional to the sound energy backscattered by the fish in a cubic meter of water; it is the backscattering cross section per unit volume

$$s_v = N_f\langle l^2 \rangle = N_f \sigma_{bs} \qquad (7.3.12)$$

where s_v has the dimensions of m^{-1}. When the volume contains a mixture of fish having different backscattering cross sections, s_v is the sum of all contributions. For example, let N_i be the number of the ith type of fish and the backscattering cross section be $\langle l^2 \rangle_i = \sigma_{bsi}$. For random phases

$$s_v = \sum_i N_i \langle l^2 \rangle_i = \sum_i N_i \sigma_{bsi} \qquad (7.3.13)$$

Expressed in decibels,

$$S_v \equiv 10 \log_{10}(s_v R_1) \qquad (7.3.14)$$

where R_1 is the reference distance, usually 1 m; S_v is called the volume backscattering strength.

7.3.3. Long Ping and Short Gate

The gate open time is very brief compared to the length of the ping T_p in the long ping–short gate mode of operation. Since the ping has a constant amplitude, the integral over T_1 to T_2 is the same as long as the ping is within the gate. Letting P_0^2 be the mean square amplitude of the ping, we write

$$\int_{T_1}^{T_2} P_0^2 \left(t - \frac{2R_j}{c} \right) dt \simeq P_0^2 T_G \qquad \text{for} \quad T_p \gg T_G \qquad (7.3.15)$$

Again, it is convenient to define an average echo

$$\langle p_e^2 \rangle = (T_G)^{-1} \int_{T_1}^{T_2} p_e^2(t)\, dt \qquad (7.3.16)$$

We make the same approximations for the average values of $\langle l^2 \rangle$ and R^2 as in Section 7.3.1 and use (7.3.3), (7.3.15), and (7.3.16) to write [after dividing both sides of (7.3.1) by T_G],

$$\langle p_e^2 \rangle \simeq s_v \frac{R_0^2}{R^4} \Delta V P_0^2 \qquad \text{for} \quad T_p \gg T_G \qquad (7.3.17)$$

where

$$\Delta V = \frac{cR^2 T_p \Psi_D}{2} \qquad (7.3.18)$$

7.3.4. Reverberation Levels

The reverberation level is the mean echo level in decibels. We use the reference pressure P_r, reference time T_r, and reference scattering distance R_1. For the case $T_p \ll T_G$ the reverberation level is

$$RL_v \equiv 10 \log_{10} \frac{\langle p_e^2 \rangle}{P_r^2} \qquad (7.3.19)$$

From (7.3.10) with $s_v = N_f \langle l^2 \rangle$ and $S_v = 10 \log_{10} s_v$, we get

$$RL_v \simeq S_v - (TL_1 + TL_2) + 10 \log_{10} \left(\frac{\Delta V}{T_G} \frac{T_r}{R_1^3} \right) + ISS \qquad (7.3.20)$$

$$ISS \equiv 10 \log_{10} \left[\int_0^{T_p} p_0^2(t) \, dt \, (P_r^2 T_r)^{-1} \right] \qquad \text{for} \quad T_p \ll T_G \quad (7.3.21)$$

where ISS is the integrated source strength, $TL_1 = 20 \log_{10} (R/R_0)$, and $TL_2 = 20 \log_{10}(R/R_1)$. When absorption and ray curvature are significant, the more general equation (3.4.5) should be used to calculate the transmission losses.

Equation (7.3.20) can be reduced to a simpler form by using (7.3.11) and combining R^2 in ΔV with the transmission loss term

$$RL_v \simeq S_v - 20 \log_{10} \frac{R}{R_1} + 10 \log_{10} \Psi_D$$

$$+ ISS + 10 \log_{10} \frac{cT_r}{2R_1} \qquad \text{for} \quad T_p \ll T_G \quad (7.3.22)$$

The reverberation level for the long ping is obtained by operating on (7.3.17).

$$RL_v \simeq S_v - (TL_1 + TL_2) + 10 \log_{10} \frac{\Delta V}{R_1^3} + SL \qquad \text{for} \quad T_p \gg T_G$$

$$(7.3.23)$$

where

$$SL \equiv 10 \log_{10} \frac{P_0^2}{P_r^2}$$

Again R^2 in ΔV can be combined with the transmission loss to give

$$RL_v \simeq S_v - 20 \log_{10} \frac{R}{R_0} + 10 \log_{10} \frac{T_p}{T_r} + 10 \log_{10}(\Psi_D)$$

$$+ SL + 10 \log_{10} \frac{cT_r}{2R_1} \qquad \text{for} \quad T_p \gg T_G \quad (7.3.24)$$

When the scatterers are uniformly distributed throughout the volume, the volume reverberation decreases as $20 \log_{10} (R/R_0)$ in both cases.

7.4. VOLUME SCATTERING IN THE OCEAN

We continue our discussion of sound scattering in the ocean by considering the bioacoustical pyramid (Fig. 1.6.1). At the base of the pyramid are phytoplankton. The phytoplankton use carbon dioxide, light, and nutrients to synthesize organic compounds. Light penetrates to about 100 to 200 m and this limits the maximum depth for phytoplankton growth. The tiny zooplankton feed on the phytoplankton. As we move up the food pyramid, we come to macroplankton and megaplankton. These are large enough to be detectable with sonars operating in the 10 to 100 kHz frequency range. The nekton, particularly the bubble-carrying species, can be detected by sonars operating below 10 kHz.

7.4.1. DEEP SCATTERING LAYERS

The macro and megaplankton may respond to light, moving up at dusk and down near daybreak. Most of our knowledge of the diel (24-hour cycle) vertical migrations of plankton is obtained by echo sounding. Condensations of typical records appear in Fig. 7.4.1. The two examples are from different parts of the world. Although the vertical migration of the main deep scattering layer (DSL) is a striking feature on the records, other groups of scatterers seem to remain at the same depths.

Figure 7.4.2 gives a day-night comparison of the volume backscattering strength for 5 kHz from surface to depth 1200 m at two locations.

Although it is easy to make acoustical observations of scattering layers, it is difficult to make direct measurements of the populations. Midwater trawls are used to sample the water above, below, and in the scattering layer. Siphonophores, copepods, pteropods, euphausiids, and small fish are collected in the deep scattering layers, but some of the plankton are so fragile that they are destroyed in the process. Sketches of zooplankton are presented in Fig. 7.4.3.

It is difficult to estimate accurately the density of fish from trawl samples because of problems in defining the swept volume of water and the efficiency of the trawl. However both observations from submersibles and trawl data suggest that the populations have a very small density. This is consistent with acoustic data because one fish having a resonant gas bubble in $1000 \, m^3$ water can give the right magnitude of the volume scattering coefficient.

7.4.2. Frequency Dependence

A great deal of data, particularly the frequency dependence of scattering layers, are obtained by using explosive sources which produce a large amount of acoustic power over a broad spectrum of sound frequencies. When the receiver is a point hydrophone, reverberating backscatter is received from a complete spherical shell. The depth resolution is improved by using a downward-beamed hydrophone with the omnidirectional source near the surface.

Studies of frequency dependence are often presented in terms of the *column strength* $S_{vc}(d)$, which is obtained by integration of the volume backscattering coefficient from the surface to depth d and taking the log.

$$S_{vc}(d) \equiv 10 \log_{10} \int_0^d s_v(z) \, dz \qquad (7.4.1)$$

Figure 7.4.4 is an example of the frequency dependence of column strength. The peak at about 3.5 kHz for the night data probably represents the resonance frequency of the swim bladder of a prominent species of fish. There appears to be another resonance peak at about 12 kHz.

The data taken during the day (Fig. 7.4.4) show that most of the scatterers are between 140 and 580 m. The strong resonance at 13 kHz and the weaker one at 7 kHz may be due to the same fish that were resonant at 3.5 kHz, within 100 m of the surface, at night. The increased resonance frequency at greater depths is explained by the fish bladder being under greater pressure and the resonance frequency following the known $\sqrt{P_A}$ pressure dependence for bubbles (6.3.9).

The resonance frequency of fish swim bladders depends also on the fish

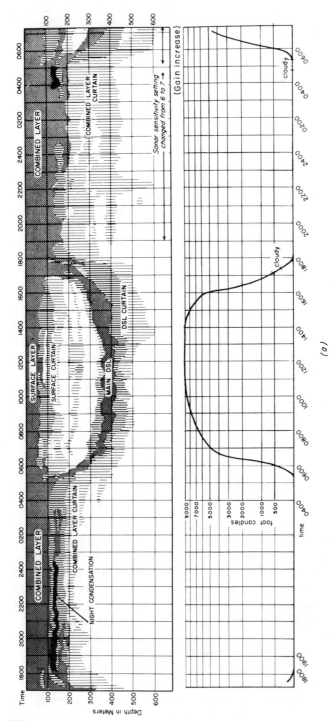

Figure 7.4.1a. *Above,* diagram of scattering layers prepared from a 37 hour long echogram recorded in the Bay of Bengal at 06°10'N, 93°07'E, beginning 1700 hours November 24 and ending 0715 hours November 26. A Simrad sonar at 30 kHz was used. Stippled patterns indicate heavy scattering; medium and light vertical lines indicate medium and light scattering, respectively. Light scattering between the surface curtain and main deep scattering layer (DSL), centered at 250 m during the day, corresponds in position to the intermediate layer seen as heavy scattering on numerous other echograms. *Below,* curve indicating intensities of incident light for the period the echogram was being recorded. Bradbury et al., 1970.

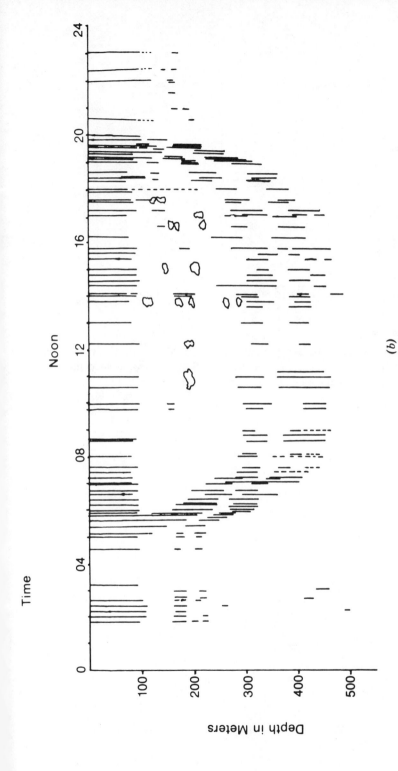

Figure 7.4.1b. The 24 hour 30 kHz DSL pattern (vertical migration) observed at the mouth of the Gulf of California, 28°N, during July 1969. A number of patchy echoes are indicated at a depth of about 200 m between 1030 and 1700 hours local time. (Dunlap, 1970.)

(b)

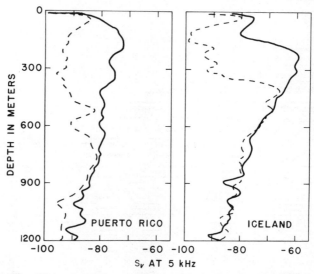

Figure 7.4.2. Volume backscattering strength S_v dB at 5 kHz versus depth for typical sites south of Iceland and north of Puerto Rico: dashed lines, day; solid lines, night. (Chapman et al., 1970.)

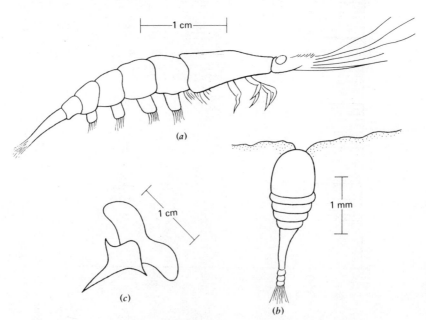

Figure 7.4.3. Zooplankton. (*a*) Euphasiid. (*b*) Copepod. (*c*) Pteropod.

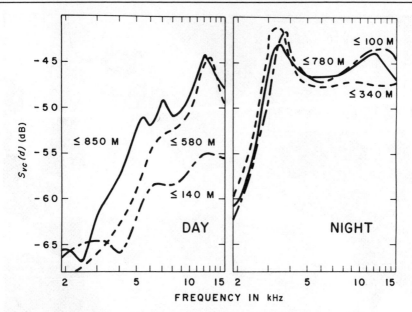

Figure 7.4.4. Column scattering strength as a function of frequency and depth, giving three different integration depths d for day and for night. Measurements at 36°41'N, 65°00'W. (Chapman, 1970.)

species. Without better depth information, from which we could get the bubble radius, it is difficult to associate resonance peaks in data such as in Fig. 7.4.4 with a particular size or species of fish.

7.4.3. Geographical Dependence

The column scattering strength is also used to describe the variation of reverberation characteristics in a frequency band with geographical position in the oceans of the world (e.g., Fig. 7.4.5). Regions of common values of column scattering strength have been called scattering "provinces."

7.4.4. Effect of Internal Waves

Many of the plankton are nearly neutrally buoyant and move about with the water mass. Probably because of temperature, salinity, chlorophyll, or oxygen preference they often appear to be stratified. Echo-sounding

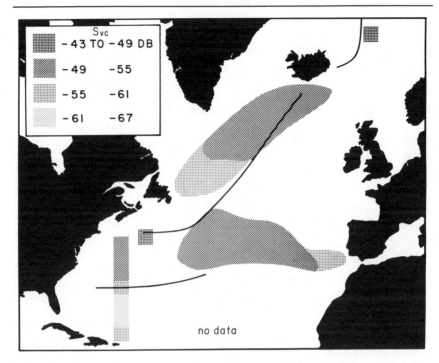

Figure 7.4.5. Nighttime values of the column scattering strength in the 1.6 to 3.2 kHz octave for the North Atlantic and adjacent seas. (Chapman, 1967.)

records taken from a stationary platform (FLIP) in the deep sea show layers oscillating up and down corresponding to passing internal waves (Fig. 7.4.6).

Because both plankton and nekton are sensitive to the presence of nutrients, food, light, and other physical-chemical characteristics of their environment, they may respond to the change of any of these. Many water properties are subject to change at the boundary of water masses; therefore the boundary is often marked also by changes of the volume scattering coefficient. Marine animals may concentrate at the interface or behave as if the interface were a barrier.

7.4.5. Near-Surface Scatterers

The near-surface region of the ocean plays host to a large number of scatterers, in addition to the transient presence of the DSLs that come up during the night. This is also the region that shows the greatest effects of

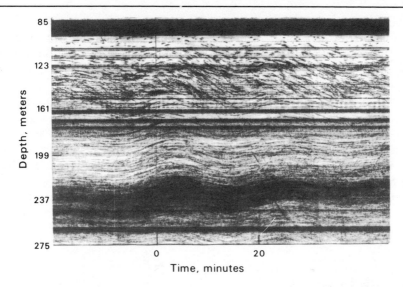

Figure 7.4.6. Internal waves. The observations were made from FLIP (floating instrument platform) in water 3600 m deep at 31°N, 120°W. The data were taken with a 6.6 ms pulse at 87.5 kHz. The echo sounder beam was 3 dB down at ±0.5°. Photograph courtesy of F. H. Fisher, Marine Physical Laboratory, University of California, San Diego. (Fisher, 1975.)

surface wave action. We divide the full-time inhabitants of the near-surface region into biological and physical scatterers.

The biological scatterers in the near surface region are predominantly of the phylum Arthropoda, subclass Copepoda. The copepods are of the order of millimeters in length. There has been very little acoustical study of their presence or extent. To make such determinations requires the use of sound frequencies of the order of hundreds of kilohertz.

One study of the near-surface scatterers (Barraclough et al., 1969) consisted of net sampling of the upper ocean to a depth of just over 100 m, and simultaneous use of a 200 kHz echo sounder. The experiment, which was conducted over a great circle route from British Columbia, Canada, to Tokyo, showed a coincidence of depth of strong echoes with depth of maximum catches that were 99% *Calanus cristatus*, a species of copepod. These plankton populations peaked at a depth of approximately 40 m during the day, where there were about 100 copepods per cubic meter, comprising a density of about 1.5 g/m^3 (wet weight).

The upper region of the ocean also contains bubbles that are not part of any zooplankton or fish and can cause substantial acoustical effects.

These bubbles, which we call "free" bubbles, can exist for a variety of reasons. They may be the vented gas of zooplankton or fish, the products of photosynthesis, or the result of breaking waves, precipitation, cosmic rays, or decaying matter. Results of an acoustical count of bubble populations are shown in Fig. 5.4.4.

The importance of near-surface bubbles is difficult to exaggerate, since these phenomena intrude on several sciences. They have been indicted as the source of water droplets produced when the bubble breaks at the surface. These droplets evaporate, become salt nuclei, and are carried aloft by the wind. The droplets formed at the ocean surface are rich in nutrients and bacteria, which have concentrated at the bubble surface during its upward trajectory; their numbers are important in sea surface chemistry. Bubbles play a major role in sonar propagation near the ocean surface.

7.5. ECHO COUNTING: FIELD ESTIMATES OF FISH DENSITIES

Echo counting procedures are based on the assumptions that a single species of scatterer dominates and that the number of echoes from the insonified volume is proportional to the density of fish. For the methods to work, echoes must not overlap. This is ensured by having narrow beam sonars and very short pings. When echoes do overlap, we use echo squared integration. The results of echo counting techniques are estimates of the number of fish per unit volume N_f and $\langle l^2 \rangle$.

We use the peak of the envelope of the echo in our analysis. Designating the peak of the echo as $|p_e|$ and the peak of the transmitted signal as $|p_0|$, (7.2.1) becomes

$$|p_e| = lD^2 |p_0| \frac{R_0}{R^2} \tag{7.5.1}$$

where D^2 is the directional response of the transducer when it is used as a source and receiver. The dependence of $|p_e|$ on the direction to the fish is contained in D^2. The range and source dependence are removed by defining the normalized echo amplitude

$$e \equiv \frac{|p_e| R^2}{R_0 |p_0|} \tag{7.5.2}$$

$$e = lD^2 \tag{7.5.3}$$

where e has the dimensions of a length. Since D^2 is a function of the direction to the fish and this is assumed to be random, D^2 is random. The scattering of sound by the fish is variable and l is random; e is the product of two quantities, both having random characteristics. Next we introduce a few concepts from probability theory.

Using the sequence of echoes in Fig. 7.5.1 as an example, the probability of e being between $(e_n + \Delta e/2)$ and $(e_n - \Delta e/2)$, that is,

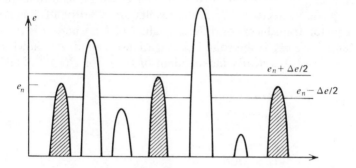

Figure 7.5.1. Normalized echo amplitudes of a sequence of echoes. Echoes that have amplitudes between $e_n + \Delta e/2$ and $e_n - \Delta e/2$ are shaded.

$\mathcal{P}(e_n + \Delta e/2 \geq e \geq e_n - \Delta e/2)$, is the ratio of the number of echoes within the region Δe to the total number of echoes. The *probability density function* (PDF) is defined as being the limit of the ratio

$$w_E(e_n) = \lim_{\Delta e \to 0} \frac{\mathcal{P}[(e_n - \Delta e/2) \leq e \leq (e_n + \Delta e/2)]}{\Delta e} \qquad (7.5.4)$$

w_E has dimensions of $(\text{length})^{-1}$. In estimating $w_E(e_n)$, measurements are made of $\mathcal{P}(e)$ for a convenient size of Δe and a set of e_n; $w_E(e_n)$ is calculated by using (7.5.4) for finite Δe. Alternatively, expressed as an integral, $\mathcal{P}(e)$ and $w_E(e)$ are

$$\mathcal{P}(e_2 \geq e' \geq e_1) = \int_{e_1}^{e_2} w_E(e') \, de' \qquad (7.5.5)$$

To simplify the problem, we assume that a single species of fish is present and all the fish have the same average backscattering cross section. We have two variables to consider, and each is random. First, the fish can be located any place in the sonar beam. Its insonification D^2 depends on its location. Second, the magnitude of the scattering factor l is random. The echo is the product of these two random functions. We assume they are independent, that is, D^2 does not affect l, and vice versa. When the probabilities of occurrence of two events are independent, the probability of both occurring is the product of the two probabilities, that is,

$$\mathcal{P}(x, y) = \mathcal{P}(x)\mathcal{P}(y) \tag{7.5.6}$$

In the fish echo problem, more than one combination of l and D^2 can give the same size of echo and these combinations add. An extended discussion of the calculation of the PDF of e, $w_E(e)$, as a function of D and l appears in Section A7.1. The results are a set of PDF curves for circular piston transducers of different values of ka, where a is the piston radius and $k = 2\pi/\lambda$ is the sound propagation constant. As shown in Fig. 7.5.2, $(ka)^2 w_E(e)$ is nearly independent of $(ka)^2$ for $e/\langle l^2 \rangle^{1/2} > 0.05$. The

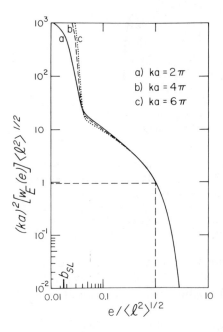

Figure 7.5.2. Probability density function of the envelope of echoes from fish: $w_E(e)$ computation are for a circular piston transducer having negligible back radiation; a = transducer radius. The mean square scattering length of the fish is $\langle l^2 \rangle = \sigma_{bs}$ where σ_{bs} is the backscattering cross section. (Peterson et al., 1976.)

very large value for smaller values of e is largely due to fish in the side lobes of the transducer. The maximum of the first side lobe b_{SL} is 0.0175. The most useful range for the interpretation of data is 0.1 to 2. In this range

$$w_E(e)_1 (k_1 a_1)^2 \langle l_1^2 \rangle^{1/2} \simeq w_E(e)_2 (k_2 a_2)^2 \langle l_2^2 \rangle^{1/2} \tag{7.5.7}$$

where the subscripts indicate two different ka values and l values.

In measuring the PDF of a set of measurements, we choose a number of amplitude cells and count the number of events that occur in each cell.

We designate $\mathcal{N}_n$ as being the number of events in the nth cell. That is, $\mathcal{N}_n$ is the number of echoes that are between $e_n \pm \Delta e/2$. From (7.5.4) or (7.5.5), the probability of an echo being in the cell $e_n \pm \Delta e/2$ is

$$\mathcal{P}\left(e_n \pm \frac{\Delta e}{2}\right) \simeq w_E(e_n)\,\Delta e$$

The number of events within a cell is proportional to the probability of an event being in the cell and the product of the rest of the factors, such as fish density N_f, volume V_G, and the number of pings M_p; $\mathcal{N}_n$ is

$$\mathcal{N}_n = N_f V_G M_p w_E(e_n)\,\Delta e \qquad (7.5.8)$$

where

$$V_G = \frac{2\pi}{3}(R_2^3 - R_1^3)$$

$$\simeq 2\pi R^2\,\Delta R$$

and the back radiation of the transducer is negligible. The product $w_E(e_n)\Delta e$ is dimensionless and if Δe is in meters, $w_E(e_n)$ has the units m^{-1}. After we remove the response of the sonar system (7.5.3), the unit of Δe is $\langle l^2 \rangle^{1/2} = \sigma_{bs}^{1/2}$, in m. Of course, Δe can be measured in arbitrary units as necessary with an uncalibrated sonar system.

Although it is better to use calibrated systems, the lack of calibration does not preclude quantitative fish density estimates when a single species of fish is present. Even if the sonar is not calibrated, its relative gain must be known for the complete set of measurements. The data processing system is shown in Fig. 7.3.1. We usually make a minor modification to the receiver and take the signal out of the receiver ahead of the gain control for the graphic display. This change prevents "knob twiddlers" from affecting the signal that is recorded on magnetic tape. We also measure the gain by injecting calibration signals into the input of the receiver and recording them on the magnetic tape.

By operating echo squared integration in parallel with echo counting, measurements of $w_E(e)$ can be used to estimate N_f and $\langle l^2 \rangle$ when the density is small enough for nonoverlapping echoes. This calibrates the echo squared integration processor. An example follows.

7.6. A FISH DENSITY PROFILE

The signal processing system in Fig. 7.3.1 has been used to process data that were taken during a night transect of Lake Michigan. The sonar operated at 50 kHz and had a transducer radius $a = \lambda$. The ping duration was 1.3×10^{-3} s. The signals were heterodyned to the 2 to 3 kHz frequency range and recorded on magnetic tape. The signals of the replayed tape were envelope detected and low pass filtered for the bandwidth of the original transmission, about 900 Hz. The envelope amplitudes were then digitized and processed by a computer. The sonar system was not calibrated, and the measurements were in arbitrary units.

Graphic records and comparisons of the echo squared integrator output with echo counts were used to identify sections of the record where echoes did not overlap. These sections were chosen for the amplitude distribution measurements. The fish are believed to be Lake Michigan alewife (*Alosa pseudoharengus*). The species is about 15 cm long and has a swim bladder that is a prolate spheroid, about 0.5 cm diameter and 2 cm long. The bubble resonance of the swim bladder is in the kilohertz range, which is much below the operating frequency of the sonar. Therefore the backscatter is probably due principally to the fish body rather than the swim bladder.

The echo amplitudes were measured by detecting the peak of the envelope. Peaks that occurred within the times for 8 to 10 m depth were analyzed. Since the times of the peaks of single events were measured, the constraint that $T_p \ll T_G$ did not apply.

The number of events within each amplitude cell is given in Fig. 7.6.1. The widths of the shaded areas are the heights (Δe) of the amplitude cells.

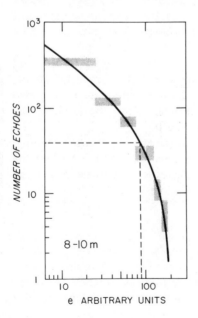

Figure 7.6.1. Amplitude distribution of the envelopes of echoes from individual fish; shaded areas are data, and widths are heights of amplitude cells, 25 arbitrary units. The heights of the shaded areas are estimates of the fluctuations of the number of echoes. Data taken in Lake Michigan for the depth interval 8 to 10 m. The solid curve is a fit of the theoretical $w_f(e)$. (Peterson et al., 1976.)

Repeated measurements of the number of events would show fluctuations of about $\mathcal{N}_n^{1/2}$ above and below the mean value $\mathcal{N}_n$. The top and bottom of a shaded area are $\mathcal{N}_n \pm \mathcal{N}_n^{1/2}$. The solid line is an overlay of the fit of the theoretical PDF, $w_E(e)$. The vertical dashed line corresponds to $e = \langle l^2 \rangle^{1/2}$ (or unity in Fig. 7.5.2). The vertical dashed line intercepts the abscissa at $e = 88$ arbitrary units. The horizontal dashed line gives $\mathcal{N}_1 = 38$ counts. These lines correspond to $w_E(e)$ for $e_1 = \langle l^2 \rangle^{1/2}$. As shown in Fig. 7.5.2 and Table A7.1.1,

$$k^2 \sigma^2 w_E(e_1) \langle l^2 \rangle^{1/2} = 0.961 \qquad \text{for} \quad e_1 = \langle l^2 \rangle^{1/2} = 88$$

We substitute these values into (7.5.8) and solve for N_f

$$N_f = \frac{\mathcal{N}_1 (ka)^2 \langle l^2 \rangle^{1/2}}{V_G M_p (0.961) \Delta e} \tag{7.5.9}$$

where M_p is the number of pings. For the data in Fig. 7.6.1, $V_G = 1022 \, \mathrm{m}^3$, $M_p = 952$, and $\Delta e = 25$. This sample of data yields $N_f = 5.7 \times 10^{-3}$ fish per cubic meter.

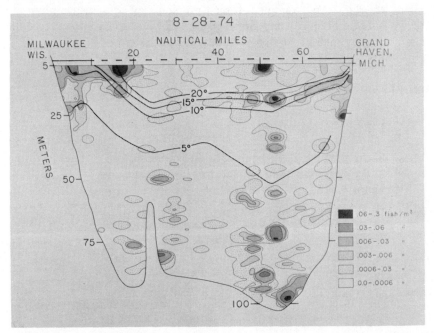

Figure 7.6.2. Cross section of fish distribution with estimates of average fish density. Horizontal bars show approximate sampled regions; taken in a single night run across Lake Michigan. (Brandt, 1975, and Peterson et al., 1976.)

For analysis of the output of the squaring and integrating circuit, it is convenient to remove the dependence on the source signal and the range from (7.3.9). We define I_E

$$I_E \equiv \left[\int_{T_1}^{T_2} p_e^2(t)\, dt \right] \left[R_0^2 \int_0^{T_p} p_0^2(t)\, dt \right]^{-1} \cdot \frac{R^4}{\Delta V} \tag{7.5.10}$$

where $\Delta V \simeq R^2 \Psi_D T_G(c/2)$; then substitute (7.3.9) into it, and get s_v of (7.3.12)

$$I_E = N_f \langle l^2 \rangle = s_v \tag{7.5.11}$$

where $\langle l^2 \rangle = \sigma_{bs}$.

Since I_E is proportional to N_f, we can use the measurement of I_E and the independent measurement of N_f to calculate the coefficient of proportionality.

The last step is to use the coefficient of proportionality to convert the measurements of I_E to fish density. This coefficient applies to all the measurements because I_E is also proportional to fish density when the echoes overlap. An example of measurements of the distribution of alewife in Lake Michigan appears in Fig. 7.6.2. Since this transect took all night, the data include hidden temporal variations.

PROBLEMS

Problems 7.1

7.1.1. Show that the peak backscattering cross section at bubble resonance is greater than the value at high frequencies $(f \gg f_R)$ by the factor $(1 + \delta_{RF}^2)/\delta_{RF}^2$.

7.1.2. A broad band sound signal is backscattered principally at 5 kHz from 100 m depth. Assuming that the principal scatterer is an air-filled, spherical swim bladder of a fish, and that the fish is neutrally buoyant, answer the following questions:
  (a) What is the equivalent radius of the bladder?
  (b) What is the estimated weight of the fish?
  (c) How are the answers changed if the swim bladder is ellipsoidal with axis ratio $a_2/a_1 = 4$?

7.1.3. Calculate the backscattering cross section of Sergestidae (see Table 7.1.1) at 30 kHz, assuming a density ratio $g = 1.03$ and an effective radius $a' = 6$ mm.

7.1.4. Calculate the target strength of a copepod of approximate equivalent radius $a' = 1$ mm when measured with sound of frequency 300 kHz. Assume that $e \approx 1.25$ and $g = 1.02$.

Problems 7.3

7.3.1. Starting from (7.3.5), show that if the scatterers are uniformly distributed in the volume ($s_v =$ constant), the reverberation intensity for a point source decays with time as t^{-2}, whereas if the scatterers are layered $[s_v = s_v(z)]$, it goes as t^{-3}.

7.3.2. Solve the following.
   (a) A scattering layer of volume backscattering strength $S_v = -70$ dB at 30 kHz is assumed to be due only to a high density of euphausiids of equivalent spherical radius 4.15 mm and backscattering cross section $\sigma_{bs} = 9.26 \times 10^{-11}$ m². What density of euphausiids is implied by this assumption?
   (b) Assuming that the euphausiids are of 10% greater radius, how many would there be?
   (c) How many of 10% lesser radius?

7.3.3. Perform the following calculations:
   (a) Given a siphonophore float of volume 4 mm³ at sea level, calculate f_R for the bubble.
   (b) Calculate the backscattering cross section at resonance for a single float.
   (c) If there are 10 resonant floats in a volume of 1000 m³, calculate the volume reverberation strength S_v.

7.3.4. A near-surface experiment at 300 kHz yields $S_v = -95$ dB. Simultaneous sampling reveals 10^7 diatoms per cubic meter and 1 copepod per cubic meter. Assuming that the diatoms have a radius of 10 μ, the copepods $a' = 1$ mm, and $e \approx 1.25$, $g = 1.02$ for both plankton, decide which scatterer is the probable cause of the S_v and how many are actually present in a cubic meter.

Problems 7.4

7.4.1. Use the data of Fig. 7.4.2 to calculate the column scattering strength at 5 kHz from surface to depth 500 m off Puerto Rico during day and night. It is suggested that 100 m increments and average values of s_v be used to replace the integration of (7.4.1).

SUGGESTED READING

R. P. Chapman, "Sound Scattering in the Ocean," in V. M. Albers, Ed., *Underwater Acoustics*, Vol. 2, Plenum Press, New York, 1967, pp. 161–183.

D. H. Cushing, *The Detection of Fish*, Pergamon Press, New York, 1973.

S. T. Forbes and O. Nakken, *Manual of Methods for Fisheries Resource Survey and Appraisal*, Part 2, *The Use of Acoustic Instruments for Fish Detection and Abundance Estimation*, Food and Agriculture Organization of the United Nations, 1972.

G. Brooke Farquaher, Ed., *Proceedings of an International Symposium on Biological Sound Scattering in the Ocean*, Maury Center for Ocean Science, Department of the Navy, Government Printing Office, Washington, D.C., 1970. Stock number: 0851-0053; price, $6.50.

R. W. G. Haslett, "Acoustic echoes from Targets Under Water," in R. W. B. Stephens, Ed., *Underwater Acoustics*, Wiley-Interscience, London, 1970.

J. B. Hersey and R. H. Backus, "Sound Scattering by Marine Organisms," in M. N. Hill, Ed., *The Sea*, Vol. 1, Wiley-Interscience, New York, 1962.

R. K. Johnson, "Sound scattering from a fluid sphere revisited," *Acoust. Soc. Am.* **61**, pp. 375–378, 1977.

B. McCartney, "Underwater Sound in Oceanography," in V. M. Albers, Ed., *Underwater Acoustics*, Vol. 2, Plenum Press, New York, 1967, pp. 185–201.

Middleton, David. "A Statistical Theory of Reverberation and Similar First-Order Scattered Fields" Parts I and II, Errata, III and IV. IEEE trans. Information Theory IT-13, p. 372–414, (1967); IT-15, p. 161–162 (1969); IT-18, pp. 35–90 (1972).

V. V. Ol'shevskii, *Characteristics of Sea Reverberation*, translation from Russian by Consultants Bureau, Plenum Press, New York, 1967.

R. J. Urick, *Principles of Underwater Sound*, 2nd ed. McGraw-Hill, New York, 1975.

D. E. Weston, "Sound propagation in the Presence of Bladder Fish," in V. M. Albers, Ed., *Underwater Acoustics*, Vol. 2, Plenum Press, New York, 1967.

EIGHT

SEISMIC AND ACOUSTICAL MEASUREMENTS OF THE SEA FLOOR

Marine geophysics is geophysics done at sea. Since many marine geophysicists learned their trade on land, they use the nomenclature that they already know. For example, "seismic" has to do with elastic waves in rocks. The marine geophysicist often calls the sound signal at the hydrophone a seismic signal because the signals have traveled through the water and the elastic material in the sea floor. Sometimes they use the term "seismic frequency range." This term refers to the frequency spectrum of signals that have penetrated deeply into the ground. In this usage the seismic frequency range is less than 100 Hz. Also, geophysicists speak of the "velocity" of propagation of the signal because rocks are anisotropic and direction of propagation is important. We use the word "velocity" when we describe the properties of the subbottom.

The marine environment has its own special problems for seagoing geophysicists. Ships rock, roll, and drift about. It is difficult to repeat a measurement because both the source and receiver move between transmissions. We meet these problems by designing "under way" experiments, to use the changing geometry as an experimental parameter.

Marine geophysical surveys require accurate navigation, and knowing where one is is constantly a problem in marine work. Satellite navigation systems have eliminated much of the guesswork.

Many individuals have contributed to the development of marine geophysics. The background of the art is presented in the chapters by Ewing on seismic refraction and reflection, Shor on techniques, Hill on the sono-radio-buoy, and Hersey on continuous reflection profiling, in *The Sea*, Vol. 3 (Hill, Ed., 1963).

8.1. THE SEA FLOOR

The sea floor begins at the water-sediment interface. In the ocean basins, the sound velocity of the sediments at the interface vary from a few percent less than the sound speed in water just above the interface to somewhat greater. Often the properties of the sediments near the interface are estimated from laboratory measurements on samples (cores). *In situ* devices to measure sound velocity are being added to deep sea corers, and in a decade we can expect to have many *in situ* sound velocity profiles of the sediments near the interface (Anderson and Hampton, pp. 357–371, in Hampton, Ed., 1974).

The sound velocity in sediments increases with depth of burial beneath the sea floor. We expect this. The weight of the overlying sediments compresses the sediments and squeezes out water. As the porosity decreases, the sound velocity increases. The temperature increases with depth of burial. The temperature gradient is caused by the flow of heat from the mantle, through the sea floor, and into the water. Since the sound velocity in the porous sediments is nearly proportional to the sound speed in water, the increase of sound speed in water due to the temperature increase causes a corresponding increase in the sound velocity in the subbottom sediments.

Samples of the sea floor from the top through to the basaltic layers have been taken all over the world by the *Glomar Challenger*. The results are in a series of volumes, *Initial Reports of the Deep Sea Drilling Project* (Government Printing Office, Washington, D.C., 1970–1976). The samples from the drill holes are "tubes of ground truth" in the sea floor.

8.2. SEDIMENTS

Marine sediments are unconsolidated; that is, the particles are not cemented or fused together. Samples feel like mud, muddy sand, sand, and so on. We use the relationships between the physical properties of sediments and the geological descriptions of sediments to interpret marine geophysical data. These relationships are partly empirical and partly theoretical. The physical properties are sound velocity and attenuation, rigidity, and density. Laboratory measurements of the sound velocity in saturated sediments show that it is constant over frequencies from 10^3 to 10^6 Hz. Field measurements in bore holes extend the constancy of sound velocity to a few hertz. At the small particle displacements used in acoustic measurements, Hooke's law applies. In viscous media, the particle velocities are included, and the sound velocity depends on the frequency. On the basis of the experimental constancy of sound velocity, we assume that Hooke's law applies to sediments.

In situ measurements show that the "unconsolidated" sediments have enough rigidity to transmit shear waves. The particle displacements for shear waves are normal to the direction of propagation. Fluids do not transmit shear waves. We assume the sediment particles have a combination of mechanical and chemical bond contacts. The assemblage of particles form a structure called the frame or skeleton.

Sketches of sediment structures appear in Fig. 8.2.1. Depending on the structure of the frame, the ability of the water to move from one pore space to another ranges from being completely trapped to flowing through a tube. Biot, in a series of papers commencing in 1941, gives the general theory for the transmission of signals in fluid-filled porous media. His theory describes propagation of a shear wave and two compressional waves. The shear wave is due to the rigidity of the frame. One compressional wave is transmitted through the fluid, the other through the frame. Both compressional waves are coupled. The theory includes the viscosity of the fluid and yields attenuation coefficients that are asymptotically proportional to f^2 at low frequency and to $f^{1/2}$ at high frequency.

When the water moves with the frame the viscosity of the water is neglected and the problem is much simpler. Using these assumptions, we give Gassmann's derivation of the bulk modulus of a fluid-filled frame in Section A8.1. The resulting equation is the same as obtained from Biot's theory for the same conditions. There is an extensive literature, and general references are listed at the end of Appendix A8.

8.2.1. Sound Velocity and Elastic Properties of Fluid-Filled Sediments

The ability of the sea floor to support objects depends on its elastic properties, that is, on the *bulk modulus E* and *modulus* of *rigidity G*. These are related to the compressional and shear wave velocities c and c_s

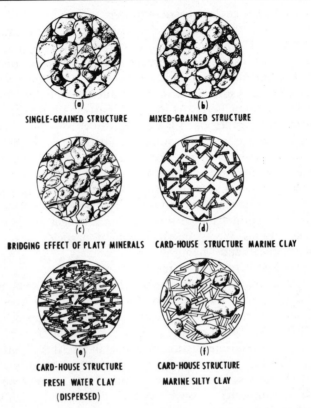

Figure 8.2.1. Sketches of common sediment structures. (Hamilton, 1970.)

by

$$c^2 = \left(E + \frac{4G}{3}\right)\Big/\rho \qquad \text{(8.2.1)}$$

$$c_s^2 = \frac{G}{\rho} \qquad \text{(8.2.2)}$$

where ρ is the density. Measurements of the compressional wave velocity are made *in situ* and in the laboratory. The sound velocity in sediments at the interface and water above the interface have about the same dependence on pressure and temperature (Laughton, 1957; Shumway, 1958; Hamilton, 1963, 1971b). Assuming a simple proportionality, the velocities in the sediment and water at depth z and *in situ* temperature and salinity are $c(z)$ and $c_w(z)$; the sound velocities at 1 atm, *in situ* salinity, and reference temperature are $c(0)$ and $c_w(0)$.

$$\frac{c(z)}{c_w(z)} = \frac{c(0)}{c_w(0)} \qquad (8.2.3)$$

Ordinarily laboratory measurements of $c(0)$ and $c_w(0)$ are made by transmitting ultrasonic pings through a few centimeters of the sample. We assume the dependence of $c(z)$ on frequency to be negligible (Horton, 1974).

Hamilton (1971a) gives laboratory values of the porosity n, density ρ, and compressional velocity c for many types of sediments and environments (Table 8.2.1). The laboratory measurements are for 23°C and 1 atm. The velocity of shear wave propagation is calculated. Laboratory measurements of c_s and G are often unreliable because the coring process disturbs the sediment structure. Hamilton's procedure for calculating E, G, σ, and c_s is based on Gassmann's theory (Section A8.1).

The conversion of the incident compressional waves into transmitted shear waves and compressional waves at the water sediment interface alters the reflection coefficient from its fluid-fluid form. Section A8.3 gives an expression for the reflection at the fluid-solid interface.

8.2.2. Gas Bubbles in Sediments and Water

Often lake bottom sediments contain much organic material, which decays to yield gases, which may in turn form bubbles within the sediments. The compressibility of the gas bubbles is orders of magnitude larger than the compressibility of the fluid and solid particles; therefore presence of the gas bubbles is the dominating effect. (This is for low frequency sound, when all the resonance effects of the bubbles are negligible.) The effective compressibility E^{-1} is

$$E^{-1} = nE_g^{-1} + (1-n)E_{sw}^{-1} \qquad (8.2.4)$$

where subscripts g and sw refer to bubble gas and the sediment-water combination. This is Wood's equation for sound transmission in a mixture. It also follows directly from (A8.1.16) by letting $E_f = 0$, $E_w = E_g$, and $E_s = E_{sw}$. For an air bubble $E_g = \gamma P_A \simeq 1.4 P_A$

When (8.2.4) is applied to lake bottom sediments, the impedance of the sediment may be found to be very much less than that of the water. Levin (1962) reports sound velocities in gas-filled mud bottom of Lake Maracaibo ranging from 0.26 to 0.7 km/s. The impedance mismatch can be so large that very little sound energy penetrates the sediment. In some

TABLE 8.2.1. AVERAGE MEASURED AND COMPUTED ELASTIC CONSTANTS, NORTH PACIFIC SEDIMENTS*

Sediment Type	Measured†			Computed			
	n	ρ	c	E	σ	G	c_s
Continental terrace (shelf and slope)							
Sand							
Coarse	38.6	2.03	1836	6.6859	0.491	0.1289	250
Fine	43.9	1.98	1742	5.6877	0.469	0.3212	382
Very fine	47.4	1.91	1711	5.1182	0.453	0.5035	503
Silty sand	52.8	1.83	1677	4.6812	0.457	0.3926	457
Sandy silt	68.3	1.56	1552	3.4152	0.461	0.2809	379
Sand-silt-clay	67.5	1.58	1578	3.5781	0.463	0.2731	409
Clayey silt	75.0	1.43	1535	3.1720	0.478	0.1427	364
Silty clay	76.0	1.42	1519	3.1476	0.480	0.1323	287
Abyssal plain (turbidite)							
Clayey silt	78.6	1.38	1535	3.0561	0.477	0.1435	312
Silty clay	85.8	1.24	1521	2.7772	0.486	0.0773	240
Clay	85.8	1.26	1505	2.7805	0.491	0.0483	196
Abyssal hill (pelagic)							
Clayey silt	76.4	1.41	1531	3.1213	0.478	0.1408	312
Silty clay	79.4	1.37	1507	3.0316	0.487	0.0795	232
Clay	77.5	1.42	1491	3.0781	0.491	0.0544	195

* *Source.* Hamilton (1971a).
† Laboratory values: 23°C, 1 atm pressure. See Hamilton (1970) for additional properties. Definitions as follows:

n = porosity (%)
ρ = density (g/cm³; Mg/m³, Note: M = 10^6)
c = compressional wave (sound) velocity (m/s)
E = bulk modulus (GN/m², Note: G = 10^9)
σ = Poisson's ratio; computed with $\sigma = (3E - \rho c^2)/(3E + \rho c^2)$
G = rigidity (shear) modulus; computed with $G = [(\rho c^2 - E)3]/4$ (GN/m²)
c_s = shear wave velocity; computed with $c_s = (G/\rho)^{1/2}$ (m/s)

areas this condition is responsible for what is called "offshore ringing" in seismic exploration (Backus, 1959).

8.2.3. ATTENUATION OF SOUND IN SEDIMENTS

We use the attenuation of sound in sediments to predict the performance of subbottom profiling systems. Often *in situ* or laboratory measurements of attenuation are made in higher frequency ranges than the frequencies used for subbottom profiling, 25 to 3500 Hz. To interpolate and extrapolate measurements, we need to know the frequency dependence of attenuation.

Experimental measurements of attenuation are presented in Fig. 8.2.2. For any given sediment type, lines can be drawn through the measurements, with the slopes nearly 1, or attenuation nearly proportional to f.

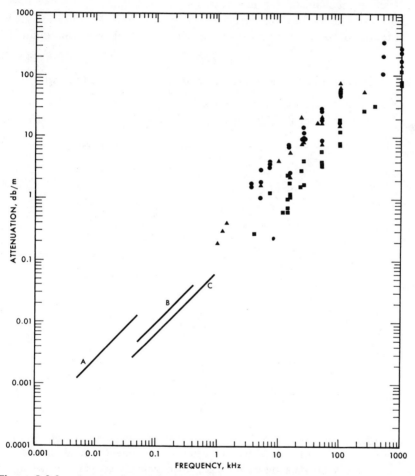

Figure 8.2.2.   Attenuation versus frequency. The measurements are for natural saturated sediments and sedimentary strata: ●, sands (all grades); ■, clayey silt, silty clay; ▲, mixed sizes (e.g., silty sand, sandy silt, sand-silt-clay); sand data at 500 and 100 kHz. Low frequency data: line *A* land, sedimentary strata; line *B* Gulf of Mexico coastal clay-sand; line *C* sea floor, reflection technique. (Hamilton, 1972.)

Since Biot's general theory (which includes viscous losses in the fluid) gives attenuation that varies with f^2 at low frequency and $f^{1/2}$ at high frequencies, other loss mechanisms may be more important in sea floor sediments than those involving viscosity. For example, as the sound wave passes groups of touching particles, the particles move and rub against each other, and losses are due to the friction of the rubbing particles. In the friction mechanism, the loss is proportional to particle displacement and does not depend on the particle velocity. Much of the literature gives the attenuation α dB per unit distance with f in kilohertz. Thus defined

$$\alpha \equiv bf^m \tag{8.2.5}$$

where b is a constant. Alternatively attenuation is expressed as $\exp(-\alpha_e x)$

α dB per unit distance $= 8.686\ \alpha_e$
$\alpha_e =$ nepers per unit distance   **(8.2.6)**

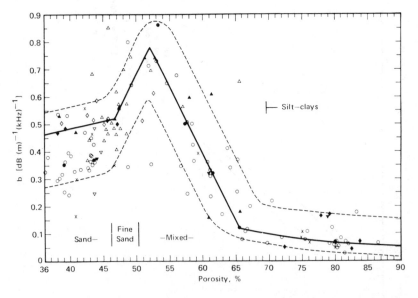

Figure 8.2.3. Attenuation coefficient b. Porosity (%) and b (in $\alpha = bf$). Dashed lines are estimates of the upper and lower limits of attenuation. Solid symbols are averaged values for data off San Diego; open symbols are data in averages; different symbols are for different sets of data: ◆ 1966–1970, ▲ –1956, ▼ –1962; ● Shumway (1960); × are selected values from literature. Regression equations for solid lines: coarse, medium, and fine sand (36 to 46.7%), $b = 0.2747 + 0.00527(n)$; very fine sand and lower porosity mixed sizes (46.7 to 52%), $b = 0.04903(n) - 1.7688$; mixed sizes (52 to 65%), $b = 3.3232 - 0.0489(n)$; silt-clays (65 to 90%), $b = 0.7602 - 0.01487(n) + 0.000078(n)^2$. (Hamilton, 1972.)

The relationship of α_e to the "damping factor" of the material Q^{-1} is in Appendix A8.2.

Most values of m range from a little less than 1 to more than 1. The rigidity of the sediments depends on porosity. Since the rigidity depends on the interparticle contacts, we expect friction losses to be larger when the rigidity is large. Comparison of the dependence of the attenuation constant b on the porosity (Fig. 8.2.3) and the rigidity (Fig. A8.1.2) confirms this relationship.

The attenuation loss in the sediments reduces the reflection coefficient for signals reflected at the water-sediment interface relative to the no-loss case. Beyond the critical angle, the loss causes a reduction of the reflected energy because the sound field penetrates the sediments in the process of total reflection. Section A8.4 gives a short derivation of the reflection coefficient.

We have barely touched the surface of a large literature on this specialized subject. Hamilton (1972) lists more than 90 references. The chapters of Hamilton, Bucker, Smith, and Stoll in Hampton, Ed., 1974 are a good place to begin. Norman Ricker has given theoretical and experimental studies of the propagation of a transient signal in a viscous medium having $\alpha \sim f^2$ (Ricker, 1953).

8.3. MARINE SEISMIC MEASUREMENTS

In our experience the main differences between working at sea and on the land are the constant motion of the sea and the acoustic transparency of the ocean. The motion of the seawater causes instruments and ships to move relative to each other and relative to the bottom. The marine geophysicists measure these displacements as part of their experiments.

Geophysicists have a structure in mind and a model for interpreting the data when they design an experiment. In analyzing the data, they adjust the parameters of the structural model until numerical calculations of the signal fit the experimental data. There are three stages of comparison: (1) adjusting the model until the numerical travel times fit the data; (2) including reflection, transmission, and spreading losses to calculate signal amplitudes; (3) calculating theoretical signals and adjusting the model until the signals fit the data.

We are primarily concerned with step 1, and we choose the simplest model that fits the data within the range of uncertainty of the measurements. As we acquire more and better data, we expect our "best fitting models" to change.

8.3.1. Wave Fronts and Ray Paths for Head Waves

Let us recall a few results of Section 3.2, which contains the essential material about ray paths. The ray paths follow Snell's law, $\sin \theta_1 / c_1 = \sin \theta_2 / c_2$. The travel time of a signal traveling along that path is a minimum and obeys Fermat's principle. Using Fig. 8.3.1, when $c_2 > c_1$ at

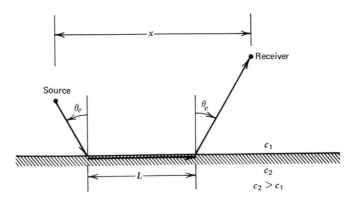

Figure 8.3.1. Critical angle ray path. The signal travels distance L in medium 2 at the sound speed c_2.

the critical angle θ_c, the refracted ray path is parallel to the interface at $\theta_1 = \theta_c$ and $\sin \theta_c = c_1 / c_2$. A signal traveling parallel to the interface at c_2 can refract into the upper medium at θ_c. The reader can show that the signal that follows the path in Fig. 8.3.1 has a minimum travel time. For $\theta_1 > \theta_c$, the reflection coefficient is complex and has an absolute value of 1. The phase of the reflected signal depends on θ_1. We call this "total reflection."

We use Huygens wavelets and wave front constructions to explain the processes involved at the critical angle. We use a homogeneous liquid half space having low sound speed over a higher sound speed. A sequence of positions of the wave front constructions appears in Fig. 8.3.2. The source is at s and the image is at s'. The sound speed in the lower medium is twice the upper and θ_c is 30°. The critical angle is drawn on all the constructions. The constructions use the same Huygens "wavelet sources" r_0, r_1, r_2, c, t, and so on. When the contacts of the wave front and the interface are within the region bounded by the critical angle, the wave front constructions look like Fig. 8.3.2a. For the contact beyond the critical angle, the initial step of the wave front construction has the appearance of Fig. 8.3.2b. The wavelets from r_0, r_1, r_2, and c form the envelope of the wave front in the medium 2. The wavelet from c intersects the interface at h_3. Wavelets from t_1 and t_2 do not add to form a wave front in medium 2. In medium 1, the wavelets add to give the reflected wave. Wavelet sources t_1 and t_2 are in the region of total reflection. The contact of the wavelet from c is at h_3. We need to consider boundary conditions at h_3 to complete the construction.

The sound pressures and displacements are equal on both sides of the interface. The sound pressure and displacement at h_3 in medium 2 causes a corresponding sound pressure and displacement at the interface in medium 1 and the Huygens wavelets radiate into medium 1. Figure 8.3.2c illustrates the construction in medium 1 for the sequence of wavelet sources c, h_1, h_2, and h_3. The line labeled *head wave* is the envelope of these wavelets. For a point source, the head wave is a conical surface. The contact h moves along the interface at c_2. The head wave travels along the critical angle into the upper medium. Many geophysicists refer to this wave as the *refracted wave* or *arrival*. It is followed by the totally reflected wave in medium 1. The complete construction is shown in Fig. 8.3.2d. The head wave is also called the *lateral wave*.

Analytical developments of the head wave are beyond the scope of our text (Cagniard, 1939; Heelan, 1953; Grant and West, 1965, Ch. 6). The following results are of interest: the energy in the head wave is due to the curvature of the wave front in medium 2.

$$\text{amplitude of head wave} \sim x^{-1/2}L^{-3/2}k^{-1} \tag{8.3.1}$$

where x and L are defined in Fig. 8.3.1. Finally, plane wave reflection and transmission coefficients are accurate for curved wave fronts at large distances from the source (i.e., $kR \gg 1$) and for incident angles that are not too close to critical angles. The conditions are set forth in Section 2.9.

8.3.2. Head Waves (Refraction Method)

Ewing and Worzel (1948) measured the first arrivals for a seismic experiment in Chesapeake Bay. The shots were fired in water about 15 m deep. The arrival that traveled in the water path had high frequency components and was separated from signals traveling deep in the ground by a high pass filter. The travel times of the first arrival versus distance are plotted in Fig. 8.3.3. The first arrivals fall along two straight lines. The receiving hydrophone was on the bottom, and the shots were fired on the bottom.

We use the "simplicity principle" to interpret the data. The simplest model and its travel times are shown in Fig. 8.3.4. We omit the thin layer

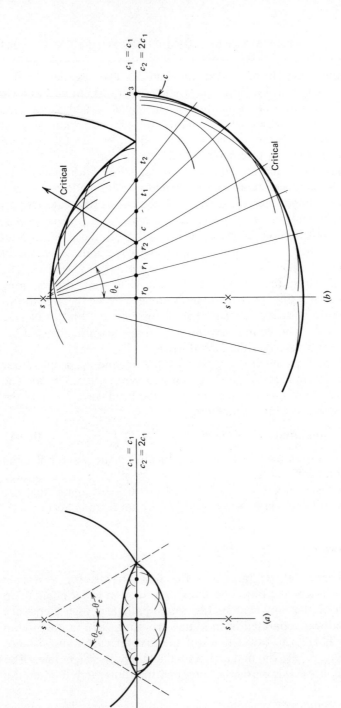

Figure 8.3.2. Huygens' construction for "head wave." (*a*) Less than critical. (*b*) Reflected and transmitted waves, head wave omitted. (*c*) Head wave. (*d*) Complete construction.

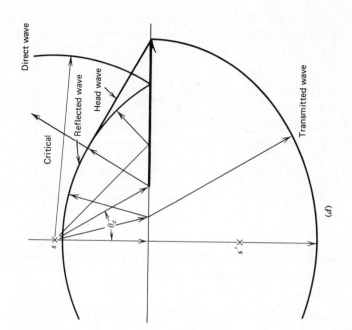

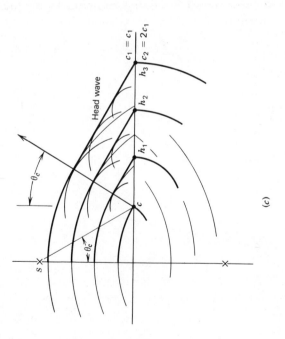

Figure 8.3.2. (*Continued*)

265

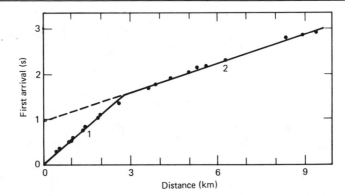

Figure 8.3.3. Seismic data, T versus distance. The data are given by Ewing and Worzel (1948, pp. 6–7), Solomon's Shoal, Chesapeake Bay. Their water travel times were changed to distance using $c = 1500$ m/s.

of water in modeling the structure of the subbottom. The sound speed in the upper layer is for the direct arrival 1 to 2.7 km. Beyond that distance, the head wave arrival 2 from medium 2 is first. From the figure, the travel time for the direct arrival is

$$\text{arrival 1:} \qquad\qquad t_d = \frac{x}{c_1} \qquad\qquad (8.3.2)$$

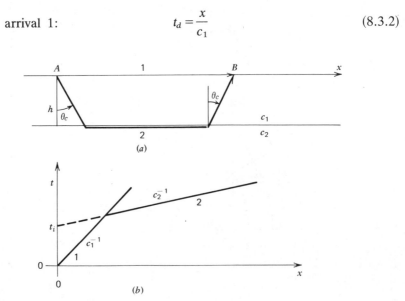

Figure 8.3.4. Direct path and head wave. (a) Ray paths for $c_2 > c_1$. (b) Line 1 is direct path, line 2 is head wave.

The travel time for the head wave is

arrival 2:
$$t_h = \frac{2h}{c_1 \cos \theta_c} + \frac{x - 2h \tan \theta_c}{c_2}$$

or

$$(8.3.3)$$

$$t_h = 2h \frac{\cos \theta_c}{c_1} + \frac{x}{c_2} \qquad \text{for} \quad x > \frac{2h}{c_1 \tan \theta_c}$$

where $\theta_c = \arcsin(c_1/c_2)$. These are the equations of two straight lines. We adjust parameters, c_1, c_2, and h to fit the data. Line 1 has the slope c_1^{-1} and line 2 has the slope c_2^{-1}. The values are

$$c_1 = 1.82 \text{ km/s}, \qquad c_2 = 4.72 \text{ km/s}$$
$$\theta_c = 22.7°$$

We use the intercept of (8.3.3) at $x = 0$, t_i to calculate h.

$$h = \frac{c_1 t_{x=0}}{2 \cos \theta_c}$$

$$(8.3.4)$$

$$t_i = 0.97 \text{ s}, \qquad h = 0.96 \text{ km}$$

Ewing and Worzel (1948) say that the value of $c_2 = 4.72$ km/s is low for "basement." They imply that 5.5 km/s is reasonable and that the bottom may be sloping downward along the profile to give a lower apparent velocity. Additional profiles are needed to determine whether this is the case.

Figure 8.3.5 presents geometry for a sloping bottom. The distances along the paths L_1, L_2, and L_3 are

$$L_1 = \frac{h}{\cos \theta_c}$$

$$L_2 = x(\cos \phi - \sin \phi \tan \theta_c) - 2h \tan \theta_c \qquad (8.3.5)$$

$$L_3 = \frac{h}{\cos \theta_c} + x \frac{\sin \phi}{\cos \theta_c}$$

The travel time for the head wave is

$$t_h = \frac{L_1 + L_3}{c_1} + \frac{L_2}{c_2} \qquad (8.3.6)$$

After substitution of (8.3.5) and Snell's law into (8.3.6), we express t_h as follows:

$$t_h = 2h \frac{\cos \theta_c}{c_1} + \left| \frac{x}{c_2} \right| \frac{\sin(\theta_c + \phi)}{\sin \theta_c} \qquad (8.3.7)$$

where $|x/c_2|$ eliminates the coordinate sign of x.

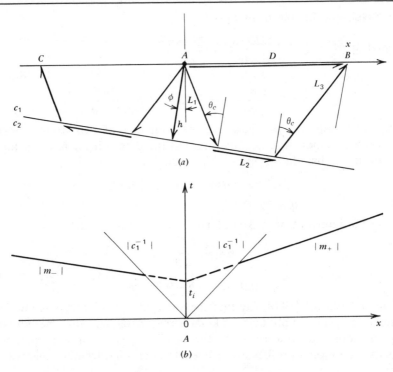

Figure 8.3.5. (a) Direct path and head wave and (b) their time-distance plot. The absolute values of the slopes are $|c_1^{-1}|$, $|m_-|$, and $|m_+|$.

We cannot evaluate the effect of the dipping interface from a single profile as shown in Fig. 8.3.3. A second profile is needed. Two procedures are commonly used. (1) The travel times are measured with B as the starting point of the profile $B–A$. Geophysicists call this a reversed profile. (2) Marine geophysicists often use A as the listening ship and B as the shooting ship: B starts shooting at large range, passes A, and continues to a large range C. We use the labels $t_h(+)$ for the shot to the right of A and $t_h(-)$ for the shot to the left of A and write

$$t_h(+) = 2h\frac{\cos\theta_c}{c_1} + \left|\frac{x}{c_2}\right|\frac{\sin(\theta_c+\phi)}{\sin\theta_c}$$

$$t_h(-) = 2h\frac{\cos\theta_c}{c_1} + \left|\frac{x}{c_2}\right|\frac{\sin(\theta_c-\phi)}{\sin\theta_c}$$

(8.3.8)

Plots of the data look like Fig. 8.3.5b. Our first step in solving for ϕ and θ_c is to use Snell's law and replace $c_2\sin\theta_c$ by c_1 in (8.3.8). The slopes of head wave arrival (Fig. 8.3.5b) are (m_+) and (m_-) and are equal to the coefficients of x in (8.3.8).

$$c_1 m_+ = \sin(\theta_c + \phi)$$
$$c_1 m_- = \sin(\theta_c - \phi)$$

(8.3.9)

The solutions are

$$2\phi = \arcsin(c_1 m_+) - \arcsin(c_1 m_-)$$
$$2\theta_c = \arcsin(c_1 m_+) + \arcsin(c_1 m_-)$$

(8.3.10)

The rest of the calculations are straightforward.

Elaborations of this type of measurement for more layers and dipping interfaces are in specialized texts.

8.3.3. Reflection Measurements

Our purpose is to measure the sound velocity and thickness of the thin layers of sediment on the sea floor. In many ocean basins, this is the first few hundred meters of sediments. The sound velocity in sediments at the water-sediment interface is less than the sound speed in water and increases with subbottom depth. The layers are very thin relative to the depth to the bottom. To make a continuous identification of the reflection

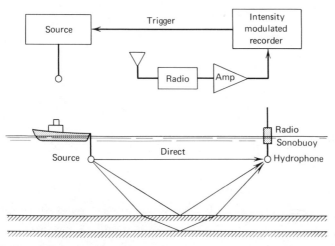

Figure 8.3.6. Sonobuoy reflection system. Radio in sonobuoy transmits signals received on its hydrophone.

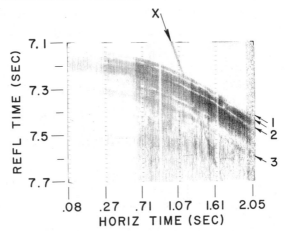

Figure 8.3.7. Hatteras abyssal plain, profile 4, 400 Hz source, 32°24′N, 71°01′W; x is the direct path. (Clay and Rona, 1965.)

from a layer, we need many reflection transmissions as the distance between the source and receiver increases.

A reflection system using a controlled source and sonobuoy receiver is illustrated in Fig. 8.3.6. We place the sonobuoy in the water, then drift or steam away from the buoy. The recorder keys the source transmissions and makes a graphic record of the signals. A reflection profile taken by

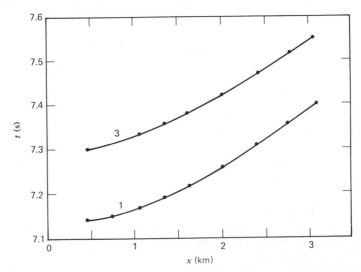

Figure 8.3.8. Time-distance graph. Hatteras abyssal plain (32°24′N, 71°01′W); points taken from data in Fig. 8.3.7.

Clay and Rona in the Hatteras abyssal plain is reproduced in Fig. 8.3.7. The record is for the travel time of the signal versus experimental time. The direct arrival, or horizontally traveling signal, gives a measure of the separation of the source and receiver. Figure 8.3.8 is our graph of the reflection time for the interfaces labeled 1 and 3. The curves are not like those for the head wave, and we need a different method of analysis.

The sound speed in the water above the bottom is known from oceanographic data. We use this in constructing a model of horizontally stratified ocean over a layered sea floor (Fig. 8.3.9). Our analysis for this

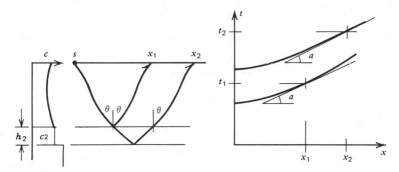

Figure 8.3.9. Ray parameter method. Travel paths in water from source s to x_1 and to x_2 have the same angles θ and ray parameter a. Slopes of time distance curves at x_1, t_1 and x_2, t_2 are a. After Bryan, in Hampton, Ed. (1974).

model uses ray traces and is known as the "ray parameter method" (Bryan, in Hampton, Ed., 1974, pp. 119–130). Recalling (3.2.2), (3.2.7) and (3.2.8), the travel time and displacement for the ray reflected at the interface 1 are

$$t_1 = 2 \int_0^{h_1} c^{-1}(1 - a^2 c^2)^{-1/2}\, dz$$

$$x_1 = 2 \int_0^{h_1} ac(1 - a^2 c^2)^{-1/2}\, dz$$

(8.3.11)

$$a \equiv \frac{\sin \theta}{c}$$

(8.3.12)

where c is a function of z and a is the *ray parameter*. The transmitted ray, having the same value of "a" returns to the surface at x_2 and at the time t_2. For a layer having thickness h_2 and sound velocity c_2, we write

$$t_2 = t_1 + 2h_2 c_2^{-1}(1 - a^2 c_2^2)^{-1/2}$$
$$x_2 = x_1 + 2ah_2 c_2(1 - a^2 c_2^2)^{-1/2}$$

(8.3.13)

The solutions of (8.3.13) for c_2 and h_2 are

$$c_2 = \left[\frac{x_2 - x_1}{a(t_2 - t_1)} \right]^{1/2}$$

$$h_2 = \frac{c_2(t_2 - t_1)(1 - a^2 c_2^2)^{1/2}}{2}$$

(8.3.14)

We next show that a is dt_1/dx. The formula for implicit differentiation is

$$\frac{dt}{dx} = \frac{\partial t}{\partial a} \frac{\partial a}{\partial x} + \frac{\partial t}{\partial x}$$

(8.3.15)

Differentiation of (8.3.11) and (8.3.12) under the integral sign yields

$$\frac{\partial t}{\partial a} = 2a \int_0^{h_1} c(1 - a^2 c^2)^{-3/2} \, dz$$

$$\frac{\partial t}{\partial x} = 0$$

(8.3.16)

$$\frac{\partial x}{\partial a} = 2 \int_0^{h_1} c(1 - a^2 c^2)^{-3/2} \, dz$$

$$\frac{dt}{dx} = \frac{\partial t}{\partial a} \left(\frac{\partial x}{\partial a} \right)^{-1} = a$$

(8.3.17)

These are measurables on the time-distance graph. We find that we can measure the slopes a to two or three significant figures. The tangent points are difficult to determine graphically. Correspondingly, $t_2 - t_1$ and $x_2 - x_1$ are uncertain.

An alternative to picking the tangent points graphically is fitting the time-distance data to a polynomial in x, then differentiating the polynomial to determine a. Since

$$t_1^2 = t_1^2(0) + \frac{x^2}{c_1^2}$$

$$t_1(0) \equiv \frac{2h_1}{c_1}$$

(8.3.18)

is exact for a layer having a constant speed c_1 and thickness h_1 we choose this form for the polynomial. The polynomials for the first and second interfaces are

$$t_1^2 = A_0 + A_1 x_1 + A_2 x_1^2 + \cdots$$

$$t_2^2 = B_0 + B_1 x_2 + B_2 x_2^2 + \cdots$$

(8.3.19)

Over a horizontally stratified bottom, the odd powers of x are zero. We also limit the expansion to second order. The ray parameters are

$$a_1 = \frac{dt_1}{dx_1} = \frac{A_2 x_1}{t_1}$$

$$a_2 = \frac{dt_2}{dx_2} = \frac{B_2 x_2}{t_2}$$

(8.3.20)

where a_1 is the value of a at x_1, and so on. In the ray parameter method, x_1 and x_2 are chosen to make $a_1 = a_2 = a$; $x_2 - x_1$ and $t_2 - t_1$ are

$$x_2 - x_1 = a \left(\frac{t_2}{B_2} - \frac{t_1}{A_2} \right)$$

$$t_2 - t_1 = \frac{B_2 x_2 - A_2 x_1}{a}$$

(8.3.21)

Recalling (8.3.14), c_2 is

$$c_2^2 = \frac{t_2/B_2 - t_1/A_2}{t_2 - t_1}$$

(8.3.22)

or

$$c_2^2 = \frac{x_2 - x_1}{B_2 x_2 - A_2 x_1}$$

(8.3.23)

and t_2 and t_1 are any pair of values for which $a_1 = a_2 = a$. We can choose a, solve (8.3.20) for x, and substitute this into (8.3.19) to obtain

$$t_1^2 = A_0 \left(1 - \frac{a^2}{A_2} \right)^{-1}$$

$$t_2^2 = B_0 \left(1 - \frac{a^2}{B_2} \right)^{-1}$$

(8.3.24)

By fitting t_1^2 and t_2^2 to the polynomial for a limited range of x, the method can be made applicable to wide angle reflection data.

The interval velocity formula, independently given by Durbaum (1954) and Dix (1955), applies to reflection data near vertical incidence. We obtain their formula by first making an analogy of (8.3.19) to (8.3.18) and defining the *apparent velocities* $\tilde{c}_1$ and $\tilde{c}_2$ and vertical incidence reflection times $t_1(0)$ and $t_2(0)$:

$$\tilde{c}_1^2 \equiv \frac{1}{A_2}, \qquad \tilde{c}_2^2 \equiv \frac{1}{B_2}$$

$$t_1^2(0) \equiv A_0, \qquad t_2^2(0) \equiv B_0$$

(8.3.25)

where $\tilde{c}_1$ and $\tilde{c}_2$ are also called the *rms velocities*.

Next we choose very small values of x_1 and x_2 in (8.3.22) and use the limiting values $t_2(0)$ and $t_1(0)$ for t_2 and t_1. On letting a tend to zero in (8.3.24) and substituting (8.3.25) into (8.3.22), the Dix–Durbaum formula is written

$$c_2^2 = \frac{t_2(0)\tilde{c}_2^2 - t_1(0)\tilde{c}_1^2}{t_2(0) - t_1(0)} \qquad (8.3.26)$$

This formula is limited to the range of x, starting at zero, that fits a second-order polynomial accurately. Clay and Rona (1965) and Shah and Levin (1973) give formulas that include terms through x^4. LePichon et al. (1968) furnish computational techniques.

The range of x for the reflection data in Figs. 8.3.7 and 8.3.8 is smaller than the water depth. We use (8.3.26) to determine the sound speed in the layer bounded by reflectors 1 and 3. A graph of $t_2^2(0)$ and $t_1^2(0)$ versus x^2 is convenient for determination of the coefficients because the intercepts are $t^2(0)$ and $t_2^2(0)$, and the slopes are $\tilde{c}_1^{-2}$ and $\tilde{c}_2^{-2}$ (Fig. 8.3.10). Our values of the coefficients are

$$t_2(0) = 7.304 \text{ s} \qquad t_1(0) = 7.134 \text{ s}$$
$$\tilde{c}_2 = 1.514 \text{ km/s} \qquad \tilde{c}_1 = 1.508 \text{ km/s}$$
$$c_2 = 1.7 \text{ km/s}$$

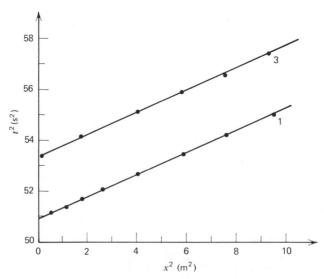

Figure 8.3.10. Plot of T^2 versus x^2 for Hatteras abyssal plain. The apparent velocities of lines 1 and 3 are 1.508 and 1.514 km/s, respectively. The reflection times at $x = 0$ are 7.134 and 7.304 s.

The estimate of c_2 is very dependent on the accuracy of $\bar{c}_1$, $\bar{c}_2$, $t_1(0)$, and $t_2(0)$. Small errors in these quantities cause large errors in c_2. In practice, (8.3.26) is useful when the second layer is larger than $h_1/10$.

Some of the major interferences or difficulties in making interval-velocity measurements are due to multiple reflections, that is, signals that have traveled from the top to the bottom of a layer and back, twice. Because of their later arrival time, these signals may be erroneously identified as the reflection from a deeper interface. Evaluation of c_2 often yields values of c_2 that are anomalous and even negative. We interpret anomalously low or high values of c_2 to mean that multiple reflections are present. Advanced methods of interpretations are given by Maynard et al., Bryan, Houtz (in Hampton, Ed., 1974), and the references therein.

8.3.4. Average Sound Velocity in Sediments

The most common use of average sediment sound velocity data is in the interpretation of seismic profiler data. The reflections from subbottom layers appear as reflection times on the records, and investigators use sound velocity curves to convert the travel times to sediment thickness. Except where there are drill hole data, measurements come from interval velocity calculations, and usually these are limited to layer thicknesses greater than one-twelfth the water depth. Houtz and his co-workers have measured and compiled thousands of individual measurements since 1968. The measurements are in card files at Lamont-Doherty Geological Observatory, Palisades, New York.

Marine geophysicists have chosen to express the dependence of sound velocity on depth of burial as a function of the one-way travel time T to that depth, measuring T from the water-sediment interface. They call the sound velocity $c(T)$ the instantaneous velocity. In Houtz's notation, $c(T)$ is

$$c(T) = V_0 + kT \tag{8.3.27}$$

The constants V_0 and k are the results of least squares fit of many interval velocity measurements to (8.3.27). Values of k range from 0.9 to 3.9 km/s^2 in various parts of the world. The intercept, V_0 ranges from 1.2 to 1.8 km/s. Houtz states that the intercept is an artifact of the distribution of sound velocities at greater depths. For thin layers near the water interface, we use Table 8.2.1 and a gradient of $1\,s^{-1}$. In the Western North Atlantic Abyssal Plains, $c(T)$ is

$$c(T) = 1.67 + 0.97T \qquad T < 0.8\,s \tag{8.3.28}$$

and the standard deviation for 33 measurements is 0.17. In the equatorial Pacific

$$c(T) = 1.46 + 3.9T \qquad T < 0.35 \text{ s} \qquad (8.3.29)$$

and the standard deviation is 0.17 for 29 measurements (Houtz, in Hampton, Ed., 1974). Their data show that different constants are required for different ocean basins and regions within the ocean basins.

The thickness of the layer h for one wave travel time T is found by integration of (8.3.27):

$$h = V_0 T + \frac{kT^2}{2} \qquad (8.3.30)$$

This equation is used to convert seismic profiles to depth sections. The equations for $c(T)$ and h are empirical and are the result of the fitting of the equation to experimental measurements. We expect the coefficients and perhaps the form of equations to change as more data are accumulated.

8.4. ECHO SOUNDING OF THE SEA FLOOR

The echo sounder measures the time for a sonic signal to go from the ship to the sea floor and back to the ship. The known sound speed is then used to convert the travel times to depths. Depth sounders range in complexity from a system having a single transducer and visual depth readout to a computer-controlled system having an array of transducers. In many of the older installations, the transducer is attached to the hull and may have a half power to half power beam width as large as $60°$. The scientific echo sounders on oceanographic ships usually have accurate internal clocks, and travel errors are less than 10^{-3} s. Many of the scientific echo sounders built in the United States display the returning signals on a graphic recorder having a 1 s sweep across 19 in. (0.47 m) paper width.

A configuration of a computer-controlled multibeam sonar system is shown in Fig. 8.4.1. A pair of arrays are crossed to give a set of pencil-like beams. The transmitting array is along the keel and has a fan-shaped beam normal to the ship. The receiving array is athwartships and has a set of beams along the ship. The effective beams are the product of the transmitting beam and the receiving beams. The effective beam widths may be of the order of $1°$. Since roll and pitch of a ship is normally greater than $1°$, this effect must be eliminated. The roll and pitch can be measured by a gyroscope and fed to the computer for corrections.

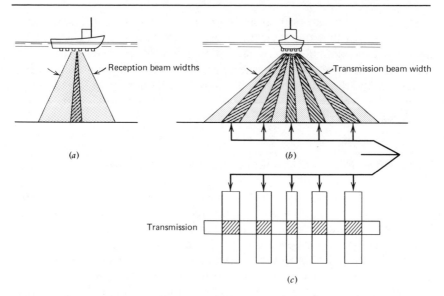

Reception beam widths

Transmission beam width

(a)

(b)

Transmission

(c)

Figure 8.4.1. Multibeam echo sounder using crossed arrays. Shaded areas are products of transmission and reception beams. (a) Side view; transmission beam is shaded. (b) Front view; reception beams are shaded. (c) Plan view.

The positions and course of the ship are fed into the computer. This, together with ray traces for the beam directions and the sound speed, is used to convert the travel times along the beams to a chart of depths of the sea floor.

Returning to the simpler single transducer system, many echo sounders give a record of "depth" versus experiment time. There are two steps in reducing the data. During the survey, the watch standers mark time on the echo sounding record, permitting the combination of time with the log of the ship's position to plot depth along the ship's track. We use the actual sound speed to convert the depth sounder reading to true depth. Most echo sounders are built to read an approximate depth by using an average sound speed (sometimes called "sounding velocity"). Some instruments are adjustable, and the user can set the sounding velocity to give true depth in a limited area and depth range. More commonly, the sounding velocity is set by the builder. In the United States, for example, many echo sounders use 4800 ft/s or 800 fm/s (1461 m/s). Some instruments use 1500 m/s. Depths based on an arbitrary sounding velocity are called "uncorrected depths." The procedure for correction is to use the echo sounder calibration to convert "uncorrected depth" to time, then to use the sound speed profile to calculate true depths.

Some charts of the bottom are corrected, and some are uncorrected. We have heard that the navigator prefers uncorrected charts because he uses the echo sounder as a navigational aid by comparing the echo sounding with depths given on the chart. Many of the hydrographic charts published by the U.S. Naval Oceanographic Office give uncorrected depths and use 4800 ft/s as the sounding velocity. Most chart makers either tell whether the depths are corrected or give the sounding velocity.

8.4.1. Broad-Beamed Echo Sounder Over a Rugged Sea Floor

Many echo sounders have sonar beam widths ranging from 10° to 60°. For example, in 3 km water depth, the 60° beam width transducer illuminates a strip 3.5 km wide under the ship. Features having dimensions larger than 3.5 km are resolved. The echo sounding in Fig. 8.4.2 was taken over a rugged bottom in the Atlantic Ocean. The sonar operated at 12 kHz, and the penetration of subbottom sediments is small. The locations of the mountains are apparent. Some overlap, and many appear to have the same profile. The overlapping echo traces and the crescent

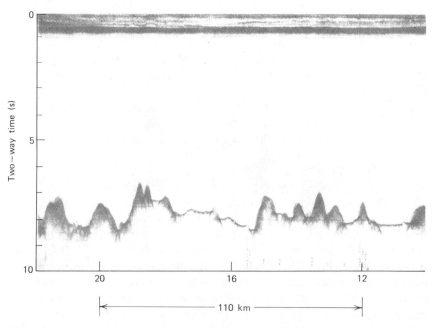

Figure 8.4.2. A 12 kHz echo sounding. Data from Lowrie and Escowitz (1969, p. 232).

shapes of the traces are due to the combination of an acoustic diffraction effect and the echo sounder display.

Recalling Fig. 2.2.4, the edge of a feature is the source of a diffracted wave. As Fig. 8.4.3 indicates, the source at 1 illuminates the feature M.

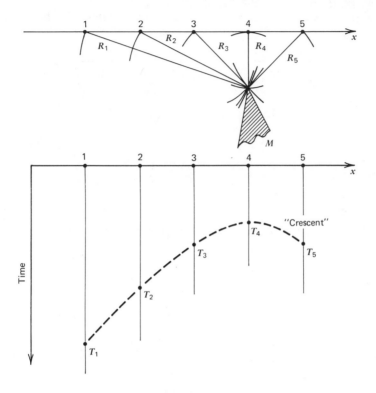

Figure 8.4.3. Diffracted arrivals and the echo sounding record; $T_1 = 2R_1/c$, etc.

The diffracted wave from the edge of M travels back to position 1 and is received at time T_1. The graphic recorder makes an intensity mark at that time on paper, as in the lower part of the figure. The transmission at position 2 gives an intensity mark at T_2, and so on. The dashed line indicates the appearance of the diffraction arrival that results from continuous operation of the echo sounder. The shape of the diffraction signal on the echo sounder record is a hyperbola. The shape of the diffraction arrival, or "crescent," is nearly the same as long as M is more acute than the crescent. The record in Fig. 8.4.3 is drawn for equal vertical and horizontal scales. The actual echo sounding record has a horizontal scale that depends on the paper speed and the ship's speed. The vertical scale depends on the depth scale. Usually the vertical scale is exaggerated relative to the horizontal scale, and the crescents are much sharper. One can make a template for the depth and vertical exaggeration to aid interpretation of the data.

Figure 8.4.4 includes a profile and the combination of reflections and

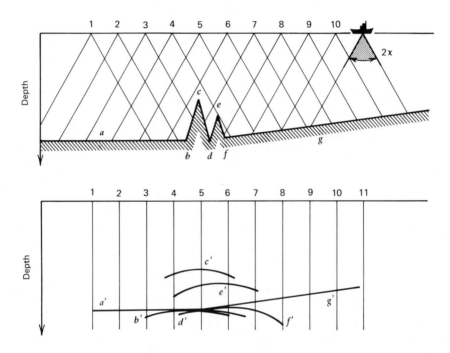

Figure 8.4.4. Profile of the bottom and echo sounding record: b, c, d, e, and f are the sources of the diffraction signals b', c', d', e', and f', a' and g' are reflections from a and g.

diffraction signals or crescents. The beam width of the echo sounder (2χ) is 60°. The length of the crescent depends on the beam width. In this sketch several features can be illuminated at the same time and give overlapping crescents. Over a sea floor having roughness in two dimensions, the features on each side of the ship also contribute their crescent arrivals.

Two alternatives may be used to "increase the resolution." The resolution of the echo sounding can be improved by using a very narrow beam width. This decreases the size of the illuminated area. The other alternative is to place the echo sounder in a deep-towed fish so that it is near the bottom. This also decreases the illuminated area. Both alternatives involve problems. On a ship-mounted system, the direction of the sonar beam has to be stabilized. In deep towing, the technology is difficult and expensive (Lonsdale et al., in Hampton, Ed., 1974, pp. 293–318).

8.4.2. Subbottom Profiling

The subbottom profiler is a powerful tool for the study of the structure of the sea floor. The subbottom profiler is simple because it is basically an echo sounder. The depth of penetration of sound signals into the subbottom depends on the frequency of the signal, the absorption coefficient of the sediments, and, of course, the signal to noise power ratio. A combined system is presented in Fig. 8.4.5. Usually the transducers of the 12 and 3.5 kHz echo sounders are mounted on the hull of the ship. The penetration of the 12 kHz is limited to a few tens of meters and does not show on the scale of the record in Fig. 8.4.5. The 3.5 kHz echo sounder penetrates to more than 100 m in areas where the sediments have small absorption coefficients. Deeper penetration requires lower frequency systems (i.e., <100 Hz), called "seismic profiling" systems.

The low frequency sources have a high output, often like an explosion, and it is inconvenient to build them into the ship. The receiving hydrophone is towed behind the ship because the ship has a high noise level in the frequency range of the signal.

Examples of impulsive sources are explosives, sparks, and the "air gun." The air gun is routinely used to make thousands of kilometers of seismic profiling. It has a large signal in the low frequency range, has reproducible signal transmissions and can transmit at about 10 s intervals. Mechanically, the air gun is simple. A chamber is filled with air at a pressure of 100 atm, then the chamber is opened very quickly, permitting the air to expand into the water. The expanding air is a small explosion. The down-traveling signals, including the surface reflection, from a 5 liter (300 in.^{3}) air gun at different depths of the air gun are shown in Fig. 8.4.6. The air initially expands, giving the signal contribution labeled 1 in Fig. 8.4.6. The air bubble collapses and expands again, giving 2, 3, and so on. The time between bubble pulses increases with the volume of air and decreases with depth. The spectral analysis of the signal for 6.1 m depth appears in Fig. 8.4.7. Sometimes combinations of air guns of several

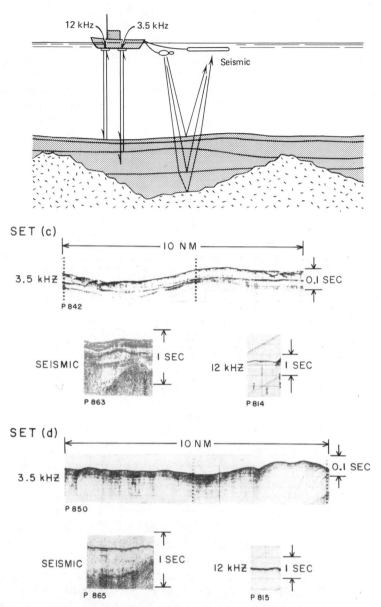

Figure 8.4.5. Echo sounding arrangement and seismic profiling records, taken along same profiles, for different frequencies. Seismic data are at about 25 Hz. Set (c) is for a smoothly rolling bottom. Sub bottom reflections appear on 3.5 kHz and seismic traces. Set (d) is for a slightly rough bottom. The 3.5 kHz signals are scattered by small features on the bottom. The bottom appears to be smooth for the seismic signal (60 m wavelength). (Lowrie and Escowitz, 1969.)

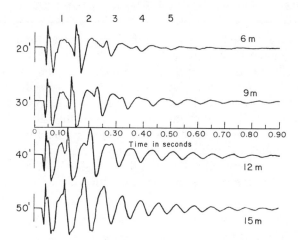

Figure 8.4.6. Signatures of standard air gun with 5 liter chamber at different depths. Air pressure is $1.2 \times 10^7 \, \text{N/m}^2$. (Mayne and Quay, 1971.)

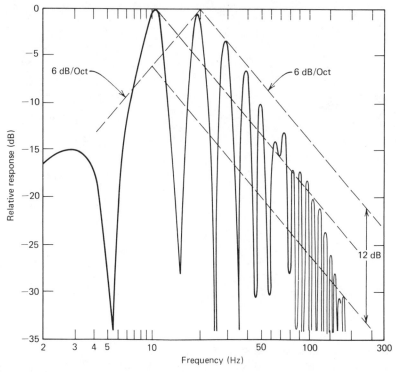

Figure 8.4.7. Spectrum analysis of raw signature of standard air gun with chamber at a depth of 6.1 m (see Fig. 8.4.6). An octave (oct) is a doubling of the frequency. (Mayne and Quay, 1971.)

283

different sizes are used to enhance the initial signal and reduce the contributions of the bubble pulses.

The widespread use of seismic profiling by the oceanographic laboratories of the world has given the distribution of sediments on the sea floor and their relationship to basic processes. The profile across the mid-Atlantic ridge shows that the sediments are thin at the ridge crest and thicken away from the crest. Sections of the profile are shown in Fig. 8.4.8. The sketches below the seismic profiles are line drawings of the sections of seismic profile and, beneath these, a simplified cross section of the structure. By the sea floor spreading hypothesis, we expect the

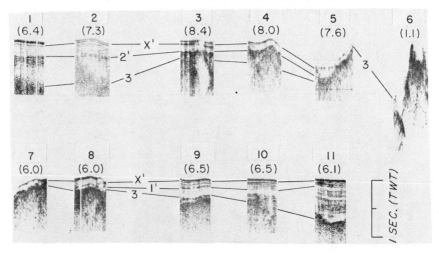

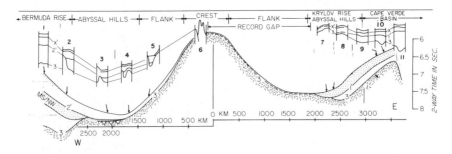

Figure 8.4.8. Transatlantic seismic profiles. Sections are aligned on returns from water-sediment interface. Two-way travel time from surface to bottom is given in parenthesis. Line drawings of corresponding seismic profile sections are shown for possible correlations of subbottom reflectors x, x', y, y', z, 1, $1'$, 2, $2'$, and 3 (acoustic basement). (Lowrie and Escowitz, 1969.)

sediments to be thin at the spreading center and to thicken with distance from the spreading center, because the age of the sea floor increases with distance from the center. These data show the effect described.

PROBLEMS

Problems 8.2

8.2.1. Calculate the sound velocity in a lake sediment. The average density of the water-filled sediment is 1.3 g/cm^3. The sediment is at 9.6 m depth. At the surface, the gas volume is 2% of the sample volume. Calculate the sound velocity in the sediment and the reflection coefficient at the water sediment interface.

8.2.2. Measure the effect of a tiny population of bubbles on the sound speed in a liquid by doing the following experiment. Close one end of a meter of water pipe, stand it on end, and fill it with water to form a resonance tube. Drive it with a small transducer and determine the resonance frequency. You may need a second transducer in the water to "hear" the resonance peak. After dropping bits of a "seltzer" tablet (the kind people take for upset stomaches and hangovers), measure the resonance frequency of the bubbly water. It is easy to obtain sound speeds as low as a few hundred meters per second.

8.2.3. Estimate the bulk modulus E for sand having 50% porosity. Assume laboratory conditions: 23°C and 1 atm pressure.

8.2.4. Estimate G for 50% porosity. Calculate the shear wave velocity. Use the sand as in Problem 8.2.3.

8.2.5. Use the estimates of E and G from Problems 8.2.3 and 8.2.4 to calculate a compressional wave velocity at laboratory conditions and 4000 m depth.

8.2.6. In an attenuating medium, the decrease of amplitude of a plane wave signal ΔA is proportional to the distance Δx and the amplitude A. Prove $A = A_0 \exp(-gx)$, where g is the proportionality constant and A_0 is the initial amplitude.

8.2.7. The intensity of Rayleigh scattered sound is proportional to $(ka)^4$, where a is the mean radius of tiny particles in the sound beam. (Rayleigh scattering which occurs when $ka \ll 1$ is discussed in Section 6.2.1.) Use the loss of amplitude from the plane wave beam to show that the attenuation coefficient is proportional to $N(ka)^4$, where N is the number of particles per unit volume. Would you expect the sound speed to be frequency dependent?

8.2.8. A passing wave causes particles to move back and forth. Assume that the energy loss is due to friction and is proportional to the distance traveled as the particles move. Consider the number of cycles in a fixed distance x and show that α is proportional to f and that Q is a constant. (See Section A8.2).

Problems 8.3.

8.3.1. Calculate the first arrival times and distance for 3.5 km of water over rock having a compressional wave velocity of 4.3 km/s. For simplicity, assume that the sound speed in water is 1.5 km/s.

8.3.2. Use the time distance data from Problem 8.3.1 for a study of the effect of errors. For each arrival toss a coin; if a head, add 0.05 s to the arrival time, and if a tail, add -0.05 s to the arrival time. Graph the data and interpret.

8.3.3. Interpret the reflection time–distance data. The interfaces are known to be horizontal.

x(km)	$t_1 s$	$t_2 s$	$t_3 s$
0	4.000	6.000	—
0.794	4.035	6.014	8.017
1.602	4.140	6.075	8.071
2.434	4.317	6.171	8.163
3.306	4.567	6.313	8.298
4.232	4.895	6.502	8.483

Apply the interval velocity method to the following combinations: t_1 and t_2 and t_3, and t_1 and t_3. *Answer.* $h_1 = 3$ km, $h_2 = 2$ km, $c_1 = 1.5$ km/s, $c_2 = 2$ km/s.

Problems 8.4

8.4.1. Use the geometry in Fig. 8.4.3 to calculate the equation of the crescents for a constant sound speed.

8.4.2 Make a template of a crescent for the scale and vertical exaggeration in Fig. 8.4.2. Use 1500 m/s for a sounding speed. Use the template to identify features which are more acute than the crescent.

8.4.3. You are to specify the equipment and ship tracks for a bathymetric survey in a rough area. The area is 10×10 km and at 3 km average depth. From a preliminary survey, the relief of the bottom is known to be about 100 m. The closest spacing between the peaks of features is about 0.5 km. Specify the maximum beam width of the transducer and spacing between sounding lines. At ship's speed of 15 km/hr, how much time will be required for the survey?

SUGGESTED READING

Appendix A8

M. B. Dobrin, *Introduction to Geophysical Prospecting, 3d Ed.*, McGraw-Hill, New York (1976).

A. R. Gregory, "Fluid Saturation Effects on Dynamic Elastic Properties of Sedimentary Rocks", *Geophysics* **41,** 895–921 (1976).

E. L. Hamilton, "Shear-Wave Velocity versus Depth in Marine Sediments: A Review", *Geophysics* **41,** 985–996 (1976).

L. L. Hampton, Ed., *Physics of Sound in Marine Sediments*, Plenum Press, New York (1974). Contains chapters on sound transmission in marine sediments. The chapters by D. T. Smith and R. D. Stoll are useful for an introduction to the seminal papers of Biot, 1941, 1956, 1962a, and 1962b.

M. N. Hill, Ed., *The Sea*, Vol. 3, *The Earth Beneath the Sea*, Wiley-Interscience, New York, 1963.

A. L. Inderbitzen, Ed. *Deep-Sea Sediments*, Physical and Mechanical Properties. Plenum Press, New York (1974).

A. W. Musgrave, Ed., *Seismic Refraction Prospecting*, Society of Exploration Geophysics, Tulsa, Okla. (1967).

R. D. Stoll, "Acoustic Waves in Ocean Sediments," *Geophysics* **42,** 715–725 (1977).

Ellis Strick, "A Predicted Pedestal Effect for Pulse Propagation in Constant—Q Solids," *Geophysics* **35,** 387–403 (1970).

M. N. Toksoz, C. H. Cheng, and A. Timur, "Velocities of Seismic Waves in Porous Rocks," *Geophysics* **41,** 621–645 (1976).

NINE

LONG RANGE SOUND TRANSMISSION IN A WAVEGUIDE

Our analysis in the preceding chapters was limited to ranges that are at most a few water depths. Direct calculation of the ray paths is satisfactory for those cases. At ranges that are many water depths, we encounter trapped waves and waveguide phenomena. Waves are completely trapped when the reflection coefficients at the upper and lower boundaries have a magnitude of unity.

We qualitatively show the transition from solutions of the previous chapters to the waveguide by considering the arrivals as the range is increased (Fig. 9.1.1). Assume a point source. At short horizontal range we receive the direct path, single reflections from the surface and bottom, and multiple reflections. At short ranges the ray paths for the reflected rays are near vertical incidence and the reflection coefficient is -1 for the surface and of the order of 0.1 for the bottom.

The contributions of the multiple reflections are small because the reflection coefficient $\Re$ at the bottom is less than unity and amplitudes of n multiple reflections are proportional to $\Re^n$. At intermediate range, we may receive the head wave and the totally reflected signals. The amplitude of the head wave decreases rapidly with increasing range, and the totally reflected signals become the most important components of the sound pressure.

To establish the criterion for a *long range* region in the waveguide, we consider the path of the ray that leaves the source at the critical angle θ_c, reflects at the bottom, reflects at the surface, and passes through the source depth. The horizontal distance between up crossings at the source depth is $2h \tan \theta_c$, where h is the thickness of the layer. At larger distances more ray paths can have total reflections. As an arbitrary criterion we use $10h \tan \theta_c$ as the transition distance to the *long range* region.

At long range the totally reflected components have had so many reflections that we find it convenient to reformulate the problem. The reformulation uses solutions of the wave equation and boundary conditions to calculate the sound pressure by the *normal mode method*.

Since much of the theoretical literature uses potential functions in this problem, we do so here to facilitate comparison with the literature. The various potential functions are defined so that differentiation operations on them yield the observables such as sound pressure p, particle displacement $\mathbf{d}$, and particle velocity $\mathbf{u}$. Recalling section A2.2, we use the gradient ∇ (A2.2.5) to define a *displacement potential* φ

$$\mathbf{d} = \nabla \varphi$$

The particle velocity is then

$$\mathbf{u} = \nabla \frac{\partial \varphi}{\partial t}$$

The substitution of $\partial \mathbf{d}/\partial t$ in (A2.2.6) gives after a little manipulation (drop the Δ)

$$p = -\rho \frac{\partial^2 \varphi}{\partial t^2}$$

The reader can continue the derivation, using φ, and obtain

$$\nabla^2 \varphi = \frac{1}{c^2} \frac{\partial^2 \varphi}{\partial t^2}$$

For the harmonic source $\exp(i\omega t)$, $p = \omega^2 \rho \varphi$ and $\nabla^2 \varphi = -(\omega^2/c^2)\varphi$.

At the interface between the media 0 and 1, the pressure is continuous or $p_0 = p_1$ and $\omega^2 \rho_0 \varphi_0 = \omega^2 \rho_1 \varphi_1$. The vertical components of the displacements are continuous at the interface and $\partial \varphi_0/\partial z = \partial \varphi_1/\partial z$. Next we show that the reflection coefficient for reflected and incident displacement potentials is $\varphi_r/\varphi_i = \mathcal{R}_{01}$, where $\mathcal{R}_{01}$ is the pressure reflection coefficient.

$$\mathcal{R}_{01} = \frac{p_r}{p_i} = \frac{\omega^2 \rho_0 \varphi_r}{\omega^2 \rho_0 \varphi_i} = \frac{\varphi_r}{\varphi_i}$$

Similarily, the ratio of the transmitted and incident displacement potentials is

$$\frac{\varphi_t}{\varphi_i} = \frac{(p_t/\rho_1)}{(p_i/\rho_0)} = \frac{\rho_0}{\rho_1}\mathcal{T}_{01}$$

The reader can extend these results to multilayer media and show that $\mathcal{R}_{on} = \varphi_r/\varphi_i$.

In our description of the fields in the waveguide we can use the term "sound pressure" interchangeably with "displacement potential" because of the proportionality of the two within a single layer.

9.1. NORMAL MODE SOLUTION IN A WAVEGUIDE

We illustrate how the method of separation of variables applies directly to normal mode solutions. After the variables are separated, we use the boundary conditions to obtain solutions. These lead to expressions of the sound transmission in terms of the "natural modes of vibration" of the bounded medium, which are called the *normal modes* of the waveguide.

The physical situation is sketched in Fig. 9.1.1. The source is at depth

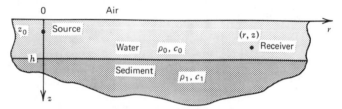

Figure 9.1.1. Layered waveguide. Source is in a water layer over half space of sediment. Source position is $(0, z_0)$, and receiver is (r, z).

z_0 and the receiver is at horizontal range r and depth z. For illustration we let the bottom impedance $\rho_1 c_1$ be infinite; then the bottom and surface are both perfect reflectors. The surface has a pressure reflection coefficient of -1 and the bottom has $+1$. Later we relax this condition and require only that the reflections be total (i.e., have the magnitude of unity at the angle of incidence).

We limit our discussion to unconsolidated sediment layers. The first approximation is to treat these layers as fluids. Eventually we incorporate an empirical loss term into the theory. The loss term includes absorption loss due to transmission in a lossy medium and losses due to imperfect reflections. The latter loss can be ascribed to reflection from a viscous medium, a rough interface, and/or a sediment that has slow shear wave velocities.

Extensions to the more general problem, including elastic media, can be found in the Suggested Readings section at the end of the chapter.

9.1.1. Separation of Variables

In a layered waveguide, the sound speed is a function of z. For sources on the z axis, the field is symmetrical about the z axis and the cylindrical wave equation is suitable

$$\frac{\partial^2 \varphi}{\partial r^2} + \frac{1}{r}\frac{\partial \varphi}{\partial r} + \frac{\partial^2 \varphi}{\partial z^2} = \frac{1}{c^2}\frac{\partial^2 \varphi}{\partial t^2} \tag{9.1.1}$$

To separate the variables, as in Section A2.3.1, we write $\varphi \sim U(r)Z(z)T(t)$. We include the constant of proportionality later. After substitution, (9.1.1) becomes

$$\frac{U''(r) + (1/r)U'(r)}{U(r)} + \frac{Z''(z)}{Z(z)} = \frac{T''(t)}{c^2 T(t)} \tag{9.1.2}$$

For the harmonic source, $T(t) = \exp(i\omega t)$, the right-hand side reduces to $-\omega^2/c^2$. The separation constants are designated $-\gamma^2$ and $-\kappa^2$. The separated equations are

$$U''(r) + \frac{U'(r)}{r} = -\kappa^2 U(r) \tag{9.1.3}$$

$$Z''(z) = -\gamma^2 Z(z) \tag{9.1.4}$$

$$k^2 \equiv \frac{\omega^2}{c^2} \tag{9.1.5}$$

$$\kappa^2 + \gamma^2 = k^2 \tag{9.1.6}$$

where we have changed notation from k_r to κ and k_z to γ to simplify the subscript notation.

Equation (9.1.3) and $U(r)$ are only functions of r, and this implies that κ, the horizontal component of the wave number, is constant throughout the waveguide. This is equivalent to Snell's law because

$$\kappa = k \sin \theta = \frac{\omega}{c} \sin \theta \tag{9.1.7}$$

where θ is the angle of incidence, ω is constant, and $(\sin \theta)/c$ is constant.

The solution of (9.1.6) for γ^2 shows that γ depends on z when c is a function of z.

9.1.2 Range Dependence

Equation (9.1.3) is Bessel's equation of the zeroth order. Its solution is the cylindrical Bessel function of the first kind $J_0(\kappa r)$ (Fig. 9.1.2). At large

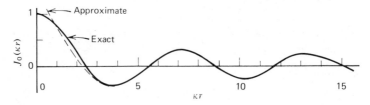

Figure 9.1.2. Range dependence of $J_0(\kappa r)$ and its approximation: solid line, $J_0(\kappa r)$; dashed line, approximation (9.1.8), merges with $J_0(\kappa r)$ for $\kappa r > 4$.

range the asymptotic expansion (Abramowitz and Stegun, 9.2.1) is convenient

$$U(r) = J_0(\kappa r)$$

$$J_0(\kappa r) \simeq \left(\frac{2}{\pi \kappa r}\right)^{1/2} \cos\left(\kappa r - \frac{\pi}{4}\right) \qquad \text{for} \quad \kappa r \gg 1 \qquad \textbf{(9.1.8)}$$

The approximation is the dashed line in Fig. 9.1.2. We use the exponential expansion of the cosine to express (9.1.8).

$$J_0(\kappa r) \simeq (2\pi \kappa r)^{-1/2} \left[\exp\left(i\kappa r - \frac{i\pi}{4}\right) + \exp\left(-i\kappa r + \frac{i\pi}{4}\right) \right] \qquad \text{for} \quad \kappa r \gg 1$$

$$\textbf{(9.1.9)}$$

When combined with the time dependence $\exp(i\omega t)$, the first term on the right represents an incoming wave and the second term is an outgoing wave. We choose the outgoing wave

$$U(r) \simeq (2\pi \kappa r)^{-1/2} \exp\left[-i\left(\kappa r - \frac{\pi}{4}\right)\right] \qquad \textbf{(9.1.10)}$$

The amplitude of $U(r)$ and φ decrease as $r^{-1/2}$. This amplitude decrease is characteristic of the waves that spread radially from a source in a plane layered waveguide. The range dependent function $U(r)$ is the same for all depths.

9.1.3. Depth Dependence in the "Ideal" Water Layer

The waveguide consisting of a water layer over an infinite half space is shown in Fig. 9.1.1. When the surface is perfectly free and the bottom is perfectly rigid we call it an "ideal water layer" because the boundaries are perfect reflectors at all angles of incidence. We use this idealization because it is analytically simple and displays the basic method.

The pressure reflection coefficient at the surface is -1 and at the bottom it is 1. The sound speed c_0 and thickness h are constant. We use these conditions and h to determine the z dependence $Z(z)$ of the pressure. A trial solution of (9.1.4) is

$$Z(z) = A \sin \gamma z + B \cos \gamma z \qquad (9.1.11)$$

Verification of the solution is left to the reader.

The next task is to choose A, B; and γ to fit the conditions at the surface and at the bottom (i.e., the boundary conditions). At the free surface p is zero for all r and t, therefore

$$Z(z) = 0 \qquad \text{at} \quad z = 0 \qquad (9.1.12)$$

which requires that $B = 0$. At the bottom, the pressure is a maximum because the bottom is perfectly rigid (let ρ_1 be infinite and $c_1 = c_0$)

$$|A| \cdot |\sin \gamma h| = |A| \qquad (9.1.13)$$

or

$$|\sin \gamma h| = 1 \qquad (9.1.14)$$

where A is a constant to be determined later. The solution of (9.1.14) is

$$\gamma h = \frac{\pi}{2}, \frac{3\pi}{2}, \frac{5\pi}{2}, \dots$$

or

$$= \left(m - \frac{1}{2}\right)\pi, \quad m = 1, 2, \dots \qquad \textbf{(9.1.15)}$$

Equation (9.1.15) is the *characteristic equation* for the ideal water layer. The conditions are satisfied for a *discrete set* of values of γ. These are called the *eigenvalues* (loosely translated from German, *eigen* means characteristic).

For the ideal water layer we have

eigenvalue: $\qquad\qquad \gamma_m \equiv \dfrac{(m - \frac{1}{2})\pi}{h} \qquad (9.1.16)$

eigenfunction: $\qquad\quad Z_m(z) \equiv \sin \gamma_m z \qquad (9.1.17)$

$$m = 1, 2, 3, \dots$$

where the subscripts refer to discrete solutions. The solutions are called modes because they are natural ways in which the system vibrates. Figure 9.1.3 gives $Z_m(z)$ for the first few modes.

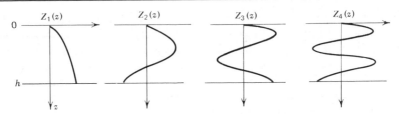

Figure 9.1.3. Depth dependence of the eigenfunctions $Z_m(z)$, where m is mode number. Waveguide has a pressure release surface over a rigid bottom.

The eigenvalues κ_m and γ_m are related by (9.1.6). Considering γ_m to be the vertical component and κ_m to be the horizontal component of k, we write, using the geometry in Fig. 9.1.4,

$$k^2 = \kappa_m^2 + \gamma_m^2$$

$$\kappa_m = k \sin \theta_m \qquad\qquad \textbf{(9.1.18)}$$

$$\gamma_m = k \cos \theta_m$$

For a given waveguide and mode m, γ_m is a constant because of (9.1.16). This means that changes of frequency $k = \omega/c$ cause corresponding changes of κ_m. Figure 9.1.4 shows γ_m and κ_m for two modes and two

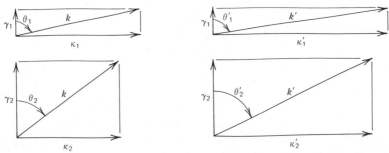

Figure 9.1.4. Eigenvalues of γ_1 and γ_2. The propagation constant $k = \omega/c$ is for one frequency and k' is for a higher frequency. The vertical components γ_1 and γ_2 do not depend on the frequency. To satisfy the conditions on γ_1 and γ_2, θ_1 changes to θ_1' for mode 1 and θ_2 changes to θ_2' for mode 2. Correspondingly, κ_m changes to κ_m'.

frequencies k and k'. Since γ_m and κ_m are the vertical and horizontal components of the propagation constant k, they are also called *propagation constants*.

The requirement that k, κ_m, and γ_m be real when combined with (9.1.18) implies that $|k| \geq |\gamma_m|$. Since γ_m increases with m, the requirement that $\gamma_m \leq k$ places an *upper limit* on m for a given $k = \omega/c$.

9.1.4. Orthogonality

We want to show that the depth [or $Z(z)$] eigenfunctions are orthogonal in the mathematical sense. We evaluate the integral

$$\int_0^h \rho_0 Z_m(z) Z_n(z)\, dz = \int_0^h \rho_0 \sin\left[\frac{(m-\tfrac{1}{2})\pi z}{h}\right] \sin\left[\frac{(n-\tfrac{1}{2})\pi z}{h}\right] dz$$

(9.1.19)

and find that it is zero when $n \neq m$. Here, ρ_0 is the density in the water layer. We define ν_m

$$\nu_m = \int_0^h \rho_0 Z_m^2(z)\, dz \tag{9.1.20}$$

Although we do not show it, ν_m is proportional to the mean energy flux passing through a vertical section (Tolstoy and Clay, 1966, p. 23). Using the delta function $\delta(m-n)$, (9.1.19) becomes

$$\int_0^h \rho_0 Z_m(z) Z_n(z)\, dz = \nu_m \delta(m-n) \tag{9.1.21}$$

This integral is a statement of the orthogonality of the eigenfunctions. We integrate from 0 to h because the sound is confined to the region between 0 and h. The range of integration can be written as $-\infty$ to ∞ with the understanding that the $Z_m(z)$ tend to zero very rapidly above and below the waveguide. When the density ρ is a function of z, as in the general case, it is necessary to include ρ in (9.1.21) for the orthogonality condition to hold.*

*The orthogonality of $Z_m(z)$ and $Z_n(z)$ can be proved by using (9.1.4) to form

$$Z_m''(z) = -\gamma_m^2 Z_m(z)$$

and a similar equation for $Z_n(z)$. Multiply $Z_m''(z)$ by $\rho Z_n(z)$ and $Z_n''(z)$ by $\rho Z_m(z)$ and take the difference

$$\rho Z_n(z) Z_m''(z) - \rho Z_m(z) Z_n''(z) = (\gamma_n^2 - \gamma_m^2)\rho Z_m(z) Z_n(z)$$

$$\rho \frac{\partial}{\partial z}[Z_n(z) Z_m'(z) - Z_m(z) Z_n'(z)] = (\gamma_n^2 - \gamma_m^2)\rho Z_m(z) Z_n(z)$$

where $\rho\varphi$ and $\partial\varphi/\partial z$ are continuous and $\rho Z_m(z)$ and $\partial Z_m(z)/\partial z$ are continuous. Also, assume that $\rho Z_m(z)$ and $\partial Z_m(z)/\partial z$ vanish at $\pm\infty$. The integration of the left side with respect to z gives an expression that vanishes, when the limits are substitued, because of the continuity of the quanitities. The integral of the right side is

$$(\gamma_n^2 - \gamma_m^2)\int_{-\infty}^{\infty} \rho Z_m(z) Z_n(z)\, dz = \nu_m\, \delta(m-n)$$

This proves the orthogonality of $Z_m(z)$ and $Z_n(z)$. Mathematicians will recognize this as an example of the Sturm-Liouville problem.

We use the orthogonality of the eigenfunctions to expand a particular dependence of pressure on z as a sum of the eigenfunctions. (This is analogous to the Fourier expansion of $f(x)$ as the sum terms F_m exp $[i2\pi mx/L]$, where F_m is the coefficient and L is the spatial period.) The function of interest is the source at $r = 0$ and $z = z_0$.

Letting the source function be $\delta(z - z_0)$ and using the analogy to the Fourier series, we write

$$\delta(z - z_0) = \sum_n A_n Z_n(z) \qquad (9.1.22)$$

where A_n are constants to be determined. We multiply both sides of (9.1.22) by $\rho_0 Z_m(z)$ and integrate over z.

$$\int_0^h \rho_0 Z_m(z) \delta(z - z_0) \, dz = \sum_n A_n \int_0^h \rho_0 Z_m(z) Z_n(z) \, dz \qquad (9.1.23)$$

The left-hand side is the delta function integration and the right-hand side contains the orthogonality integral

$$\rho_0 Z_m(z_0) = A_m \nu_m \qquad (9.1.24)$$

$$A_m = \frac{\rho_0 Z_m(z_0)}{\nu_m} \qquad (9.1.25)$$

where ρ_0 and $Z_m(z_0)$ are the density and the evaluation of the eigenfunction at the source depth, respectively.

9.1.5. Sound Pressure

The solution of the wave equation with boundary conditions is the product of $Z_m(z)$, $U_m(r)$, $T(t)$, and A_m. The sound pressure is the sum over all modes, and is a function of ω and t

$$p = a_0 \exp\left[i\left(\omega t + \frac{\pi}{4}\right)\right] \sum_m \frac{\rho_0 Z_m(z_0) Z_m(z) \exp(-i\kappa_m r)}{\nu_m (2\pi \kappa_m r)^{1/2}} \qquad (9.1.26)$$

where

$$k^2 = \kappa_m^2 + \gamma_m^2 = \frac{\omega^2}{c^2}, \qquad (9.1.27)$$

and a_0 is a constant that depends on the source power. Evaluations that include the source power require much more analysis, and we give the result derived in Tolstoy and Clay 1966, pp. 81–84. By comparison of our

(9.1.26) and their (3.122), we find

$$a_0 = (2\pi)^{3/2} \frac{\rho}{\rho_0} (\rho_0 c_0 \Pi)^{1/2} \qquad (9.1.28)$$

where Π is source power and ρ and ρ_0 are ambient densities at the receiver and source, respectively. On combining the mode dependence of A_m and a_0 into a *source excitation* function for each mode q_m, we find

$$q_m \equiv \frac{2\pi(\rho_0 c_0 \Pi)^{1/2}}{[\nu_m \sqrt{\kappa_m}]} \qquad \textbf{(9.1.29)}$$

where c_0 is the sound speed at the source depth. In units, q_m has the units of $m^{5/2}/s^2$, $\rho q_m r^{-1/2}$ has units of $(kg\ m\ sec^{-2})\ m^{-2}$ or pressure in pascals.

$$p = \rho \exp\left[i\left(\omega t + \frac{\pi}{4}\right)\right] r^{-1/2} \sum_m q_m Z_m(z_0) Z_m(z) \exp(-i\kappa_m r) \quad \textbf{(9.1.30)}$$

These expressions are valid for any waveguide even though we derived them for simplified conditions. The normal mode expressions of the sound pressure (9.1.26) or (9.1.30) are especially convenient for calculating the sound field as a function of depth and range. A similar calculation using ray tracing techniques would require a complete set of ray traces for each point in space.

9.2. "REAL" WAVEGUIDES

This section is titled "Real Waveguides" because the theoretical models are close to reality. The most important requirement is that the waveguide have horizontal stratification or parallel interfaces between the layers. This is realizable because many areas of the ocean have nearly horizontal stratifications over ranges of hundreds of kilometers. The structure beneath the ocean bottom is variable; however in many areas it also has nearly horizontal stratifications. In going from an "ideal" to a "real" waveguide, the major changes are to have many layers, to let the sound speed and density changes at the interface be finite, and to allow absorption losses.

The basic method demonstrated for the ideal water layer can be applied to any number of layers. It becomes clumsy, however, because each layer introduces a new set of constants to be fitted at the boundaries. Instead, we use Tolstoy's characteristic equation method to compute the propagation constants because it reduces the general problem to the simplicity of the ideal water layer (Tolstoy and Clay, 1966, p. 35).

9.2.1. CHARACTERISTIC EQUATION METHOD
FOR COMPUTING PROPAGATION CONSTANTS

As shown in Fig. 9.2.1 the waveguide is assumed to have the upper and lower reflection coefficients $\mathcal{R}_u$ and $\mathcal{R}_l$ and thickness h. The sound speed

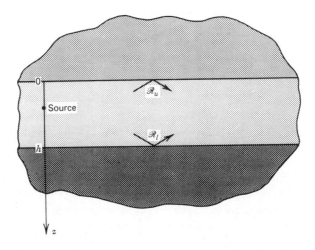

Figure 9.2.1. Source in a layered waveguide. Section containing source has constant sound speed c_0; $\mathcal{R}_u$ and $\mathcal{R}_l$ are the plane wave reflection coefficients from the top and bottom of the waveguide.

is constant within h. This is general because $\mathcal{R}_u$ and $\mathcal{R}_l$ include the reflections from layered half spaces above and below the section. (The reflection coefficient at a layered half space is given in Section 2.10.)

Within the guide we write $Z(z)$ as being the sum of the exponentials

$$Z(z) = Ae^{i\gamma z} + Be^{-i\gamma z} \tag{9.2.1}$$

where A and B are constants. Considering the time dependence $\exp(+i\omega t)$ and z positive downward, the first term on the right represents an uptraveling wave $Z\uparrow$ and the second is a down-traveling wave $Z\downarrow$.

$$Z\uparrow = Ae^{i\gamma z} \tag{9.2.2}$$

$$Z\downarrow = Be^{-i\gamma z} \tag{9.2.3}$$

$$Z(z) = Z\uparrow + Z\downarrow$$
$$Z(z) = Ae^{i\gamma z} + Be^{-i\gamma z} \tag{9.2.4}$$

At the upper interface $z = 0$, $Z\uparrow$ reflects and becomes $\mathcal{R}_u A$. Therefore

$$Z(0) = Z\uparrow + \mathcal{R}_u Z\uparrow|_{z=0} \tag{9.2.5}$$

$$= A(1 + \mathcal{R}_u) \tag{9.2.6}$$

This is equal to $Z(0)$ as given by (9.2.4) and at $z = 0$

$$A + B = A + A\mathcal{R}_u \tag{9.2.7}$$

$$\frac{B}{A} = \mathcal{R}_u \tag{9.2.8}$$

We now evaluate $Z(z)$ at the lower interface $z = h$. After reflection the uptraveling component becomes $\mathcal{R}_l B e^{-\gamma z}$. At $z = h$ (9.2.4) *is*

$$Z(h) = A e^{i\gamma h} + B e^{-i\gamma h} \tag{9.2.9}$$

$$= \mathcal{R}_l Z\downarrow + Z\downarrow|_{z=h}$$

$$= B(\mathcal{R}_l + 1) e^{-i\gamma h} \tag{9.2.10}$$

The pair of equations (9.2.9) and (9.2.10) are combined to give

$$A e^{i\gamma h} = B \mathcal{R}_l e^{-i\gamma h} \tag{9.2.11}$$

The substitution of (9.2.8) gives

$$\mathcal{R}_u \mathcal{R}_l e^{-i2\gamma h} = 1 \tag{9.2.12}$$

This is the *characteristic equation*. It places conditions on "allowed" values of γh for a set of waveguide parameters.

Recalling the discussion in Section 2.9 for reflections beyond the critical angle, we write $\mathcal{R}_u$ and $\mathcal{R}_l$ as having a magnitude unity and phase angles $2\Phi_u$ and $2\Phi_l$ respectively, (2.9.11). For a source in the zeroth layer (Fig. 9.2.1) the pressure reflection coefficients are

$$\mathcal{R}_u = \exp(i2\Phi_u)$$
$$\mathcal{R}_l = \exp(i2\Phi_l) \tag{9.2.13}$$

where

$$\Phi_l = \arctan \frac{b_1 \rho_0 c_0}{\rho_1 c_1 \cos \theta_0}$$

$$b_1 = \left[\left(\frac{c_1}{c_0} \right)^2 \sin^2 \theta_0 - 1 \right]^{1/2}$$

The substitution of these into (9.2.12) gives

$$\exp[i2(\Phi_u + \Phi_l - \gamma h)] = 1 \tag{9.2.14}$$

This is true when

$$2(\Phi_u + \Phi_l - \gamma h) = 2n\pi \qquad (9.2.15)$$

where $n = 0, \pm 1, \ldots$. For convenience we rearrange (9.2.15) and write it as follows

$$\gamma_m h - \Phi_u - \Phi_l = (m - 1)\pi \qquad \textbf{(9.2.16)}$$

where $m = 1, 2, 3, \ldots$ and γ_m is the value of vertical component of the wave number that satisfies (9.2.16)

The signs of Φ_u and Φ_l are based on the requirement that the function decreases exponentially in the second medium (Section A2.3.3), and $(m - 1)$ is chosen to make the first mode $m = 1$. Equation (9.2.16) is another form of the characteristic equation. In a multilayered half space, the phase depends on partial reflections from all layers down to the bottom. Formulas for computing the angle Φ_l are in Section 2.10.2.

As an example, the ideal waveguide having air at the surface and a rigid bottom has $\Phi_u = \pi/2$ [$\exp (i\pi) = -1$] and $\Phi_l = 0$. Substitution gives

$$\gamma_m h = (m - \tfrac{1}{2})\pi \qquad (9.2.17)$$

This is the same result given in (9.1.15).

For simplicity in considering a few general limitations on the number of modes, we use the waveguide in Fig. 9.1.1 and let the sediment have a sound velocity c_1. For trapped modes all angles of incidence must be greater than the critical angle

$$\sin \theta_m \geq \frac{c_0}{c_1} \qquad (9.2.18)$$

where θ_m is the angle of incidence for the wave number of the mth mode (Fig. 9.1.4). Since

$$\kappa_m = k \sin \theta_m = \frac{\omega}{c_0} \sin \theta_m \qquad (9.2.19)$$

we can write an inequality for κ_m by replacing $\sin \theta_m$ by (9.2.18)

$$\kappa_m \geq \frac{\omega}{c_1} \qquad \textbf{(9.2.20)}$$

Using $\gamma_m^2 = k^2 - \kappa_m^2$, the inequality for κ_m gives an upper limit for γ_m

$$\gamma_m \leq \left[k^2 - \left(\frac{\omega}{c_1}\right)^2 \right]^{1/2} \qquad \text{or} \qquad \gamma_m \leq \frac{\omega}{c_0}\left[1 - \left(\frac{c_0^2}{c_1^2}\right) \right]^{1/2}$$

$$\textbf{(9.2.21)}$$

So far we have shown that the maximum value of γ_m is limited by frequency ω, the ratio c_0/c_1, and c_0.

Solving (9.2.16) for γ_m gives

$$\gamma_m = \frac{(m-1)\pi + \Phi_u + \Phi_l}{h} \tag{9.2.22}$$

We combine (9.2.21), (9.2.22), and $\omega/c = 2\pi/\lambda$ and rearrange the resulting inequality to obtain

$$(m-1)\pi + \Phi_u + \Phi_l \leq \frac{2\pi h}{\lambda}\left[1 - \left(\frac{c_0}{c_1}\right)^2\right]^{1/2} \tag{9.2.23}$$

This shows that the maximum number of modes primarily depends on two factors h/λ and c_0/c_1. The inequality also sets a lower limit for h/λ, thereby giving the low frequency "cutoff" for each mode.

9.2.2. Attenuation

As the sound travels in the waveguide it reflects many times at the upper and lower boundaries. Since reflections are seldom perfect, there is a small loss ϵ at each reflection and

$$|\mathcal{R}| \approx 1 - \epsilon, \qquad \epsilon \ll 1 \tag{9.2.24}$$

For the mth mode in a waveguide, a wave front has the angle of incidence θ_m and the horizontal distance between successive reflections is $2h \tan \theta_m$. Therefore the change of amplitude $|\Delta p|$ along the horizontal direction due to reflection losses is

$$|\Delta p| = -|p|\left(\frac{\epsilon}{2h \tan \theta_m}\right)\Delta r \tag{9.2.25}$$

Integration yields

$$|p| = |p_0| \exp(-\delta'_m r) \tag{9.2.26}$$

where

$$\delta'_m \equiv \frac{\epsilon}{2h \tan \theta_m} \tag{9.2.27}$$

represents the attenuation coefficient for imperfect reflection. This is an example of attenuation caused by losses at the boundaries.

In addition there are losses due to absorption in the medium. Since the absorption losses in sediments are orders of magnitude larger than the losses in water, modes having large penetration into the sediments have high attenuations. For a given mode the proportion of losses in the water column, sediments, and reflections is constant. We designate δ''_m as the attenuation coefficient for losses in the medium including the water and the sediment column. The total attenuation coefficient δ_m is

$$\delta_m = \delta'_m + \delta''_m \qquad (9.2.28)$$

Including the attenuation for each mode, (9.1.30) becomes

$$p = \rho \exp\left[i\left(\omega t + \frac{\pi}{4}\right)\right] r^{-1/2} \sum_m q_m Z_m(z_0) Z_m(z) \exp\left(-\delta_m r - i\kappa_m r\right)$$

$$(9.2.29)$$

Theoretical estimates of δ_m require much detailed knowledge of the media. Usually the δ_m are determined by fitting (9.2.29) to experimental data.

9.3. SOUND TRANSMISSION IN A WAVEGUIDE: NUMERICAL EXAMPLE AND EXPERIMENT

To keep the example simple, we limit the number of modes that are excited. On the basis of (9.2.23), we choose an example in which the water depth is a few λ at the most. In deep ocean basins (3500 m) this corresponds to frequencies of the order of 1 Hz and less. At convenient experimental frequencies of about 100 Hz, the water depth of the example would be less than 35 m.

Extensive measurements of the geophysical structure and sound transmission have been made in the shallow water south of the Long Island, New York. The water depth in the test area is 22.6 m. The waveguide and its parameters are shown in Fig. 9.3.1.

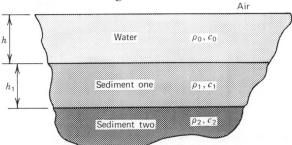

Figure 9.3.1. Water layer over a sediment column, south of Long Island, New York. Parameters are $\rho_0 = 1033$ kg/m³, $c_0 = 1508$ m/s, $h = 22.6$ m; $\rho_1 = 2\rho_0$, $c_1 = 1.12c_0$, $h_1 = 0.9h$; $\rho_2 = 2\rho_0$, $c_2 = 1.24c_0$. Data from Tolstoy and Clay, p. 133 (1966).

The sound transmission measurements were made by towing a source slowly away from the receiver. The source was driven by an oscillator and power amplifier so that it transmitted a CW signal. The sound pressure was received at a hydrophone in the water, amplified, and filtered. The rms sound pressure was recorded on a strip chart recorder. As Fig. 9.3.2

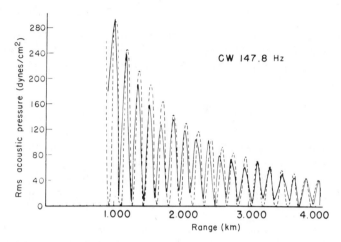

Figure 9.3.2. Sound transmission in a shallow water waveguide. Data taken in the waveguide of Fig. 9.3.1: solid line, experimental data, dashed line, a theoretical calculation. Source depth was approximately 10 m, and receiver was at 19.8 m. Frequency was 147.8 Hz. Clay (1964).

indicates, the rms amplitude of the received signal has a regular variation from maximum to minimum as a function of range. The variations are due to interferences between the sound pressures traveling in different modes.

The data in Fig. 9.3.2 are one example of a typical sound transmission measurement. For this run the depths of the source and receiver were chosen to give nearly equal amplitudes of sound pressure in the first two modes and almost no energy in the higher modes. A summary of the research is given in Tolstoy and Clay (1966, Ch. 4).

Our comparison of theory and experiment can include the calculation of the interference wavelength of the fluctuation. This is a good comparison because, as we show shortly, the interference wavelength depends on accurate determinations of the eigenvalues γ_m. Instead of calculating the rms sound pressure, we calculate the mean square sound pressure, which is simpler to do. The absolute square of (9.2.29) is

$$\langle pp^* \rangle = \frac{\rho^2}{r} \left\{ \sum_m q_m^2 Z_m^2(z_0) Z_m^2(z) \exp(-2\delta_m r) \right.$$

$$+ \sum_{m \neq n} q_m q_n Z_m(z_0) Z_m(z) Z_n(z_0) Z_n(z) \exp[i(\kappa_n - \kappa_m)r]$$

$$\left. \exp[-(\delta_m + \delta_n)r] \right\} \qquad \textbf{(9.3.1)}$$

The first summation on the right represents the attenuation of the signal as a function of range. The second summation contains the interference terms. The spatial frequency of the interference is $\kappa_n - \kappa_m$. The interference wavelength Λ_{nm} is

$$\Lambda_{nm} \equiv \left| \frac{2\pi}{\kappa_n - \kappa_m} \right| \qquad \textbf{(9.3.2)}$$

In this experiment most of the sound transmission is in the first two modes and there is one term for the mode interferences. Here it is easy to interpret the data. When many modes are present, the pressure signal has a complicated appearance and it is difficult to determine either Λ_{nm} or $\kappa_n - \kappa_m$ by inspection.

To analyze $\langle pp^* \rangle$ when many modes are present we suggest four steps: (1) multiply $\langle pp^* \rangle$ by $r \exp(\delta r)$, where δ is an estimate of the average attenuation coefficient, to compensate for cylindrical spreading and attenuation; (2) select a portion of the record and prepare it for "time series" or frequency analysis by removing the mean and tapering the ends to zero; (3) do an FFT analysis to determine the wave numbers and amplitudes of the mode interference terms (the length of the section should be more than 5 times the longest interference wavelength); (4) compare the spectrum of the experimental $\kappa_n - \kappa_m$ with the theoretical $\kappa_n - \kappa_m$.

The basic steps in making a numerical calculation are (1) determine the the eigenvalues, (2) use the eigenvalues to determine the eigenfunctions and mode excitations, and (3) estimate the mode attenuations. Section 9.3.1 gives numerical examples for each of these steps.

9.3.1. Determination of the Propagation Constants κ_m and γ_m

To calculate the propagation constants or eigenvalues γ_m and κ_m for the waveguide, we use the characteristic equation (9.2.16)

$$\gamma_m h - \Phi_u - \Phi_l = (m-1)\pi \qquad \textbf{(9.3.4)}$$

For the water layer over a sediment column as in Fig. 9.3.1, we consider first Φ_u, then Φ_l, and then calculate γ_m. Recalling that the phase shift at total reflection 2Φ is defined as

$$\mathcal{R} = e^{i2\Phi} \tag{9.3.5}$$

$\mathcal{R}$ at the water-air interface is -1 and $\Phi_u = \pi/2$.

The reflection from the entire sediment column is involved in $\mathcal{R}_l$. Because of the relatively high attenuation of sound waves that travel deep in the section compared to signals in the water layer, we can neglect the structure beneath the water-sediment interface for signals near the experimental frequency of 150 Hz. The approximate structure consists of the water layer ρ_0, c_0 over a sediment half space ρ_1, c_1.

Using the half space approximation, Φ_l is given by (2.9.12) for θ_m greater than the critical angle

$$\Phi_l = \arctan\left(\frac{\rho_0 c_0 b_1}{\rho_1 c_1 \cos \theta_m}\right) \tag{9.3.6}$$

$$b_1 = \left[\left(\frac{c_1}{c_0}\right)^2 \sin^2 \theta_m - 1\right]^{1/2} \tag{9.3.7}$$

We use (2.9.5), (9.3.5), and (9.3.6) to calculate $\mathcal{R}_l$ for the water-sediment interface in Fig. 9.3.1; $\mathcal{R}_l$ appears in Fig. 9.3.3.

After substitution of $\pi/2$ for Φ_u, solution of (9.3.4) for γh gives

$$\gamma_m h = (m - \tfrac{1}{2})\pi + \Phi_l \tag{9.3.8}$$

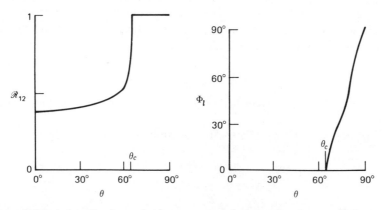

Figure 9.3.3. Amplitude and phase terms of the reflection coefficient as a function of the angle of incidence θ. Calculations are for the water-sediment interface and parameters in Fig. 9.3.1. $\rho_0 = 1033$ kg/m³, $c_0 = 1508$ m/s, $h = 22.6$ m; $\rho_1 = 2\rho_0$, $c_1 = 1.12 c_0$.

We evaluate this equation by expressing κ_m and γ_m as functions of ω and θ_m,

$$\kappa_m = k \sin \theta_m = \frac{\omega}{c_0} \sin \theta_m$$

$$\gamma_m = k \cos \theta_m = \frac{\omega}{c_0} \cos \theta_m$$

(9.3.9)

The substitution of (9.3.9) in (9.3.8) and solution for ω gives

$$\omega = \frac{c_0[(m - \frac{1}{2})\pi + \Phi_l]}{h \cos \theta_m}$$

(9.3.10)

We seek the propagation constant κ_m at our experimental frequency. First we use (9.3.6) and (9.3.7) to calculate Φ_l for a set of θ_m, where θ_m varies from θ_c to values near $\pi/2$. This gives us a graph such as Fig. 9.3.3 for the sediment. Then, for a given mode, we use (9.3.10) and the values of Φ_l and θ_m to calculate a set of values of ω and use (9.3.9) to calculate a set of values of κ_m. We then have a table or graph of ω vs. κ_m. We interpolate to determine κ_m at our experimental frequency. We can obtain γ_m from (9.3.9) or (9.1.18).

Numerical Example. The structure of the water layer and first two subbottom sediment layers are shown in Fig. 9.3.1. The source frequency is 147.8 Hz.

First use 147.8 Hz to estimate the number of modes that will be trapped in the water layer at this frequency. The sound waves are totally reflected at the 0–1 interface. The limiting sound wave is reflected at or slightly greater than critical angle. At the critical angle of incidence, the phase term Φ_l is zero and (9.3.10) reduces to

$$f = \frac{c_0(m - \frac{1}{2})}{2h \cos \theta_c}$$

This expression also gives the "cutoff" frequency for a given mode number. At $f = 147.8$ Hz, $c_o = 1508$ m/s, $c_1 = 1689$ m/s, $h = 22.6$ m, and $\theta_c = \arcsin(1/1.12) = 63.23°$, the upper limit on m is

$$m \le \frac{f}{c_0}(2h \cos \theta_c) + \frac{1}{2}$$

$$m \le 2.50$$

The highest mode number for a water layer over a half space of sediment (in our approximation we ignore the second sediment) is 2. The inclusion of deeper sediment layers (having larger sound velocities) allows the ray paths to have smaller angles of incidence and still be totally reflected. For the same frequency, γ_m becomes larger as the angle of incidence decreases, and correspondingly m can be larger. Both by choice of source depth and the larger attenuation of the higher modes, the transmissions of sound pressures in modes 3, 4, ... are negligibly small in our example.

We give an example of a calculation of ω using the characteristic equation. From Fig. 9.3.3, and for $\theta = 80°$, $\Phi_l = 50°$. Substitution of these values in (9.3.10) gives, for $m = 1$,

$$\omega = 1508\left[\frac{\pi/2 + 50°(\pi/180°)}{22.6\cos 80°}\right]$$

$$\omega = 939 \text{ rad/s}$$

$$\kappa_1 = \frac{939}{1508}\sin 80° = 0.614 \text{ m}^{-1}$$

Application of this procedure over the range $\theta = 63.23°$ to $88°$ for modes 1 and 2 gives values of ω versus κ_1 and κ_2. Figure 9.3.4 plots the dependence of ω on the propagation constants.

For the first mode, ω must be greater than 233 rad/s if a solution is to exist. The second mode has a minimum ω of 699 rad/s. These are the "cutoff" frequencies for the first two modes.

The experimental measurement of p was made at 147.8 Hz or 928.7 rad/s. For our comparison of theory and experiment, we first compare the theoretical Λ_{12} with the measurement. The values of κ_1 and κ_2 at 928.7 rad/s are

$$\kappa_1 = 0.607 \text{ m}^{-1}$$

$$\kappa_2 = 0.573 \text{ m}^{-1}$$

$$\kappa_1 - \kappa_2 = 0.034 \text{ m}^{-1}$$

$$\Lambda_{12} = 185 \text{ m}$$

From Fig. 9.3.2, the average interference wavelength is 177 m. The error is 5%, which is within the limits of the measurements. For example, about one percent decrease of depth reduces Λ_{12} to the observed value, the water depths are constant to within 2% over the range.

The values of γ_m and θ_m for $\omega = 928.7$ rad/s are calculated by using (9.3.9).

$$\gamma_1 = 0.104 \text{ m}^{-1}, \qquad \theta_1 = 80.3°$$

$$\gamma_2 = 0.226 \text{ m}^{-1}, \qquad \theta_2 = 68.5°$$

9.3.2. Eigenfunctions and Mode Excitation

Continuing our numerical example, the boundary condition at $z = 0$ is a pressure release surface and $Z_m(0) = 0$. Therefore in (9.2.1) $A = -B$ and,

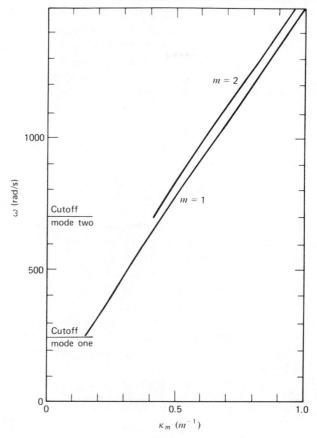

Figure 9.3.4. Propagation constants and frequency for the first two modes. Waveguide is a water layer over an infinite sediment half space with parameters of Fig 9.3.1 and Fig. 9.3.3.

considering only z dependence, $p_i = \rho_i Z_m(z)$. In the first layer

$$p_0 = \rho_0 \sin\gamma_m z \qquad 0 \leqslant z \leqslant h \qquad (9.3.11)$$

At the bottom of the water layer, the sound pressure is continuous across the boundary. Beneath the interface, the sound pressure is exponentially damped because the incident sound wave is totally reflected. From (A2.3.34), the sound pressure in the lower medium is $p_1 = \rho_1 Z_m(z)$

where $Z_m(z) = a_1 \exp\left[-b_m(z-h)\right], \quad z \geqslant h \qquad (9.3.12)$

$$b_m \equiv (\kappa_m^2 - k_1^2)^{1/2} \qquad (9.3.13)$$

$$k_1 = \frac{\omega}{c_1} \quad \text{and} \quad a_1 = \text{constant}$$

Using the continuity of sound pressure across the boundary, $Z_m(z)$ for the lower medium, by solving for a_1 at $z = h$, is

$$Z_m(z) = (\rho_0/\rho_1)\sin \gamma_m h \exp[-b_m(z-h)] \quad \text{for} \quad z \geq h \quad (9.3.14)$$

We use (9.3.13), the values of κ_1 and κ_2, and $c_1 = 1.12c_0$ to calculate b_1 and b_2

$$b_1 = 0.2561 \text{ m}^{-1} \quad \text{and} \quad b_2 = 0.1614 \text{ m}^{-1}$$

Graphs of $Z_1(z)$ and $Z_2(z)$ appear in Fig. 9.3.5. The eigenfunctions are

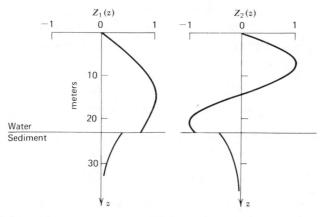

Figure 9.3.5. First two eigenfunctions for the waveguide in Fig. 9.3.1. The continuity of pressure gives equal $\rho Z_m(z)$ on each side of the interface where $\rho_1 = 2\rho_0$. The slopes $\partial Z_m(z)/\partial z$ are equal on each side. Both eigenfunctions are negligible at the depth of the lower sediment layer, $z = 43$ m.

negligible at the depth of the sediment layer that we neglected. This justifies our simplification of the waveguide.

The experiments were made with an uncalibrated source. The power was not known. Replacing the power and other constants by a constant a, q_m in (9.1.29) is

$$q_m = \frac{a}{v_m\sqrt{\kappa_m}} \quad (9.3.15)$$

Next we calculate v_m. From (9.1.20), since ρ is a function of z

$$v_m = \int_0^\infty \rho Z_m^2(z)\, dz$$

and the substitution of (9.3.11) and (9.3.14) in ν_m gives

$$\nu_m = \int_0^h \rho_0 \sin^2 \gamma_m z \, dz + \int_h^\infty \rho_1 \left(\frac{\rho_0}{\rho_1} \sin \gamma_m h\right)^2 \exp\left[-2b_m(z-h)\right] dz \tag{9.3.16}$$

$$\nu_m = \rho_0 \left(\frac{h}{2} - \frac{\sin 2\gamma_m h}{4\gamma_m}\right) + \frac{\rho_0^2}{2b_m \rho_1} \sin^2 \gamma_m h$$

Assuming $\rho_0 = 1033 kg/m^3$, $\rho_1 = 2\rho_0$, numerical evaluation gives $\nu_1 = 1.5 \times 10^4$, and $\nu_2 = 1.4 \times 10^4$. Values of the excitation function are

$$q_1 = 8.8a \times 10^{-5}$$
$$q_2 = 9.5a \times 10^{-5}$$

The amplitude for each mode depends on source and receiver depths. The data in Fig. 9.3.2 were taken at $z_0 \approx 10$ m and $z = 19.8$ m. From (9.3.11) or Fig. 9.3.5,

$$Z_1(10) = 0.86, \qquad Z_1(19.8) = 0.88$$
$$Z_2(10) = 0.77, \qquad Z_2(19.8) = -0.97$$

On combining these with q_m, we obtain

$$q_1 Z_1(z_0) Z_1(z) = 6.6a \times 10^{-5}$$
$$q_2 Z_2(z_0) Z_2(z) = -7.1a \times 10^{-5}$$

The transmission in the two modes is nearly equal. For convenience we combine the z dependence in one term

$$P_m(\omega) \equiv \rho q_m Z_m(z_0) Z_m(z) \exp\left(\frac{i\pi}{4}\right) \tag{9.3.17}$$

The numerical values of $P_1(\omega)$ and $P_2(\omega)$ give

$$P_1(\omega) \simeq -P_2(\omega)$$

9.3.3. Attenuation Coefficients

The attenuation is due to cylindrical spreading, absorption losses, and imperfect reflection at the boundary. We use the field measurements and the theoretical expression for p to estimate the attenuation coefficients δ_m. Using the results of the preceding section, (9.2.29) reduces to

$$p = P_1(\omega) r^{-1/2} \exp(i\omega t)[\exp(-\delta_1 r - i\kappa_1 r) - \exp(-\delta_2 r - i\kappa_2 r)] \tag{9.3.18}$$

The absolute square of (9.3.18) gives

$$\langle pp^* \rangle = \frac{|P_1|^2}{r} \{ \exp(-2\delta_1 r) + \exp(-2\delta_2 r)$$

$$-2\exp(-\delta_1 r - \delta_2 r) \cos[(\kappa_1 - \kappa_2)r] \} \qquad (9.3.19)$$

Continuing the numerical example, δ_1 and δ_2 are determined by trying numerical values of δ_1 and δ_2 until (9.3.19) fits the data in Fig. 9.3.2.

By inspecting Fig. 9.3.2, we notice that the minima of the sound pressure are a very small fraction of the maxima over most of the range. Even if the minima were zero, the ambient noise precludes recording the nulls. To obtain these deep minima, (9.3.19) must be near zero when $\cos[(\kappa_1 - \kappa_2)r] = 1$. If the attenuation coefficients were equal ($\delta_1 = \delta_2$), the minima would be zero. Therefore we assume that they are equal.

By taking the ratio of the pressure maxima (Fig. 9.3.2), at 1 and 4 km and doing the same in (9.3.18), we obtain the sediment attenuation rates

$$\delta_1 \approx \delta_2 \approx 4.3 \times 10^{-4} \text{ m}^{-1}$$
$$\approx 3.7 \text{ dB/km}$$

Comparison with Fig. 3.3.1 reveals that the absorption loss in water is a negligible part of the total attenuation.

The theoretical rms sound pressure curve was calculated by using the foregoing values of δ_1 and δ_2. We also made a small adjustment of $\kappa_1 - \kappa_2$ from the computed value of 0.034 to 0.0355 m^{-1}. We can do this by decreasing the average depth over the transmission path from 22.6 to 22.4 m. The proportionality constant $|P_1(\omega)|$ is chosen to match the first peak at about 1000 m. The theoretical sound pressure is the dashed line in Fig. 9.3.2.

9.4. DISPERSION: PHASE AND GROUP VELOCITIES

9.4.1. Phase Velocity

The phase velocity is the speed at which a constant phase surface appears to move along a given direction. In a waveguide we measure the phase velocity along the horizontal direction r. Recalling the equation for the sound pressure (9.3.18), we define the sound pressure for the mth mode as p_m and write

$$p_m \equiv P_m(\omega) r^{-1/2} \exp[i(\omega t - \kappa_m r) - \delta_m r] \qquad \textbf{(9.4.1)}$$

From $\omega\Delta t - \kappa_m \Delta r$, the *phase* of the pressure signal moves at the velocity ω/κ_m. We use the symbol v_r for the phase velocity along the r direction. Since the phase velocity depends on the mode, we have

$$v_{r,m} = \frac{\omega}{\kappa_m} \tag{9.4.2}$$

It is easy to obtain the phase velocity from the graph of the eigenvalues (Fig. 9.3.4) because $v_{r,m}$ is the ratio ω/κ_m. The maximum value of $v_{r,m}$ is c_1, the "free" speed in the sediment, and its minimum value is c_0, the "free" speed in the water. Figure 9.4.1 is a graph of $v_{r,m}$. When the phase velocity depends on frequency, the effect is called *dispersion*.

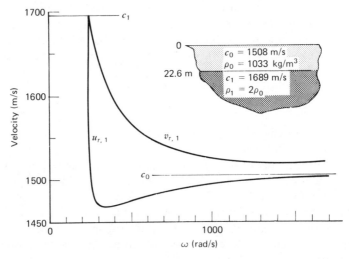

Figure 9.4.1. Phase and group velocities for the first mode; waveguide is the water over first sediment layer as in Fig. 9.3.1. The eigenvalues are given in Fig. 9.3.4.

Dispersion in a waveguide is the result of the geometry and the parameters of the waveguide. This is *geometrical dispersion* in contrast to *intrinsic dispersion*, which is due to the properties of the material such as for relaxation phenomena, (Section A3.2) and bubbly liquids (Section A6.2).

9.4.2. Group Velocity

The group velocity is the speed at which energy is transported in the waveguide. A heuristic way to develop the concept of a group velocity is to consider the pressure signal due to an impulsive source, We use the Fourier transformation and source spectrum $S(\omega)$ to write [using (9.4.1), (4.2.2), and $\omega = 2\pi f$, the transformation gives $P_m(t)$ from $P_m(\omega, t)$]

$$p_m(t) = r^{-1/2} \int_{-\infty}^{\infty} P_m(\omega) \exp\left[i(\omega t - \kappa_m r) - \delta_m r\right] df \qquad (9.4.3)$$

This is a difficult integral to evaluate in closed form, and much analytical effort has been devoted to approximate evaluations. Numerical evaluations are made by making a table of values for all the functions in $P_m(\omega)$, then doing the Fourier transformation numerically.

We can learn much about the solution by noticing that the function $P_m(\omega)$ has a relatively slow dependence on frequency as compared to $\exp\left[i(\omega t - \kappa_m r)\right]$. An old method of approximation (stationary phase) is based on the observation that if $P_m(\omega)$ varies relatively slowly, the oscillations of the integrand due to $\exp\left[i(\omega t - \kappa_m r)\right]$ effectively cancel each other on integration, except where $\omega t - \kappa_m r$ is constant. These values of $\omega t - \kappa_m r$ are known as stationary values. Differentiating with respect to ω gives

$$\frac{d}{d\omega}(\omega t - \kappa_m r) = 0$$

$$\frac{d\omega}{d\kappa_m} = \frac{r}{t} \qquad (9.4.4)$$

where t is the time required for the component of signal having frequency ω to travel the distance r. The ratio r/t is called the *group velocity* for the mode $u_{r,m}$

$$u_{r,m} = \frac{d\omega}{d\kappa_m} \qquad (9.4.5)$$

and $u_{r,m}$ is the velocity at which energy travels in the waveguide. In the numerical example the group velocities are the slopes of the curves of ω versus κ_m in Fig. 9.3.4.

In a graph of $u_{r,m}$ for the first mode (Fig. 9.4.1), $u_{r,1}$ has a minimum at 350 rad/s. The minimum is called the Airy phase. For frequencies below the minimum, the group velocity increases rapidly to c_1 at the mode cutoff frequency 233 rad/s. Waves are not trapped below the cutoff frequency. This portion of the group velocity curve is associated with what is called the *ground wave* because it is closely related to the sediment sound velocity c_1.

For frequencies above the Airy phase or minimum of $u_{r,m}$, the group velocity slowly approaches c_0. The portion of the group velocity curve is associated with a phenomenon designated the *water wave* or *rider* because it is mainly a function of the water depth and sound speed in water c_0.

9.4.3. Impulsive Source in a Waveguide

We can use the group velocity curve in Fig. 9.4.1 to show many important features of the transient arrival from an explosion in a waveguide. We can think of the source as transmitting many wavelets having different peak frequencies. At the instant of the explosion, the peaks of all the wavelets add together to make an impulsive function. An example is sketched in Fig. 9.4.2a, where we have chosen a low frequency wavelet g in the ground wave portion of $u_{r,m}$, a wavelet A for the frequency at the Airy phase, and a high frequency wavelet w for the water wave. The sum of the three wavelets has the appearance of an impulse.

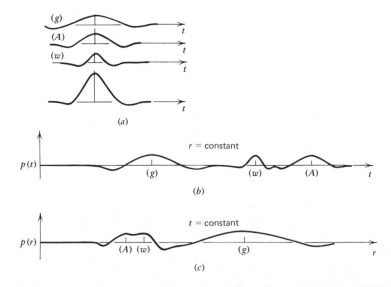

(a)

(b)

(c)

Figure 9.4.2. Wavelets in a waveguide. (a) An impulsive wavelet is formed of the sum of wavelets having different peak frequencies: g is a "ground" wavelet having a frequency below the group velocity minimum, A is a wavelet having a peak frequency at the minimum or Airy phase, w is a "water" wavelet having a frequency above the minimum. (b) Sound pressure $p(t)$ observed at a large distance from the source when all wavelets are initiated at the same time. (c) Distance dependence of the sound pressure $p(r)$ at an instant of time.

The three wavelets are transmitted into the waveguide at the same time. Because all have different group velocities, they pass by the distant hydrophone at different times, $t = r/u_{r,m}$. The arrivals versus time and the sound pressure versus distance (at constant time) are sketched in (Figs. 9.4.2b and 9.4.2c, respectively.

An actual explosion has many more components than are indicated in Fig. 9.4.2. On applying the same principle over the whole frequency spectrum, we obtain the pressure signal in Fig. 9.4.3a. For clarity, the water wave is drawn separately (Fig. 9.4.3b). The approximate periods of the wavelets are indicated at their arrival times. Much more analysis is needed to calculate the amplitudes of the wavelets. This is for the first mode. The other modes have similar signals.

It is easy to use an explosion in a waveguide to measure the group velocity as a function of frequency. At a large distance from the source, the signal is spread out and has the appearance of a slowly changing sinusoidal wave form. The peak to peak time difference is used to estimate the frequency (of the wavelet), and the group velocity is estimated from the arrival time of the wavelet.

As an example, we use Fig. 9.4.3 to calculate a few data points on the group velocity curve in Fig. 9.4.1. The highest frequency component of the water wave (Fig. 9.4.3b), arrives at 2.663 s. The range is 4.00 km.

$$u_{r,1} \simeq \frac{4000}{2.663}$$

$$\simeq 1500 \text{ m/s}$$

The periods are about 4 ms and

$$\omega \simeq \frac{2\pi}{4 \times 10^{-3}} = 1570 \text{ rad/s}$$

This group velocity is in the high frequency branch of the curve in Fig. 9.4.1.

The tail of the water wave has a period of about 18 ms and arrives at about 2.71 s. The group velocity is 1476 m/s and the frequency is 349 rad/s. The group velocity of the tail is the minimum value of $u_{r,1}$ at 349 rad/s. This is the Airy phase.

In Fig. 9.4.3a the low frequency ground wave is seen to arrive at 2.38 s. It has a periodicity of about 25 ms. This corresponds to a group velocity of about 1688 m/s and an angular frequency of about 251 rad/s, which is approximately the maximum velocity at mode cutoff as seen in Fig. 9.4.1.

This is a simple example of a basic technique for measuring the dependence of group velocity on frequency. It has also been applied to

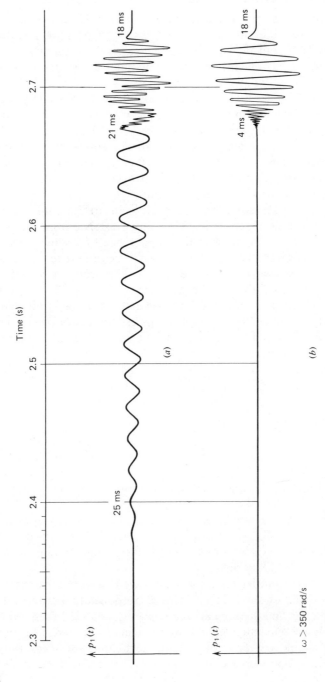

Figure 9.4.3. First mode of a distant signal due to an impulsive source in a waveguide. Sketch based on the group velocity curve (Fig. 9.4.1). The distance is 4 km and the approximate periods of the arrivals are indicated. (*a*) Total pressure signal. (*b*) High frequency part of the arrival, $\omega > 350$ rad/s. This arrival is called the water wave.

layers of ice on water, the surface waves from earthquakes, and gravity waves on water. Spectrum analyzers that have been designed to study transient phenomena and speech are used to analyze dispersed signals electronically. The older manual and numerical procedures are described in detail by Pekeris (1948).

PROBLEMS

9.1. Perform the following. (Use $\rho_1 = \infty$, $c_1 = c_0$, $\mathcal{R}_{01} = 1$)
   (a) Derive expressions for the eigenfunctions and eigenvalues of a 5 m thick layer of water over a perfectly rigid half space.
   (b) Determine the cutoff frequencies for the first three modes. What are the directions of the propagation constant k for each of these modes at the cutoff frequency?
   (c) Calculate the (three) mode interference wavelengths for a 600 Hz transmission frequency.

9.2. Use the expressions for the eigenvalues in Problem 9.1 to calculate the phase and group velocities as functions of frequency for the first mode.

9.3. An impulsive source is fired in the waveguide of Problem 9.1. Use the group velocity curves from Problem 9.2. to sketch the signal at a 10 km range. How does this result for a perfectly rigid bottom differ from the signal over a bottom having finite ρ_1 and c_1?

9.4. Often lake floors have sediments that are filled with bubbles from the decay of organic material. Approximate this liquid layer as having pressure release surfaces at the top and the bottom; air over water $c_0 = 1500$ m/s, $\rho_0 = 1000$ kg/m^3 of thickness h, over gas $c_1 = 0$ m/s and $\rho_1 = 1100$ kg/m^3.
   (a) Use the boundary conditions to determine the eigenfunctions.
   (b) For $h = 5$ m, derive expressions for the eigenvalues or propagation constants for the first four modes.
   (c) Calculate the cutoff frequencies for these modes.
   (d) Calculate γ_m, κ_m, and k for a 300 Hz signal frequency for the first four modes.

9.5. For the waveguide in Problem 9.4, assume that $c_1 = 150$ m/s.
   (a) Write an expression for the reflection coefficient at the bottom and use it to estimate the reflection loss ϵ for the first mode. *Hint.* Choose a 300 Hz signal and calculate θ_0 for the first mode and write $|\mathcal{R}_{01}| = 1 - \epsilon$.
   (b) Estimate the attenuation coefficient δ_1'.

9.6. An ideal water layer has a depth of 15 m over a rigid bottom.

  (a)   Calculate the eigenfunctions for the first two modes.

  (b)   Determine the cutoff frequencies.

  (c)   Calculate the propagation constants for 100 Hz transmission frequency.

  (d)   At what depth would you place a 100 Hz source to excite only the first mode?

  (e)   What modes would be excited by a pair of equal 100 Hz sources at 5 and 15 m depths when they are in phase?

  (f)   When the sources e have a relative phase shift of π?

9.7. The waveguide consists of a water layer over a sediment half space. The parameters are c_0, ρ_0, h, and $c_1 = 1.5c_0$, $\rho_1 = 2\rho_0$. Use the "dimensionless frequency" (fh/c_0).

  (a)   Calculate the eigenvalues for the first and second modes and graph them.

  (b)   Calculate and graph the phase v_1/c_0 and group u_1/c_0 velocities as a function of fh/c_0. You can compare your results with those of Tolstoy and Clay (1966, p. 108).

  (c)   For $c_0 = 1500$ m/s, $\rho_0 = 1000$ kg/m^3, and $h = 15$ m, calculate the cutoff frequencies for the first and second modes.

9.8. For advanced students. Synthetic seismograms are calculated by combining theoretical expressions for sound transmission, source functions, and the Fourier transformation. The problem is to calculate $p(t)$ for a layered wave guide and an impulsive source function like an explosion. For simplicity use one mode. To test your program we suggest using the wave guide of problem 9.7. For $p(t)$ computations use $h = 100$ m and r in the range of 1 to 2 km. Suggestions: Use the convolution expression (A 4.4.7) for your computations. Identify the shot spectrum $S(f)$ as being $X(f)$ and the wave guide transmission function $P_m(\omega) \, r^{-1/2} \exp[-i\kappa_m r - \delta_m r]$ as being $F(f)$. For κ_m, it may be easier to calculate a table of f as a function of κ_m and interpolate to evaluate κ_m for each f. To use the inverse FFT (A 4.4.2), construct sets of S_m and F_m for $m > N/2$ using (A 4.4.4) and (A 4.4.5).

SUGGESTED READING

Appendix A9

I. Tolstoy and C. S. Clay, *Ocean Acoustics: Theory and Experiment in Underwater Sound*, McGraw-Hill, New York, 1966.

L. M. Brekhovskikh, *Waves in Layered Media*, Academic Press, New York, 1960.

I. Tolstoy, *Wave Propagation*, McGraw-Hill, New York, 1973.

A. O. Williams, Jr, "Comments on Propagation of Normal Mode in the Parabolic Approximation, [Suzanne T. McDaniel, *J. Acoust Soc Am.* **57**, 307 (1975)]," *J Acoust Soc Am.* **58**, 1320–21 (1975).

TEN

SCATTER AND REFLECTION OF SOUND AT ROUGH SURFACES

The real sea floor is rough and the sea surface has waves. Sound signals are imperfectly reflected and partly scattered by rough surfaces. The gross magnitude of these effects depends on the rms roughness of the surface σ relative to the wavelength of the signal. If σ is a small fraction of λ, the surface appears to be smooth, as indicated by comparing the 3.5 kHz echo soundings and the seismic profile in Fig. 8.4.5. The 3.5 kHz sound has a wavelength of 0.43 m; the many hyperbolas appear to be due to small features at the water-sediment interface. The seismic data (λ ca 60 m) show an undulating smooth interface. The deeper signals appear to have interfering segments. We believe that these are scattered at a "rough basement." To go beyond a crude assessment of smoothness and roughness, we need a scattering theory to interpret data.

Two theoretical starting places are Rayleigh's formulation in *The Theory of Sound* and a mathematical version of Huygens' principle. Rayleigh's formulation is for the scattering of an incident plane wave on a sinusoidally corrugated surface. Extension of this approach to the random surface was given by Marsh (1961). We use the Huygens' principle approach with the Helmholtz–Kirchhoff integral formulation because it is easy to apply to realistic sonar situations.

319

In this chapter we regard the sonar as being a system for measuring (or sampling) the scattering characteristics of small areas of the sea floor and sea surface. We treat two classes of measurement: (1) the features are isolated and simple, permitting scattered sound signals to be resolved and identified, and (2) the features are numerous causing the scattered sound signals to overlap so that they cannot be resolved. The first class includes the diffraction of waves at a half plane and other simple objects. Trorey (1970) and Hilterman (1970) showed the applicability of these simple diffraction problems to the interpretation of seismic reflection measurements. For the second class, Eckart (1953) gave a way to calculate the mean values and the statistics of scattering of sonar signals at a randomly rough surface.

We put the analytical developments of the theory in Appendix A10 because the manipulations are very tedious. This chapter very broadly discusses the theory and experimental results. The comparisons are important because they demonstrate the usefulness of the theoretical results with the simplifying approximations. The material in Appendix A4, and in particular the Fourier integral transformations, are used in this chapter.

10.1   HUYGENS' PRINCIPLE AND THE HELMHOLTZ–KIRCHHOFF INTEGRAL

Our discussion rests very strongly on the qualitative description of diffraction effects given in Chapter 2. The mathematical formulation of Huygens' principle yields the Helmholtz-Kirchhoff (H-K) integral. The H–K integral relates the wave field U on the scattering surface to the field $U(Q)$ at a point Q. We use the abstract term "field" instead of "sound pressure" because it is convenient to suppress the time dependence of the source emission and the sound pressure amplitude. The field at the surface is proportional to the source pressure. When the source has time dependence $\exp(i\omega t)$, the field at the surface and the field at Q have the same time dependence. After using the first two approximations in Section A10.2, and factoring out the $\exp(i\omega t)$, we find that U at the surface is approximately $\mathcal{R}U_s$, where U_s is the incident wave field and $\mathcal{R}$ is the pressure reflection coefficient. The H–K integral is

$$U(Q) \simeq \frac{1}{4\pi} \int_S \mathcal{R} \frac{\partial}{\partial n} \frac{U_s e^{-ikR}}{R} \, dS \qquad (10.1.1)$$

where $\exp(-ikR)/R$ is the "point source" or Green's function for the Huygens' wavelets; R is the distance from dS to Q; $\partial/\partial n$ is the derivative along the normal to the surface and $\partial/\partial n(\ )$ is evaluated at dS. In writing (10.1.1), we assume that the source and receiver are in the same medium. Since U_s has the form of an expanding wave front, it has the dimensions of m^{-1} and is generally a function of frequency.

10.2. DIFFRACTION OF AN IMPULSIVE SIGNAL AT A SEGMENT OF A PLANE

Explosions and air guns are common sound sources. The pressure signals from these sources have a very short duration and a broad frequency spectrum. For a medium with a constant sound speed, the pressure wave from the source spreads as a very thin spherical shell of disturbance as shown in Fig. 2.2.4. As the wave front expands, it contacts and then sweeps over the half plane, generating a moving line of sources of Huygens wavelets. The successive positions of the outgoing wave front and the Huygens wavelets are shown. We assume the receiver is at the source position 0. The first wavelets to arrive at 0 form the reflection wave front R. Wavelets from the edge at B form the diffraction wave front D and arrive later. The sequence of arrivals gives a particular pressure signal, and repeated emissions from the source give the same pressure signal. It is obvious that the pressure signal depends on the positions of the source and receiver, the scatterer, and the amplitude and time dependence of the source emission. The transmission is linear because the amplitude of the pressure signal is proportional to the amplitude of the emission of the source.

The derivation of an analytical expression for $p(t)$ uses the source emission, the H–K integral, and the Fourier transformation. We use the input-output relation for an electrical filter as an analogy to the problem we wish to solve. Recalling (4.2.2), we can write the output $h(t)$ for a source having the time dependence $g(t)$ and the spectrum $G(f)$ and a filter with response $F(f)$:

$$h(t) = \int_{-\infty}^{\infty} G(f)F(f) \exp (i 2\pi ft)\, df$$

The analogy for $G(f)$ is the emission spectrum of the acoustic source $P_G(f)$ at the distance R_0 from the source. We imagine that the source is driving the acoustic equivalent of an electrical filter. Continuing the analogy, $U(Q)$, the acoustic system response, in (10.1.1) gives the frequency dependence of the field at Q. $U(Q)$ *is the equivalent of the electrical filter* $F(f)$. We replace $G(f)$ by $P_G(f)$ and $F(f)$ by $U(Q)$; the output pressure of the acoustic system then is

$$p(t, Q) = R_0 \int_{-\infty}^{\infty} P_G(f) U(Q) \exp(i2\pi ft) \, df \qquad \textbf{(10.2.1)}$$

where $U(Q)$ depends on frequency. Here $P_G(f)$ has the units pascals per hertz. This expression is valid for any transient source when the Fourier integral exists. Similarly, we use (A4.2.2) to write the expression for the pressure caused by a periodic source:

$$p(t, Q) = R_0 \sum_{m=-\infty}^{\infty} P_{Gm} U_m(Q) \exp(i2\pi mf_1 t) \qquad \textbf{(10.2.2)}$$

where f_1 is the fundamental frequency; P_{Gm} is the amplitude coefficient (Pa) of the source signal at R_0 for the frequency mf_1; $U_m(Q)$ is acoustic system response for mf_1. In most applications (10.2.2) would be the FFT (A4.1.4); the limits are $(1 - N/2)$ to $N/2$. We remind the reader that the CW source has a single amplitude coefficient P_{G1}, which is the rms pressure amplitude at separation R_0 from the source. The coefficient P_{G1} is equivalent to P_0 for CW sources elsewhere in this book.

The delta function $\delta(t)$ type of impulsive source is particularly useful for this problem because its spectrum is constant as a function of frequency (Section A4.3). We let the emission spectrum of the impulsive source at separation R_0 be P_δ, move P_δ out of the integral, and write the sound pressure as follows:

$$p_\delta(t, Q) = R_0 P_\delta \int_{-\infty}^{\infty} U(Q) \exp(i2\pi ft) \, df \qquad (10.2.3)$$

For convenience, we define the integral as

$$u(Q) \equiv \int_{-\infty}^{\infty} U(Q) \exp(i2\pi ft) \, df \qquad \textbf{(10.2.4)}$$

and the Fourier transformation

$$U(Q) = \int_{-\infty}^{\infty} u(Q) \exp(-i2\pi ft) \, dt \qquad (10.2.5)$$

where we use the notation of Chapter 4 and indicate the time dependent function by a lowercase letter and the spectrum by the uppercase letter. The substitution of (10.2.4) in (10.2.3) gives

$$p_\delta(t, Q) = R_0 P_\delta u(Q) \qquad (10.2.6)$$

where $p_\delta(t, Q)$ has the units of pascals when R_o is in meters, P_δ is in pascals per hertz, and $u(Q)$ is in (meter seconds)$^{-1}$. These expressions,

(10.2.1)–(10.2.6), are general and apply to any situation that can be described by the Fourier integral or series.

10.2.1. Integration for an Impulsive Source

We simplify the general problem as follows. The source and receiver are at the same position; the medium has a constant sound speed and the ray paths are straight lines; the scattering object is a segment of a plane. These simplifications are reasonable approximations for many marine seismic experiments because the source and receiver are close together; ray paths within 30° of vertical are nearly straight; and the sea floor can be represented by segments of planes that have various shapes, slopes, and reflectivities.

The integration for an impulsive $\delta(t)$ type of source involves a sequence of transformations. The first step is to transform the H–K integral from an integral over the area of the segment of a plane to a line integral around the boundary of the segment. Next we insert $U(Q)$ into (10.2.4) and obtain an expression for $u(Q)$. A change of variable facilitates the identification of the time dependent expression within the double integral as being $u(Q)$. The expression for $u(Q)$ is simple and easy to evaluate.

When the source and receiver are at Q, as shown in Fig. 10.2.1a, U_s and (10.1.1) become (for $KZ>1$, see (2.9.13).

$$U_s = \frac{e^{-ikR}}{R}$$

$$U(Q) = \frac{\mathscr{R}}{4\pi} \int_S \frac{\partial}{\partial n} \left[\frac{\exp\left(-2i\omega \dfrac{R}{c}\right)}{R^2} \right] dS \qquad (10.2.7)$$

where $\omega = 2\pi f$. The scattering surface is a segment of the x–y plane. The normal derivative, $\partial/\partial z$, can be expressed as

$$\frac{\partial}{\partial n} = \frac{\partial}{\partial z} = \frac{-z}{R} \frac{\partial}{\partial R}$$

Then (10.2.7) is

$$U(Q) = -\frac{z\mathscr{R}}{4\pi} \int_S \frac{1}{R} \frac{\partial}{\partial R} \left[\frac{\exp\left(-2i\omega \dfrac{R}{c}\right)}{R^2} \right] dS \qquad (10.2.8)$$

The plane of the segment is in the x–y plane, and we express dS as $r\, dr\, d\theta$ (Fig. 10.2.1a). Using $R^2 = r^2 + z^2$, $r\, dr = R\, dR$, and new limits, the integral

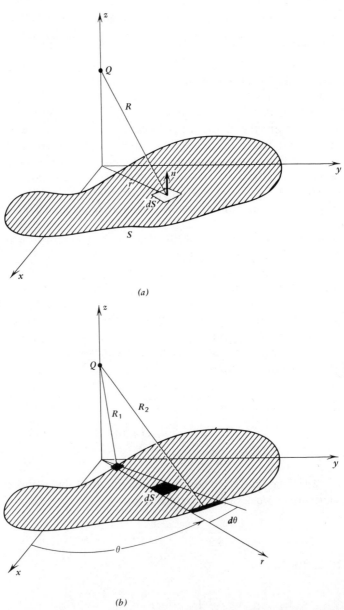

(a)

(b)

Figure 10.2.1. Geometry for boundary diffraction waves. For a given θ, the radius vector from the origin intersects the boundary of S once at its near and far sides. Source and receiver at Q. a) surface integral. b) line integral. c) source and receiver over S.

324

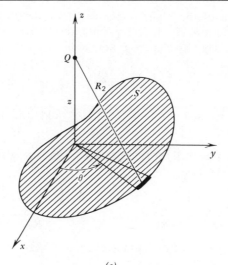

(c)

Figure 10.2.1. (*Continued*)

becomes

$$\int_S (\ )dS \rightarrow \int_{\theta_1}^{\theta_2} \int_{R_1}^{R_2} (\ )R\,dR\,d\theta$$

After substitution, for dS, the integrand is an exact differential, and integration yields

$$U(Q) = \frac{z\mathcal{R}}{4\pi} \int_{\theta_1}^{\theta_2} \left[\frac{\exp\left(-2i\omega\dfrac{R_1}{c}\right)}{R_1^2} - \frac{\exp\left(-2i\omega\dfrac{R_2}{c}\right)}{R_2^2} \right] d\theta \qquad \textbf{(10.2.9)}$$

where R_1 and R_2 are the distances from the source to the near and far boundaries at the angle θ (Fig. 10.2.1b). The remaining integral is along the boundary, and the surface integral has been changed into a line integral along the boundary. Sometimes this is called a boundary diffraction wave.

We use the Fourier transformation (10.2.4) to calculate the impulse response $u(Q)$:

$$u(Q) = \frac{z\mathcal{R}}{4\pi} \int_{-\infty}^{\infty} \int_{\theta_1}^{\theta_2} \left[\frac{\exp\left(-2i\omega\dfrac{R_1}{c}\right)}{R_1^2} - \frac{\exp\left(-2i\omega\dfrac{R_2}{c}\right)}{R_2^2} \right] \exp\left(i\omega t\right) df\,d\theta$$

$$(10.2.10)$$

If the segment of the plane contains the origin, then $R_1 = z$, and R_1 does not depend on θ, Fig. 10.2.1c. The integral over $d\theta$ yields the factor 2π. The integral over df has the following form of the delta function:

$$\delta\left(t - \frac{2z}{c}\right) = \int_{-\infty}^{\infty} \exp\left[i\omega\left(t - \frac{2z}{c}\right)\right] df$$

Evaluation of the first term of (10.2.10) yields the specular reflection, $\mathcal{R}/(2z)\ \delta(t - 2z/c)$.

The remaining integral represents the boundary wave. The complete expression for $u(Q)$ is

$$u(Q) = \frac{\mathcal{R}\delta\left(t - \dfrac{2z}{c}\right)}{2z} - \frac{z\mathcal{R}}{4\pi} \int_{-\infty}^{\infty} df \int_{\theta_1}^{\theta_2} \frac{\exp\left(i\omega t - 2i\omega \dfrac{R_2}{c}\right)}{R_2^2}\, d\theta \quad \textbf{(10.2.11)}$$

The integral appears often, and we define $\mathcal{D}(t)$ as follows to obtain Trorey's (1970) form:

$$\mathcal{D}(t) \equiv \frac{c^2}{4} \int_{-\infty}^{\infty} df \int_{\theta_1}^{\theta_2} \frac{\exp\left(i\omega t - 2i\omega \dfrac{R}{c}\right)}{R^2}\, d\theta \quad (10.2.12)$$

On making the changes of variable,

$$t' \equiv \frac{2R}{c}$$
$$(10.2.13)$$
$$d\theta = \frac{d\theta}{dt'}\, dt'$$

We obtain

$$\mathcal{D}(t) = \int_{-\infty}^{\infty} df \int_{t_1}^{t_2} \left\{\frac{1}{t'^2}\frac{d\theta}{dt'}\right\} \exp\left[i\omega(t - t')\right] dt' \quad (10.2.14)$$

where t_1 and t_2 are the travel times corresponding to the angles θ_1 and θ_2. By comparing (10.2.14) and the complete expression of the Fourier transformation [replace $G(f)$ by its integral in (4.1.5) and (4.1.6)], we see that

$$g(t) = \int_{-\infty}^{\infty} df \int_{-\infty}^{\infty} g(t') \exp\left[i\omega(t - t')\right] dt'$$

it is evident that the expression within curly brackets is $\mathcal{D}(t)$. Dropping the primes yields

$$\mathcal{D}(t) = \frac{1}{t^2}\frac{d\theta}{dt} \quad \text{for} \quad t_1 < t < t_2$$
$$= 0 \quad \text{otherwise} \quad \textbf{(10.2.15)}$$

When the angle θ to a point on the boundary of S is expressed as a function of time, we can calculate the time dependence of the boundary wave. As a function of θ, t must be single valued. If the boundary of S has a shape such that t is multiple valued, the integral can be broken into single-valued segments. For S containing the origin, (10.2.11) becomes

$$u(Q) = \mathscr{R}\left[\frac{\delta\left(t - \dfrac{2z}{c}\right)}{2z} - \frac{z}{\pi c^2}\,\mathscr{D}_2(t)\right] \qquad (10.2.16)$$

where $\mathscr{D}_2(t)$ indicates evaluation for R_2.

When the segment of the plane does not contain the origin (Fig. 10.2.1b), the first term yields $\mathscr{D}_1(t)$ for R_1 and the second $\mathscr{D}_2(t)$:

$$u(Q) = \frac{\mathscr{R}z}{\pi c^2}[\mathscr{D}_1(t) - \mathscr{D}_2(t)] \qquad (10.2.17)$$

The specular reflection is missing, and the integral over $d\theta$ has two boundary waves. The boundary wave from the nearer edge (R_1) has a positive sign, and the boundary wave from the far edge (R_2) has a negative sign. The sound pressure at Q is

$$p_\delta(t, Q) = P_\delta R_0 u(Q) \qquad (10.2.18)$$

10.2.3. Linear Plane Strips

Models of fairly complicated bottoms can be constructed of sectors of planes. The prototype of a sector is the straight edge of a half plane (Fig. 10.2.2). The source transmits an impulsive signal and the wave front spreads spherically. The contact of the wave front and edge of the plane b is a moving source of Huygens wavelets as θ goes from 0 to $\pm\pi/2$; $u(Q)$ is the signal. The boundary wave starts to arrive at T_1 and tails off as $b \to \infty$. If the edge is finite, the boundary wave lasts until θ reaches the corner, and then the boundary wave follows the new edge. The evaluation of $\mathscr{D}_1(t)$ is given in Section A.10.3. The final expressions are

$$\mathscr{D}_1(t) = 2T_x[t(t^2 + T_x^2 - T_1^2)(t^2 - T_1^2)^{1/2}]^{-1} \quad \text{for} \quad t \geqslant T_1 \qquad (10.2.19)$$

$$\mathscr{D}_1(t) = 0 \quad \text{for} \quad t < T_1$$

$$T_x \equiv \frac{2x_1}{c}$$

$$T_z \equiv \frac{2z}{c} \qquad (10.2.20)$$

$$T_1^2 \equiv T_x^2 + T_z^2$$

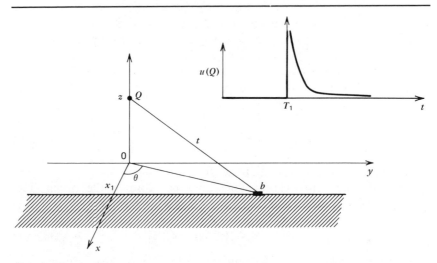

Figure 10.2.2. Boundary wave at a half plane. Source is offset from plane, t is two-way travel time from Q to b; $u(Q)$ is signal for impulsive source $\delta(t)$. Initial part of boundary wave arrives at T_1.

The value of $\mathcal{D}_1(t)$ at $t = T_1$ is infinite. We handle this infinity and the source function infinity by letting Δ be a short finite time:

$$\delta(t) = \frac{1}{\Delta} \qquad 0 \le t \le \Delta$$

$$\delta(t) = 0 \qquad \text{otherwise}$$

(10.2.21)

In the equations (A10.3.7) to (A10.3.11) we average $\mathcal{D}_1(t)$ over Δ and obtain

$$\mathcal{D}_1(t) = 2(T_1^2 \Delta)^{-1} \arctan\left[\frac{(2T_1\Delta)^{1/2}}{T_x}\right] \quad \text{for} \quad T_1 \le t \le T_1 + \Delta \quad \textbf{(10.2.22)}$$

where Δ is the time step of the numerical computation. The boundary signal for an edge is calculated with (10.2.19) and (10.2.22). Similar equations are written for other edges of the plane. The transient pressure follows by the substitution of the expressions for $\mathcal{D}_1(t)$ and $\mathcal{D}_2(t)$ in (10.2.16), (10.2.17), and (10.2.18).

In Fig. 10.2.3a a source and receiver are to the left of a half plane and the boundary wave is a positive impulsive function. We use an impulsive function having peak $1/\Delta$ and width Δ in numerical evaluations. The long tail is the boundary wave as θ goes from 0° to 90°. Moving the source to a position over the half plane produces a specular reflection and changes the sign of the boundary wave. If the source is moved to a position over

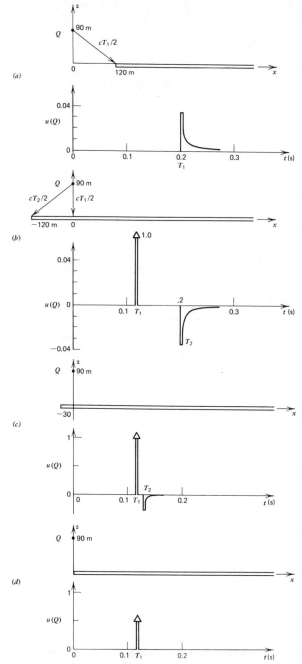

Figure 10.2.3. Boundary diffraction waves $u(Q)$ for four geometries. Time step for calculation $\Delta = 0.005$ s, $u(Q)$ is normalized to image reflection amplitude, specular reflection shown by double shafted arrow, $c = 1500$ m/s.

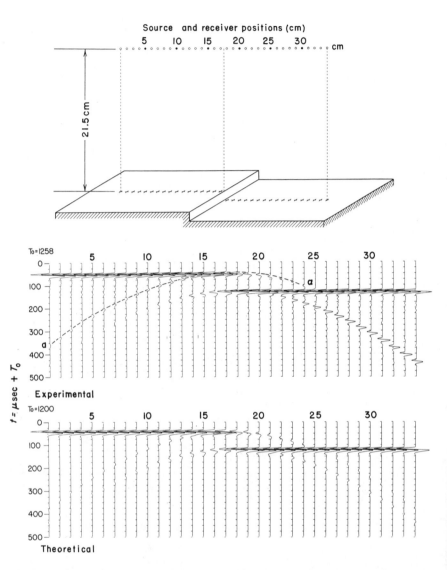

Figure 10.2.4. Reflection and diffraction waves. Source was a spark and receiver was a small microphone. The experiments were made in air. Pressure-time graphs shown for 34 adjacent source-receiver positions. T_0 is an arbitrary delay time so that trace begins just prior to arrival of reflection. $a ---- a$ indicates diffractions from top edge of the step. Data and theoretical traces from Hilterman (1970).

the edge, the image reflection and boundary wave merge. The boundary wave loses its tail because θ is constant and $\mathscr{D}_1(t)$ tends to one-half of the image reflection. The net signal is an impulse having half the amplitude of the image.

The experimental measurements of Hilterman (1970) show the relative amplitudes and sign of the boundary wave (Fig. 10.2.4). Figure 10.2.5 compares the experimental values of boundary wave amplitudes with those from (10.2.19) to (10.2.22). We believe that the comparisons are good enough to warrant using the theory.

We can model linear features on the sea floor by connecting segments of plane strips. Each strip has its own coordinate axis, because x and y are in the plane and z is perpendicular to the plane. The details are given in Section A10.4. The magnitude of the boundary wave depends on the slope of the strip because x_1 and z depend on the slope. The magnitude of the boundary wave also depends on the reflection coefficients for each strip. The boundary waves due to the edges of each strip are added. For example, at the intersection of two half planes, the boundary wave from the plane containing the origin is negative (Fig. 10.2.3b) and the boundary wave from the other half plane is positive (Fig. 10.2.3a). The net signal depends on the magnitude of the change of slope from one half plane to the next. The sign depends on the direction of the slope change. The boundary waves for several changes of slope are shown in Fig. 10.2.6. The signal from a model of a bottom is shown in Fig. 10.2.7.

Changes of reflection coefficients from strip to strip also yield boundary waves.

10.2.4. $p(t, Q)$ for Other Source Emissions

We use $u(Q)$ and $p_\delta(t, Q)$ as the expressions for the $\delta(t)$ source. For source emissions other than the $\delta(t)$, we can use $u(Q)$ and either the spectrum of the source emission $P_G(f)$ or the source's time dependent function $p_g(t)$ to calculate $p(t, Q)$. We suggest two numerical procedures, and the choice depends upon computational convenience. Both procedures are related and give the same result.

We assume that $u(Q)$ is known at a set of sample times $0, \Delta, 2\Delta, \ldots,$ $(N-1)\Delta$. We use Fig. 10.2.7 as an example and require that Δ be small enough to display the arrivals from each of the boundaries. The highest frequency of the source emission spectrum f_h gives an additional condition, $\Delta < 1/(2f_h)$. The choice for N should be large enough to isolate the transient from its periodic repetition in the FFT (see Fig. A4.1.2 for an

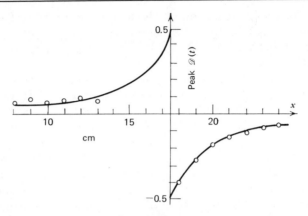

Figure 10.2.5. Comparison of $\mathcal{D}(t)$ and Hilterman's measurements. Amplitudes are for the diffraction at a, read from Hilterman's wave forms in Fig. 10.2.4. Amplitudes are relative to the image reflection. For this calculation $\Delta = 3.6\ \mu\text{s}$.

example). We use (A4.1.3) to calculate $U_m(Q)$. The finite sum from $1 - N/2$ to $N/2$ in (10.2.2) gives

$$p_n(Q) = R_0 \sum_{m=1-N/2}^{N/2} P_{Gm} U_m(Q) \exp\left(\frac{i2\pi mn}{N}\right) \quad \textbf{(10.2.23)}$$

where $p_n(Q)$ is the value of $p(t, Q)$ at $t = n\Delta$.

Another method of calculating $p_n(Q)$ uses the convolution. We sample the source emission $p_g(t)$ to obtain p_{gn}. The convolution expression (A4.4.8) is

$$h(t) = \int_{-\infty}^{\infty} f(\tau) x(t - \tau)\, d\tau \quad (10.2.24)$$

and the equivalent summation is

$$h_n = \sum_{j=-\infty}^{\infty} f_j x_{n-j} \quad (10.2.25)$$

Letting p_{gn} be f_n and $u_n(Q)$ be x_n, we obtain for the convolution expression

$$p_n(Q) = R_0 \sum_{j=-\infty}^{\infty} p_{gj} u_{n-j}(Q) \quad \textbf{(10.2.26)}$$

The convolution expression is particularly convenient when either the source emission or $u(Q)$ has a short duration.

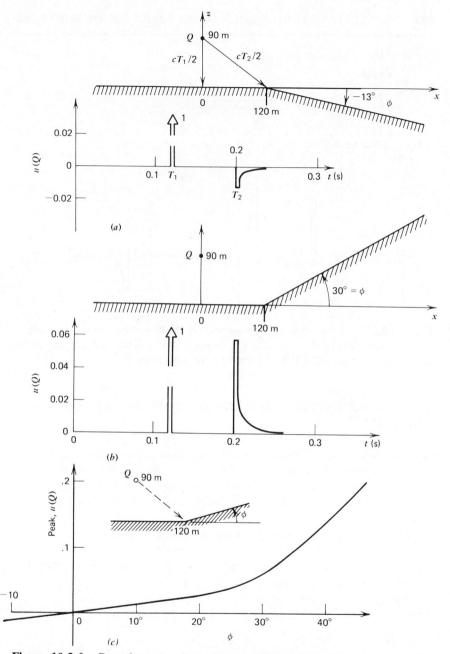

Figure 10.2.6. Boundary wave from changes of slope. Time step for this calculation $\Delta = 0.005$ s, $u(Q)$ is normalized to amplitude at T_1, $\mathscr{R} = 1$. (a), (b) are for different slope changes and show wave forms; (c) dependence of $u(Q)$ on ϕ at T_2; $c = 1500$ m/s.

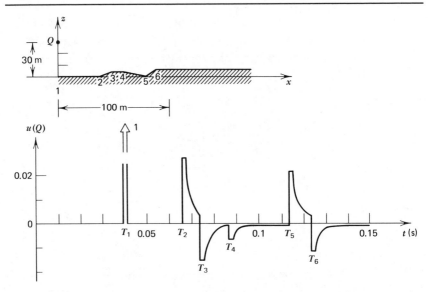

Figure 10.2.7. Signal from a bottom approximated by plane strips: $\Delta = 0.005$ s, $c = 1500$ m/s, $\mathscr{R} = 1$; T_1, T_2, ..., are two-way travel times for the ray paths $Q \to 1$, $Q \to 2$, ... $u(Q)$ is normalized by the amplitude at T_1.

10.3 DOPPLER EFFECTS FOR MOVING OBJECTS, SEA SURFACE, AND SHIPS

Motions of a source, receiver or scattering object change the frequency of returned signals. We assume that the water is still and that the propagating signal moves with sound speed c regardless of the velocities of the source or receiver or objects. A brief review of the Doppler effect follows:

Initially, we consider motions along the x axis. The source moving with velocity v_s transmits either a continuous wave or a very long ping of frequency f_s. When the source advances on its own waves, they are shortened and the wavelength along x is $\lambda = (c - v_s)/f_s$.

Assume the receiver has velocity v_r away from the source. In a unit time it detects $f_r = (c - v_r)/\lambda$ crests. The frequency at the receiver is

$$f_r = \frac{f_s(c - v_r)}{(c - v_s)} \tag{10.3.1}$$

Now consider a fixed source radiating frequency f_s. An object moving away with velocity v_o senses a different number of crests per second.

Therefore the frequency of sound observed (and scattered) by a moving object is $f_o = f_s(c - v_o)/c$. From the point of view of a fixed receiver at the source location, the scattering object becomes a moving source and the frequency at the receiver is

$$f_r = \frac{f_s(c - v_o)}{(c + v_o)} \tag{10.3.2}$$

where the object is moving away from the source and receiver.

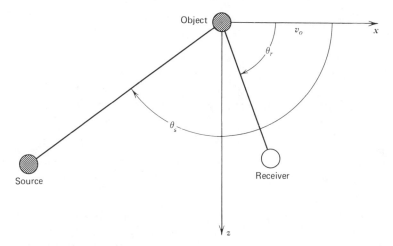

Figure 10.3.1. Doppler effect for an object moving along the x coordinate. Source and receiver are in the x–z plane.

In other directions, Fig. 10.3.1, the velocities in (10.3.2) become the components of v_o along the directions to the source and receiver

$$f_r = \frac{f_s(c + v_o \cos \theta_s)}{(c - v_o \cos \theta_r)} \tag{10.3.3}$$

The Doppler shift is positive or negative, depending on the geometry and the direction of motion.

10.3.1. Sea Surface

Signals scattered at the moving sea surface are shifted in frequency. The amount of the frequency shift is given by (10.3.3). We assume that the sonar insonifies a wave crest that is moving away at velocity v (Fig.

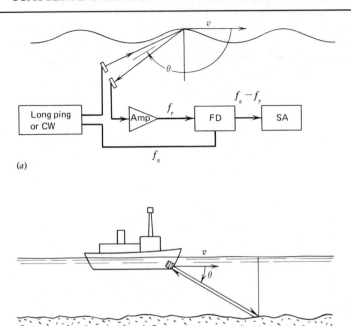

Figure 10.3.2. Applications of Doppler effect. (a) Measurement of velocity of a feature (crest) of a surface wave: FD gives frequency difference, SA is a spectrum analyzer. (b) Measurement of ship velocity. Sonar system is similar to the one in (a).

10.3.2a). For $\theta_s = +\theta_r = \theta$, f_r is

$$f_r = \frac{f_s(c + v \cos \theta)}{c - v \cos \theta} \tag{10.3.4}$$

We can use the Doppler shift to measure the component of velocity of the wave along the axis of the sonar system.

The spectrum of heights of gravity waves is very peaked. For simplicity, we represent the long-crested train of surface waves by a traveling sine wave, $\sin[2\pi f_w(t - x/v)]$. From Section 1.4 we know that the phase velocities v and f_w of gravity waves are related by

$$v \simeq \frac{g}{2\pi f_w} \tag{10.3.5}$$

Measurements of Doppler shift can be used to estimate f_w. Incoming waves cause a positive Doppler shift, and outgoing waves cause a negative shift. Standing waves are represented as the sum of a pair of waves traveling in opposite directions. Correspondingly, plus and minus Doppler shifts are expected for standing waves.

Nominally the Doppler shift is zero in the specular direction $\theta_s = \pi - \theta_r$. However some of the scattered components can reach the receiver from outer parts of the insonified area where $\theta_s \neq \pi - \theta_r$, and these *do* show a Doppler shift. Roderick and Cron (1970) give comparisons of theoretical and experimental Doppler shifts.

10.3.2. Doppler Navigation

Doppler navigation systems use the frequency shift of backscattered sound signals to measure the velocity of the ship relative to the bottom or stationary objects within the water. Operation of the system is sketched in Fig. 10.3.2*b*. The frequency of the backscattered signal is

$$f_r = \frac{f_s(c + v \cos \theta)}{c - v \cos \theta}$$

$$\Delta f \approx \frac{2vf_s \cos \theta}{c} \tag{10.3.6}$$

Sometimes people use combinations of sonars that look forward, backward, and to each side to measure the speed and direction of the ship. The velocity information is used to interpolate positions between fixes.

10.3.3 Uncertainty of Position and Velocity Measurements

The use of the same ping to measure both the Doppler frequency shift and distance to a moving object leads to a fundamental uncertainty of position and velocity. This uncertainty is referred to as "ambiguity" in the literature. We recall that a ping having a frequency f_0 and duration ΔT has a spectrum $P(f)$ (using the Fourier transformation).

$$P(f) \sim \frac{\sin[\pi(f - f_0)\Delta T]}{\pi(f - f_0)\Delta T} \tag{10.3.7}$$

The first null occurs at the frequency difference $f - f_0 = \Delta f$, and

$$\Delta f \approx \frac{1}{\Delta T} \tag{10.3.8}$$

To measure the Doppler frequency shift very accurately, we need a very narrow spectral width; this requires a long ping. This is contrary to the condition for accuracy of position, which calls for a short ping and a large spectral width. The magnitude of Δf in terms of ΔT for other kinds of signal depends on the design of the signal. We recommend Woodward (1964) as a starting place for supplementary reading.

10.4. AVERAGE REFLECTION COEFFICIENT FOR RANDOMLY ROUGH SURFACES

The preceding sections gave calculations of the signal for specific models of the bottom or sea surface. To determine an average effect of surface waves or a rough bottom on signals, we can calculate the reflection from several models and average them. Eckart's procedure (1953) is to use statistical descriptions of the rough surface to calculate statistical estimates of the signal. The rough surface is described by statistical quantities such as the rms roughness σ, the PDF, and the spatial correlation function of surface heights. We describe these quantities and how they enter into the acoustical problem.

10.4.1. Average or Coherent Reflection Coefficient

The average or coherent reflection coefficient is based on measurements of many repeated signal transmissions, as in Fig. 10.4.1. Because of ship

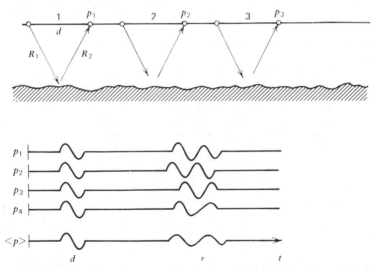

Figure 10.4.1. Average transient signal: d is direct arrival, r is reflection.

motion, the source and receiver are translated relative to the bottom between each transmission. The signals are p_1, p_2, and so on. The average signal is obtained by first aligning the signals as shown. Then the instantaneous values of the signals are averaged to form $\langle p \rangle$ (Fig. 10.4.1). This process, called "signal stacking," "signal averaging," and "coherent processing," occurs to some extent when the eye averages the adjacent traces of a graphic recorder display.

The *coherent reflection coefficient* $\langle \mathscr{R} \rangle$ is the reflection coefficient of $\langle p \rangle$. For source and receiver distances R_1 and R_2 and a source p_0 at R_0, the solution of the image reflection equation (3.2.30) for the reflection at a plane interface gives

$$\mathscr{R} = p \frac{(R_1 + R_2)}{p_0 R_0}$$

Correspondingly, measurements $\langle p \rangle$ yield $\langle \mathscr{R} \rangle$

$$\langle \mathscr{R} \rangle \equiv \langle p \rangle \frac{(R_1 + R_2)}{p_0 R_0} \qquad \textbf{(10.4.1)}$$

where $\langle \mathscr{R} \rangle$, the result of a measurement of $\langle p \rangle$, depends on the roughness as well as the impedance contrast at the bottom (i.e., the reflection coefficient $\mathscr{R}_{12}$). If subbottom reflections from thin layers are included in $\langle p \rangle$, then $\langle \mathscr{R} \rangle$ includes subbottom reflections.

We construct a theoretical relationship between $\langle \mathscr{R} \rangle$ and a statistical model of the bottom by using a simplified description of the reflection at a rough bottom. In Fig. 10.4.2 the bottom has a mean level $\zeta = 0$. The

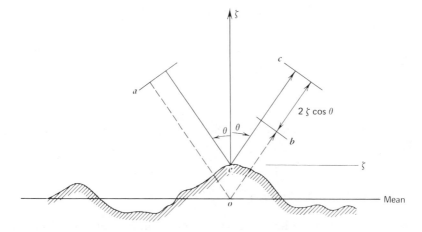

Figure 10.4.2.   Travel path at a local area.

signal has frequency ω and wave number $k = \omega/c$. The initial position of the incident wave front is a. The path a–o–b gives the distance for a reflection at the mean level. During the same time, the portion of the wave front reflected at ζ gets to c. We can ignore the source and receiver distances because they are common to both paths. The phase change of the signal relative to $\zeta = 0$ is $2k\zeta \cos \theta$.

The bottom has many different elevations, and the signal is the sum of the effects of them all. The probability of occurrence of an elevation between ζ and $\zeta + d\zeta$ is $w_a(\zeta)\,d\zeta$, where $w_a(\zeta)$ is the PDF of the rough surface. The effect of ζ on the signal is to alter the phase by $-2k\zeta \cos \theta$. To get the mean value of the effect, we multiply the effect $\exp(-2ik\zeta \cos \theta)$ by $w_a(\zeta)$ and integrate over all values of ζ

$$\langle \mathfrak{R} \rangle = \mathfrak{R} \int_{-\infty}^{\infty} \exp(-2ik\zeta \cos \theta) w_a(\zeta)\,d\zeta \tag{10.4.2}$$

Since $k = \omega/c$, we see that $\langle \mathfrak{R} \rangle$ is frequency dependent. The equation has the form of the Fourier integral transformation with ω replaced by $2k \cos \theta$. Its inverse is the acoustical estimate of the PDF.

$$w_a(\zeta) = \frac{1}{\pi \mathfrak{R}} \int_{-\infty}^{\infty} \langle \mathfrak{R} \rangle \exp(2ik\zeta \cos \theta)\,d(k \cos \theta) \tag{10.4.3}$$

Measurements of $\langle \mathfrak{R} \rangle$ over a wide range of frequency, and (10.4.3) can be used to calculate $w_a(\zeta)$.

The first guess about a random surface is to assume that it has a Gaussian PDF because many natural surfaces have a nearly Gaussian PDF

$$w_g(\zeta) = \sigma^{-1}(2\pi)^{-1/2} \exp\left(\frac{-\zeta^2}{2\sigma^2}\right) \tag{10.4.4}$$

where σ is the rms roughness. The substitution into (10.4.2) and evaluation of the infinite integral gives

$$\frac{\langle \mathfrak{R} \rangle_G}{\mathfrak{R}} = \exp(-2k^2\sigma^2 \cos^2 \theta) \tag{10.4.5}$$

$\langle \mathfrak{R} \rangle_G$ is the coherent reflection coefficient for a surface having a Gaussian PDF.

10.4.2. Laboratory Measurements of $\langle \mathfrak{R} \rangle$ and $w_a(\zeta)$

Laboratory measurements of $\langle \mathfrak{R} \rangle$ were made in the water-filled tank in Fig. 10.4.3. The water surface was roughened by blowing wind over it.

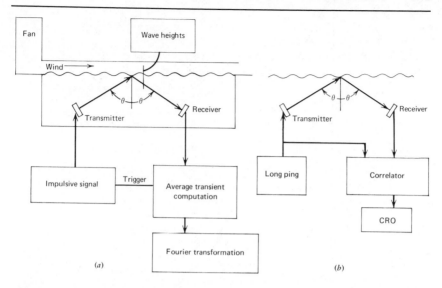

Figure 10.4.3. Laboratory tank and acoustic systems, where waves are produced by wind and wave heights are measured by a wave staff. (a) Simplified representation of the Spindel-Schultheiss system. A computer was used to process the data (1972). (b) Simplified diagram of Clay *et al.* (1973). An anechoic water tank was used. Data were processed in an analogue correlator.

Each transmission insonified a slightly different surface because the wind-driven waves moved through the area. The method represented in Fig. 10.4.3a was used by Spindel and Schultheiss (1972) to measure the average transient. They used the Fourier transformation of $\langle p \rangle$ to calculate the average transmission (transfer) function. We have used their results to calculate $\langle \mathcal{R} \rangle$ in Fig. 10.4.4a. The deviation of $\langle \mathcal{R} \rangle$ from $\langle \mathcal{R} \rangle_G$ is small.

Clay, Medwin, and Wright measured average signal by another method in which the transmitted signal was a long ping. This was then correlated with the received signal in the correlator. The correlation for each transmission was stored and averaged over many transmissions, and $\langle \mathcal{R} \rangle$ was proportional to the correlation. Since $\mathcal{R} = -1$ at the air–water interface, $|\langle \mathcal{R} \rangle|$ tended to unity as σ approached zero. A typical $|\langle \mathcal{R} \rangle|$ is given in Fig. 10.4.4b. For comparison, we show the average reflection for a Gaussian PDF (10.4.5).

Next we use (10.4.3) and measurements of $\langle \mathcal{R} \rangle$ to estimate $w_a(\zeta)$. Figure 10.4.5 shows the results for the two experiments illustrated in Fig. 10.4.4. The acoustic PDFs are compared to wave staff measurements of the PDFs of the surface wave heights $w_{ws}(\zeta)$. For the larger roughness,

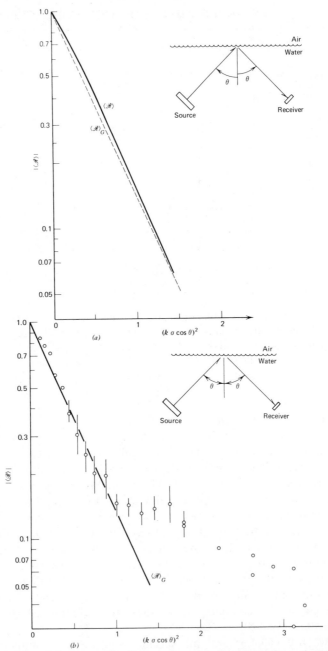

Figure 10.4.4. Coherent reflection at air-water interface $\mathcal{R} = -1$. Theoretical value for Gaussian surface $\langle \mathcal{R} \rangle_G$ is shown for comparison. (*a*) $\langle \mathcal{R} \rangle$ calculated from the average transfer function measurements of Spindel and Schultheiss (1972), where $\sigma = 0.14$ cm and $\theta = 45°$. Wave staff measurements of the PDF of the surface appear in Fig. 10.4.5*a*. (*b*) Downwind data from long ping measurements, $\sigma = 0.41$ cm, $\theta = 45°$, Clay *et al.* (1973). Wave staff measurements appear in Fig. 10.4.5*b*.

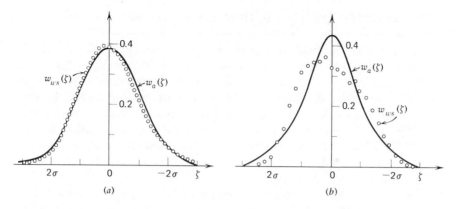

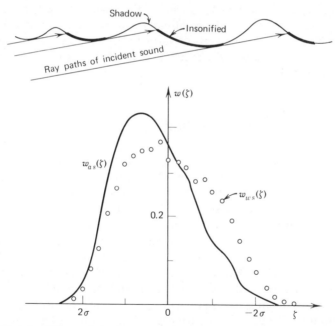

Figure 10.4.5. PDF for acoustic $w_a(\zeta)$ ——— and wave staff measurements $w_{ws}(\zeta)$ ($\circ\circ\circ\circ\circ$) of the windblown surface. ζ is positive downward. (*a*) $\sigma = 0.14$ cm and $\theta = 45°$. Near-Gaussian surface. Data from Spindel and Schultheiss (1972). (*b*) $\sigma = 0.41$ cm, and $\theta = 45°$ Skewed and peaked PDF. Data from Clay *et al.* (1973).

Figure 10.4.6. Effect of shadows on the PDF, here w_{ws} is wave staff measurement $\sigma = 0.41$ cm and w_{as} is shadowed PDF (Wagner, Eq. 40, 1967). Curve is from Clay *et al.* (1973), shadowing parameter $B = 0.3$.

$w_a(\zeta)$ is much more peaked than the $w_{ws}(\zeta)$. The examples show that the estimates of $w_a(\zeta)$ depend on the roughness of the surface. There are several hypotheses that can account for the differences between the acoustic estimates of the PDF and wave staff measurements. In Fig. 10.4.6, shadows in the geometrical optics approximation alter the apparent PDF for the acoustical signal because surfaces in the shadows do not contribute. The result of a calculation of the shadowed PDF from $w_{ws}(\zeta)$ is given in Fig. 10.4.6. Aside from the skew, $w_{as}(\zeta)$ has about the same width and height as $w_a(\zeta)$.

Another hypothesis requires corrections to the Kirchhoff method. De-Santo and Shisha (1974) show that higher order terms in the basic scattering calculation also contribute to the average reflection. At low wind speeds and small roughness, the surfaces in model tanks have the appearance of long sinusoidal corrugations at small roughness. At high wind speeds and large roughness, the wave profiles have sharp crests and rounded bottoms. The diffractions from these can contribute to the coherence. These hypotheses depend on the shape of the surface.

We believe that acoustical measurements of $\langle \mathcal{R} \rangle$ are a practical way to estimate the statistical roughness of surfaces. In the range of $k\sigma \cos \theta < 1$, Figs. 10.4.4a and b show that excellent estimates of σ can be made with the assumption that $w(\zeta)$ is approximately Gaussian.

10.5. MEAN SQUARED PRESSURE

A common method of processing sonar signals is to square (or take the absolute value of) the signals, then pass them through a low pass filter. Often signals are displayed on an intensity-modulated oscilloscope or graphic recorder. The purpose of the system is to record or display travel times of the arrivals. The low pass filter is chosen to pass the envelope of the initial ping. Figure 10.5.1 is a diagram of the system and waveforms.

Our knowledge of the frequency and phase of the signal is lost in the squaring process. The usual reasons given for using these systems are that the phase is unimportant and that the arrivals have random phase and add as the sums of squares. There are costs and benefits for this process.

The "information cost" of squaring the signal is loss of knowledge of its frequency and phase. However, the comparative "cost" of recording all the information is considerable. For example, assume that a 200 kHz sonar transmits a 10^{-3} s ping and the backscattered signal lasts 1 s. The bandwidth of the signal is about 1 kHz and is centered on 200 kHz (i.e., 199.5 to 200.5 kHz). The minimum sampling frequency is 401 kHz, and we

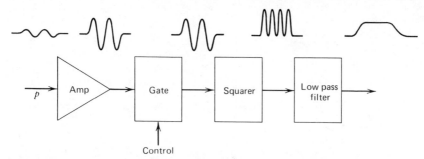

Figure 10.5.1. Mean square processing system and associated wave forms. The gate passes signal for a selected time interval and its main function is to eliminate extraneous arrivals. The functions are done by analogue or digital methods.

must record at least 4.01×10^5 data points per ping. By contrast, the output of the squaring system in Fig. 10.5.1 is in the band 0 to 1 kHz, and the minimum sampling frequency is 2 kHz. Now we record only 2×10^3 data points per ping. The benefits are reduction of the amount of stored information and, usually, simplification of the equipment.

10.5.1. Scattering Function

It is traditional to express the results of scattering experiments by means of a scattering function. We give an operational definition that is valid for sonar measurements. The geometry is presented in Fig. 10.5.2. The source is directional and insonifies an area A at an average distance R_1. The average incident sound signal at the bottom is p_1. Assuming that the scattered sound p is proportional to the incident sound

$$\int_{T_2} p^2 \, dt = \left(\int_{T_1} p_1^2 \, dt \right) \frac{A\mathscr{S}}{R_2^2} \tag{10.5.1}$$

where R_2 is the average distance to the receiver. The integration limits are T_1 for the duration of the incident signal and T_2 for the duration of all arrivals from A; $\mathscr{S}$ is the scattering function and depends on a combination of properties of the bottom and the measuring system.

We estimate the average p_1 in the insonified area by letting

$$p_1 = \frac{p_0 R_0}{R_1} \tag{10.5.2}$$

$$\int_{T_2} p^2 \, dt = \left(R_0^2 \int p_0^2 \, dt \right) \frac{A\mathscr{S}}{R_1^2 R_2^2} \tag{10.5.3}$$

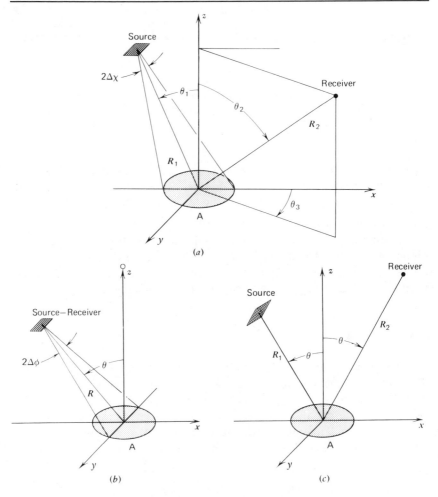

Figure 10.5.2. Scattering geometries. (*a*) Bistatic. (*b*) Monostatic. (*c*) Specular. Source is in the x–z plane; beam widths $2\Delta\chi$ and $2\Delta\phi$ shown separately in (*a*) and (*b*).

where p_0 is the average signal at R_0 in the beam of the transducer. The insonified area A is

$$A \equiv \frac{\pi\Delta\chi\Delta\phi R_1^2}{\cos\theta} \tag{10.5.4}$$

where $\Delta\chi$ and $\Delta\phi$ are the e^{-1} half beam widths of the directional source.
In the backscattering direction, $R_1 = R_2 = R$ and (10.5.3) reduces to

$$\int_{T_2} p^2\, dt = \left(R_0^2 \int_{T_1} p_0^2\, dt \right) \cdot \frac{A\mathscr{S}}{R^4} \tag{10.5.5}$$

The substitution of A in (10.5.3) and solution for $\mathscr{S}$ gives

$$\mathscr{S} = \frac{\displaystyle\int_{T_2} p^2\, dt}{R_0^2 \displaystyle\int_{T_1} p_0^2\, dt} \frac{R_2^2 \cos\theta_1}{\pi \Delta\chi \Delta\phi} \tag{10.5.6}$$

where $\mathscr{S}$ is a dimensionless function. These formulas, which are for straight ray paths, can be modified for curved ray paths by using ray tracing calculations to determine the angle of incidence at the bottom, A, and the spreading loss.

In this section we use a directional source to limit the insonified area because it is a CW theory. The dimensions of the insonified area are within the integral expression for $\mathscr{S}$ (A10.5.24). An alternative way to select an area for measurements is to use an impulsive source and to gate the received signal to process the sound scattered by a portion of the surface. These two ways of measuring the scattering of sound from a small area have different theoretical descriptions because of the following: the directional transducer insonifies a small area, and the directionality of the whole scattered sound field depends on the shape and dimensions of the insonified area. The impulsive source insonifies the area and the sound is scattered. How the sound signal is processed within the receiver does not affect the scattered sound field. For an impulsive source and gated receiver we suggest using the theory given in Section 10.2.

10.5.2. Experiment and Theory

Interpretations of $\mathscr{S}$ depend on theoretical or empirical connections between $\int p^2\, dt$ and the rough interface. Three procedures are employed. (1) We can use measurements to make a catalog of $\mathscr{S}$ as a function of interface type and sonar system. (2) Using Section 10.2, we can calculate the impulse response of likely models of the interface. (3) Following Eckart, we can square the Helmholtz–Kirchhoff integral for a CW source and evaluate $\int p^2\, dt$ and $\mathscr{S}$. Actually, we combine all three approaches by using the last two to guide the organization of the catalog. Experimental data show the kinds of feature needed in theoretical studies.

We use laboratory experiments in our comparisons of experiment and theory because we need tight controls on experiments to make crucial comparisons to theory. The theory in Sections A10.5 and A10.6 shows that $\mathscr{S}$ is dependent on the beam width. Two steps in this test are to measure $\mathscr{S}$ as a function of beam width and to compare measurements and theory. The latter is important because we want quantitative verification of the beam width effect.

The experiments were made in a laboratory tank (Fig. 10.5.3). The surface of the water was roughened by blowing air over it. The wave heights were measured by a pair of wave staffs. The spatial correlation

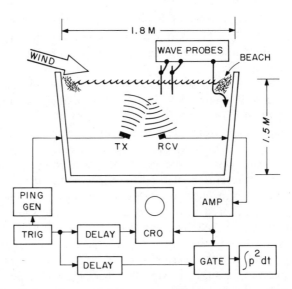

Figure 10.5.3. Laboratory sound scattering experiment: ping, 200 kHz carrier, 35 μs duration; pulse repetition frequency (Prf) 100 Hz. Cathode ray oscilloscope was intensity-modulated and photographed on a slowly moving film (similar to a graphic recorder). The surface was roughened by blowing wind over it, and heights were measured by wave probes. Multiple probes used for spatial correlations. (Clay and Sandness, 1971.)

function was measured by averaging the products of the wave heights in an analogue computer for each spatial lag. Acoustic scattering measurements were made over a range of 30 to 100 cm depth at vertical incidence. The received signals were gated to eliminate the direct signal and scattered signals from the walls of the tank. The set of records in Fig. 10.5.4 were made to show the fluctuations of the signals as the windblown waves moved over the insonified area. With increasing roughness, the signal lasts longer. The maximum duration of the backscattered sound corresponds to the travel time to the limits of the insonified area.

The signals were processed by an absolute value and integration circuit. Integral p^2 was approximated by squaring the average measurement of

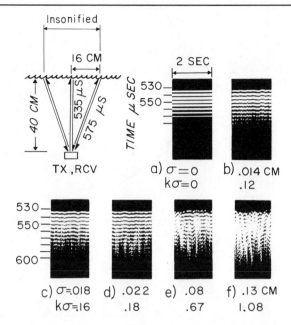

Figure 10.5.4. Signals reflected at a windblown surface. The display is intensity-modulated; white area are positive phases of signals. Signal frequency is 200 kHz, and ping length is 35 μs; σ is rms roughness. (Clay and Sandness, 1971.)

integral $|p|$. The approximation was tested in a few runs, and the difference was insignificant. The biggest problem in the measurements was a slowly drifting calibration that was attributed to the growth of tiny bubbles on the transducers.

Measurements of $\mathcal{S}$ are given in Fig. 10.5.5. Each data point is the average of about 8000 transmissions. The spatial correlation functions of the surface are in Fig. 10.5.6. The basis for the theoretical calculations requires explanation.

In (A10.5.24) we derive an expression for the scattering function $\mathcal{S}$

$$\mathcal{S} = \frac{B_1}{\pi} \int\limits_{-\infty}^{\infty}\!\!\int \mathcal{G} W \, d\xi \, d\eta \tag{10.5.7}$$

where B_1 is a constant and $\mathcal{G}$ is a function of the sonar system and geometry (A10.5.21). The scattering function is very dependent on the beam widths of the sonar. For $\theta_1 = \theta_2$, $\mathcal{G}$ has the form of an error function $\exp(-x^2)$ and tends to zero outside the insonified area. The W is the characteristic function of the bivariate probability density of the

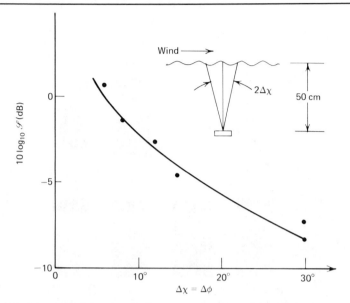

Figure 10.5.5. Dependence of $\mathscr{S}$ on beam width. (*a*) Solid data points are experimental. Data taken at vertical incidence. Measured roughness $\sigma = 0.14$ cm. Frequency 200 kHz, $k\sigma = 1.25$, and $\Delta\phi = \Delta\chi$. The spatial correlation function is in Fig. 10.5.7. (*b*) Solid line is a theoretical line, (A10.6.19): $\sigma = 0.15$ cm, $a_1 = 0.33$ cm^{-1}, $b_1 = 0.08$ cm^{-1}. Data from Clay and Sandness (1971).

surface (A10.5.17). For a normal or Gaussian surface, we write

$$\mathscr{W} = \exp\left[-4\gamma^2\sigma^2(1-C)\right] \tag{10.5.8}$$

$$\gamma \equiv \frac{k(\cos\theta_1 + \cos\theta_2)}{2} \tag{10.5.9}$$

where C is the spatial correlation function. We use (10.5.8) and the measurements of C to calculate $\mathscr{W}$, Fig. 10.5.6*b*. The values of C for ξ greater than 2 cm are unimportant because $\mathscr{W}$ is so small. The value of $\mathscr{S}$ depends on the beam width (Fig. 10.5.6*d*). We use a linear approximation to C. The corresponding components of $\mathscr{W}$ are in Fig. 10.5.6*c*. The theoretical curve appears in Fig. 10.5.5. The sphericity of the incident wave causes the dependence of $\mathscr{S}$ on beam width.

10.5.3. Mean Square Reflection Coefficient and Bottom Loss

Marine acousticians often treat $\int p^2 \, dt$ measurements in the specular direction as if the sound were *reflected* at the bottom. The process is

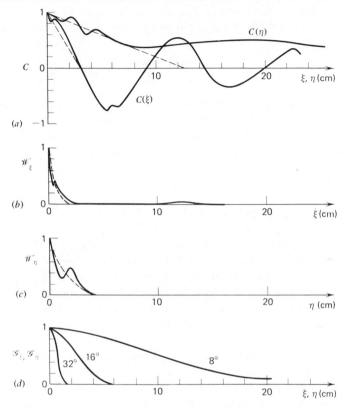

Figure 10.5.6. Functions in the integrand of $\mathcal{S}$. (a) Measured spatial correlation function for downwind $C(\xi)$ and crosswind $C(\eta)$. Dashed lines are linear approximations. (Clay and Sandness, 1971.) (b) Solid line is calculations of ξ component $\mathcal{W}_\xi$ using bivariate Gaussian PDF. Dashed line is the estimate of $\mathcal{W}_\xi$ for linear approximation. (c) The η component $\mathcal{W}_\eta$; notation follows (b). (d) Sonar function $\mathcal{G} = \mathcal{G}_\xi \mathcal{G}_\eta$ in specular direction, $\theta_1 = \theta_2$ or $\alpha = \beta = 0$. Since transducer is circular, $\mathcal{G}_\xi = \mathcal{G}_\eta$; half beam widths $\Delta\chi = \Delta\phi$ are labeled.

called specular scatter because it describes scattering measurements in the specular (mirror) direction ($\theta_1 = \theta_2$). Using the image reflection equation as a model, the mean square reflection coefficient and "bottom loss" are

$$\langle \mathcal{R}^2 \rangle \equiv \frac{\left(\int p^2 \, dt \right)(R_1 + R_2)^2}{R_0^2 D^2 \int p_0^2 \, dt} \qquad (10.5.10)$$

$$\text{bottom loss} \equiv -10 \log_{10} \langle \mathcal{R}^2 \rangle$$

where D is source directivity and p_0 is the axial signal at R_0. Comparison of (10.5.3) and (10.5.10) gives, for measurements along the axial direction $(D = 1)$,

$$\langle \mathcal{R}^2 \rangle = \frac{A(R_1 + R_2)^2 \mathcal{S}}{R_1^2 R_2^2} \qquad (10.5.11)$$

Here $\mathcal{S}$ has a complicated dependence on beam width geometry and the spatial correlation function. Our numerical studies of $\mathcal{S}$ [see (A10.6.21)] show that values of $\mathcal{S}$ tend to increase at small beam widths. Numerical evaluations of $\langle \mathcal{R}^2 \rangle$ reveal that $\langle \mathcal{R}^2 \rangle$ tends to decrease at small beam widths.

10.6. REMOTE SENSING OF THE SEA FLOOR

Sonars are one part of a remote sensing system. The rest of the system includes the data processing, the display, and ourselves. We are part of the system because we look at the output, think about it, and make our guess ("interpretation"). Our task is easy when the sonars have very high resolution and the displays have very little distortion. The task is difficult, even impossible, when the echoes merge together and none of the features are resolved.

Bottom-scanning sonars measure the amount of backscattered sound from each area increment ΔA. The simplest version consists of a one-channel sonar. As Fig. 10.6.1a indicates, the sonar transmits a short ping and insonifies a wedge of the bottom. The backscattered sound and the range are a function of the travel time. Over a rough bottom the signal decays as in Fig. 10.6.1b. The small hill at 0.3 s scatters a bit more strongly and has a shadow (no return) behind it. This figure also illustrates a basic design problem in scanning sonar. The signal changes by about 80 dB. This change is compensated by including a time variable gain (TVG) in the receiver. The quality of the output is very sensitive to the adjustment of the TVG.

Since the single-channel sonar scans a single wedge, it is turned or moved to scan the bottom. Two common methods are the side scan and the polar scan. For the side scan mode, the acoustic beam is directed perpendicular to the direction of travel. The ship moves at constant velocity, and each sonar ping insonifies a slightly different wedge of bottom. The backscattered sound is recorded on the same kind of graphic recorder as is used for echo sounders. The "depth" becomes slant range from ship. The display or "sonograph" is a rectangular coordinate map of slant range and distance along the track.

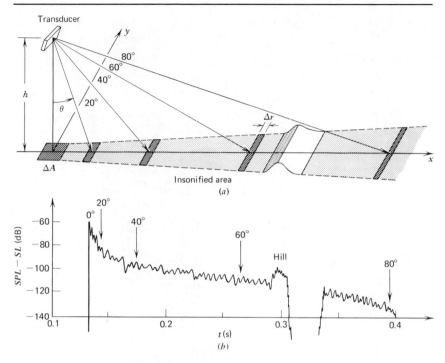

Figure 10.6.1. Geometry for bottom-scanning and backscattered sound. (a) Bottom-scanning sonar. Sonar transducer insonifies the bottom with a short ping: Δr is effective length of unresolved area ΔA and is approximately $\Delta r \approx cT_p/(\sin \theta)$ for $\theta > 45°$, where T_p is the duration of the ping. Width of ΔA is given by half beam width $\Delta \phi$ and the slant range. (b) Relative backscattered sound level $SPL-SL$ decibels. Sketch is made for a transducer 100 m above the bottom. Near side of hill has a larger scattering function, and far side is in a shadow.

An example of a side-scanning record appears in Fig. 10.6.2. The sonar was towed at constant distance from the shore along the course marked on the chart. The shore and the edge of the channel where the depth increases from 10 to 30 ft are evident on the sonograph. There is a small amount of distortion, and the sonograph needs to be stretched to bring the features into exact correspondence with the chart.

Three problems in making side-scanning sonar surveys are: keeping the ship moving along a known course at constant velocity, adjusting the paper speed of the graphic recorder to ensure that distance along the track has the same scale as maximum range to each side, and keeping the sonar transducer at a "good" height above the bottom (about one-fifth the maximum range).

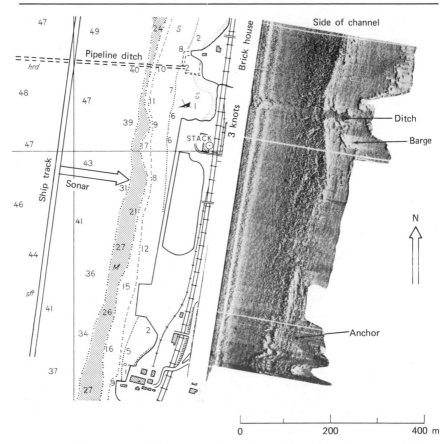

Figure 10.6.2. Comparison of a sonograph and corresponding chart of the Hudson River near Dobbs Ferry, New York. The side-scanning sonar towed about 250 m offshore and along the ship's track shown on chart. Dark means a large amount of backscattered sound and light means a small amount of backscattered sound. Chart is shaded between 20 and 10 ft contours to show side of Channel. Other features are a ditch in the bottom, a sunken barge, and anchors for small boats in a marina. For perfect match, the sonograph needs to be stretched in the north–south direction.

Polar scans are made from a stationary (or very slowly moving platform). The transducer is mounted to permit mechanical rotation to scan the area around it (Fig. 10.6.3a). The transducer is suspended on rods and is rotated from above. The data are recorded on a standard graphic recorder. Each azimuth is recorded for several seconds and the transducer is moved to the next azimuth. After return to the laboratory, the graphic

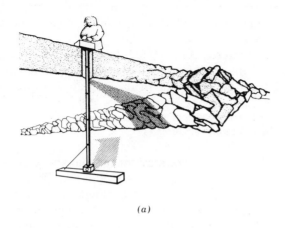

(a)

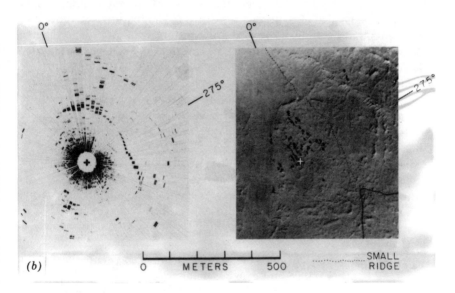

(b)

0   METERS   500   SMALL
                       RIDGE

Figure 10.6.3. Polar scan of the underside of sea ice. (*a*) Sonar transducer is suspended on rods and is turned to make the polar scan. Pressure ridges give a large amount of backscattered sound and show as dark. The ice between pressure ridges scatters very little sound back to the transducer. The sonar was 23.5 m below the ice. (*b*) Sonar map made at the 1972 AIDJEX site. Aerial photograph courtesy of Austin Kovacs, U.S. Army Cold Regions Research and Engineering Laboratory, Hanover, N.H.

records are photographed, enlarged, cut into wedges, and mounted to make the polar map of scatterers.

The scattering features (Fig. 10.6.3*b*) are downward projections of pressure ridges of the pack ice. An extremely small amount of sound is scattered back from the underside of the ice between pressure ridges.

10.6.1. Small Scale Bottom Features

Most photographs of the sea floor are taken from above at distances of 3 to 5 m. The area is about 2×3 m, or the size of a bedsheet. Features that range from a few centimeters to a few meters are recognizable in the photographs. For comparison, a sonar having a half beam width of 0.5° insonifies about 2 m width at 100 m range. The range increment is about 0.75 m for a 10^{-3} s ping. The sonar and bottom photographs complement each other in the size of features they display. The features we can see in bottom photographs are not resolved by bottom scanning sonars having medium range and directionality (ca 100 m and 0.5° half beam width).

There is a close relationship between the kind of feature and the bottom environment. For example, currents at the water-sediment interface can cause ripple features resembling those in Fig. 10.6.4. Sometimes small ripples such as these appear on larger dunelike features. The latter may have dimensions of tens of meters. Strong currents may wash the sediment away, and the bottom then appears to be covered with gravel and small boulders. Quite often the small boulders are manganese nodules. The currents are very small in the deep abyssal plains, where the bottom is covered by a gentle rain of sediments. Animals burrow in and make tracks on the otherwise flat bottom. Heezen and Hollister (1971) have used bottom photographs to describe the sea floor and the processes that shape it.

The shallow water around the edges of the continents has reefs, submerged beaches, cliffs, and canyons, similar to features we find above the water line. Unfortunately (for photographers) in many of the coastal areas, the water is much less transparent than it is in the deep sea, and bottom photography is very limited. In these areas sonar appears to be the more effective method of studying the morphology of the bottom.

10.6.2. Interpretation

The amount of sound scattered back by each of the unresolved areas depends on the composition of the bottom and the local roughness of the

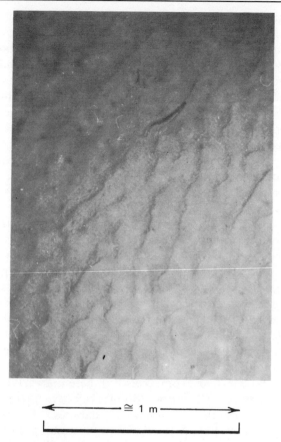

≅ 1 m

Figure 10.6.4. Bottom photograph, at 30°14′N and 78°07′W on the Blake Plateau. Crest-to-crest distances of the ripples were about 17 cm. Photograph courtesy of B. G. Hurdle, Naval Research Laboratory, Washington D.C.

interface. Usually this is insufficient to determine what is there because many combinations of shapes and compositions of the bottom can cause the scattering function to be small or large. We need more information, such as that provided by samples and photographs of the bottom. The plausibility of an interpretation can be tested by using scattering theory to calculate theoretical backscattering functions for the assumed material and shape of bottom.

We have made many numerical evaluations of the scattering function $\mathscr{S}_b$ (A10.6.21) for assumed correlation functions and profiles of the bottom. We have found that $\mathscr{S}_b$ is quite different for two classes of surface: gently curving surfaces, and ridged surfaces that have abrupt changes of slope.

Gently curving surfaces have spatial correlation functions that are described by $C = 1 - a_2\xi^2 - b_2\eta^2$ for small values of ξ and η. As Fig. 10.6.5a demonstrates, the scattering coefficients are large near vertical incidence and decrease to very small values for θ_1 greater than 45°. Since the angle of incidence is greater than 45° over most of the bottom, the backscattered sound is extremely small over a gently curving bottom.

A very, very small amount of backscattered sound indicates some combination of the following: (a) the bottom is smooth relative to the wavelength of sound or $k\sigma < 1$, (b) the surface has gentle curves even though $k\sigma$ is larger than 1, (c) the reflection coefficient is small [we can expect this when the sound speed in the sediment is a little less than the sound speed in water], (d) the area is in the shadow of a large feature.

Angular features can be modeled by intersecting planes. The spatial correlation function for these features is approximately $C \simeq 1 - a_1|\xi| - b_1|\eta|$, and the sound is scattered over a wide range of incident angles (Fig. 10.6.5b). The two curves are for two orientations of the features. The upper curve is for the simulation of sand ripples having their ridge crests perpendicular to the sonar beam. The lower curve is for ridges parallel to the sonar beam.

A large backscattered sound signal indicates a combination of some of the following options: (a) the bottom has irregular objects and features such as rocks, boulders, and coral reefs, (b) there is gravel on the bottom; (c) there are ripples on the bottom. Ripples can have a large backscattering coefficient because they often have angular crests. For 100 kHz sonar signals, σ as small as 1 cm is large enough to strongly scatter the sound (Fig. 10.6.5b). A fourth possibility is that an upward slope of the bottom decreases the angle of incidence and increases the amount of backscattered sound. This effect delineates the edge of the channel in Fig. 10.6.2.

Since features of so many different kinds can cause the backscattering coefficient to be large, additional information is needed. "Ground truth" in the form of bottom photographs and samples of the bottom are helpful. With this information an interpretation can be made for a small area and then the side-scanning sonar can be used to extend the interpretations over much larger areas.

10.6.3. Sonar Survey

Often a bottom-scanning sonar survey enables us to describe the characteristics of features on the bottom much better than is possible by using a

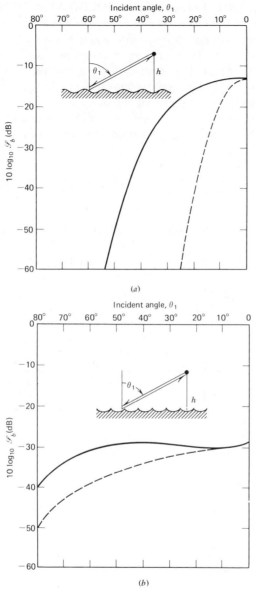

Figure 10.6.5. Scattering functions, calculated using (A10.6.21). Sonar transducer is at fixed height, $h = 50$ m, above the bottom and is tilted for the different angles of incidence. Sonar parameters: half beam widths $\Delta\phi = \Delta\chi = 0.5°$, frequency $= 100$ kHz. We let $c_1 = c_2 = 1500$ m/s and $\rho_2 = 2\rho_1$ for the bottom material. The rms roughness is $\sigma = 0.01$ m for both surfaces. (a) Gently curving surface: $C = 1 - a_2\xi^2 - b_2\eta^2$ solid line, wave crests perpendicular to sonar beam, $a_2 = 400$ m^{-2} and $b_2 = 49$ m^{-2}; dashed line, wave crests parallel to sonar beam, $a_2 = 49$ m^{-2} and $b_2 = 400$ m^{-2}; (b) Ridged surface: $C = 1 - a_1|\xi| - b_1|\eta|$, solid line, ridge crests perpendicular to sonar beam, $a_1 = 20$ m^{-1} and $b_1 = 7$ m^{-1}; dashed line, ridge crests parallel to sonar beam, $a_1 = 7$ m^{-1} and $b_1 = 20$ m^{-1}.

conventional echo sounding survey. The usual reason cited is that echo sounding surveys are made on a grid of sounding lines that are spaced hundreds of meters apart. If the features have smaller dimensions, they are sampled too sparsely for reconstruction. A scanning sonar survey can fill in the locations and shapes of features between sounding lines. Sometimes there are surprises when seemingly unrelated mounds and valleys on the echo sounding record are mapped by using a side-scanning

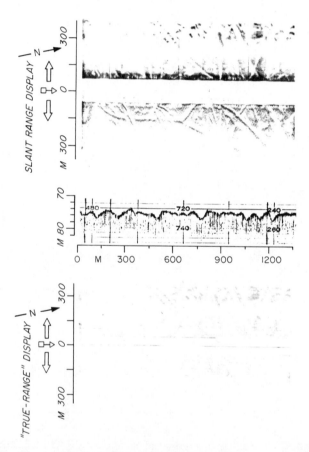

Figure 10.6.6. Sonographs and echo sounding record from Lake Superior. Dual side-scan sonar displays sonographs from both sides of the ship. Upper sonograph is slant range display. A short section of the echo sounder record is in the middle. "True-range" display appears on the bottom. Far sides of grooves scatter more sound back to receivers and are dark on sonograph. Bottoms of grooves scatter little sound back and may be shadowed. (Berkson and Clay, 1973.)

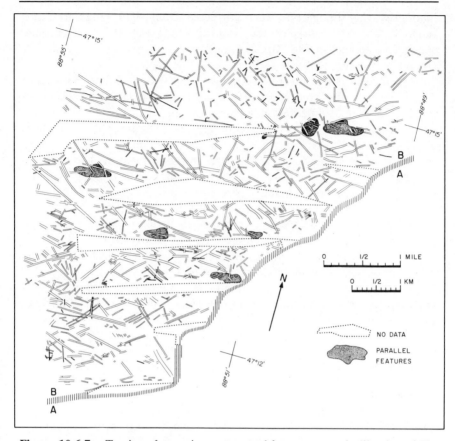

Figure 10.6.7.   Tracing of a mosiac constructed from sonographs like that of Fig.
10.6.6, indicating grooves on a region of the bottom of Lake Superior. Sonographs map a ribbon of the bottom to about 450 m to each side of the ship's track.
No-data regions are where sonographs for adjacent tracks do not touch or
overlap. The water depths north of the hashed line (region B) are greater than
50 m. Region A, not shown, becomes shallow toward the south-east. The grooves
are not found in the A region. (Berkson and Clay, 1973.)

sonar. A study of the microphysiography of Lake Superior serves as an
example.

A dual 30 kHz sonar was used to survey the right and left sides of the
ship simultaneously. Figure 5.3.4 depicts one of the transducers. For the
initial recordings, the paper speed of the graphic recorder was proportional to the ship's speed and was chosen to have the same linear scale
along the ship's track as the maximum slant range. After returning to the

laboratory the sonographs were corrected from slant range to "true-range." A short section of the original sonograph and the corrected sonograph appear in Fig. 10.6.6. The correction to true-range straightened the crossing features.

Vertical echo sounding records show depressions of 2 to 5 m depth. The bottom in the area consists of glacial till and lacustrine clay. The present rate of sedimentation is small. Using Fig. 10.6.6 as a typical sonograph for the area, we interpret the linear features to be grooves. The far wall of the groove has a smaller angle of incidence, has a larger scattering function, and appears as dark on the sonograph. The near wall and deep part of the groove may be in a shadow and is light on the sonograph. The echo sounding record shows a depth profile. A mosaic is constructed by laying the sonographs along the ship's track. A tracing of the mosaic (Fig. 10.6.7) shows that the small depressions on the echo sounding record are part of a large system of grooves.

The grooves are attributed to the plowing action of icebergs as they drifted about and their points dragged on the bottom (Berkson and Clay, 1973). Presumably the icebergs could have come from the icesheet on the northern shore of Lake Superior in late glacial time, about 10,000 years ago.

PROBLEMS

Problems 10.2

10.2.1. Calculate the diffraction waves when the surface *is a sector of a disk*. The source and receiver are at the same position on the z axis. The center of the whole disk is $(0, 0, 0)$, the sector of the disk is bounded by r_1, r_2, and the radial lines are along θ_1 and θ_2. *Answer:*

$$U(Q) \sim \frac{\Re(\theta_2 - \theta_1)z}{4\pi} \left[\frac{\exp(-2i\omega R_1/c)}{R_1^2} - \frac{\exp(-2i\omega R_2/c)}{R_2^2} \right]$$

$$R_1^2 = r_1^2 + z^2 \qquad R_2^2 = r_2^2 + z^2$$

10.2.2. We assume the source emission is $g(t)$ where

$$g(t) = 0 \qquad \text{for} \quad t < 0$$
$$g(t) = g(t) \qquad \text{for} \quad t \geq 0$$

and the Fourier integral of $g(t)$ is $G(\omega)$. Calculate $p(t)$ of the waves diffracted at the sector of a disk in Problem 10.2.1.

Answer:

$$p(Q) \sim \frac{\Re(\theta_2 - \theta_1)z}{4\pi} \left[\frac{g_1(t)}{R_1^2} - \frac{g_2(t)}{R_2^2} \right]$$

where

$$g_n(t) = 0 \quad \text{for} \quad t < \frac{2R_n}{c}$$

$$g_n(t) = g\left(t - \frac{2R_n}{c} \right) \quad \text{for} \quad t \geq \frac{2R_n}{c}$$

10.2.3. Calculate the axial reflection from a circular disk having radius r_2 (let $r_1 = 0$). Compare the result to the image procedure.
Answer:

$$p(Q) \sim \frac{\Re}{2z} \left[g_1(t) - \frac{g_2(t)z^2}{R_2^2} \right]$$

$$g_1(t) = g\left(t - \frac{2z}{c} \right)$$

10.2.4. (For advanced students.) Sometimes measurements of the sea floor are made by using a source at the surface and a receiving hydrophone that lies beneath the source and very close to the bottom. To a good approximation the incident wave front is nearly plane in the local region of the receiver and sea floor. Using the approximation that the incident signal is a vertically traveling plane wave, calculate the reflected signal and boundary diffraction wave at the receiver. The segment of the plane is in the x–y plane.

Problems 10.3

10.3.1. A 100 kHz sonar insonifies a fish. The transmitter and receiver are side by side. The fish is swimming toward the sonar at 12 cm/s. Calculate the frequency shift of the scattered signal.

10.3.2. A "deep water gravity" wave is measured by a 150 kHz sonar. The source and receiver are side by side. As shown in Fig. 10.3.1, the crest of the wave is moving along the x direction and the sonar is in the x–z plane. The angle of incidence of the sonar is 60°. The frequency shift is -1370 Hz.

(*a*) What is the velocity of the wave along the direction of the sonar?

(*b*) If the direction of the wave is along the direction of the sonar, what is the frequency of the wave?

(*c*) If the direction of the surface wave is 30° relative to the $x-z$ plane and the sonar is in the $x-z$ plane?

10.3.3. Show that the signal frequency f_s is scattered into preferred directions when the surface is corrugated. *Answer:*

$$(f_s/c)(\sin \theta_1 - \sin \theta_2) = n/\Lambda,$$

where Λ is the wavelength of the corrugation and n is the order and an integer.

10.3.4. A 30 kHz backscatter sonar is incident at θ on a long train of surface waves having 10 cm wavelength. Calculate the directions for maximum scatter. Can this method be used for spectral analysis of the waves? See Liebermann, *J. Acoust. Soc. Am.*, **35**, 932 (1963).

Problems 10.4

10.4.1. Calculate the coherent transmission coefficients for a rough interface. The sound speeds are c for the upper medium and c' for the lower medium. Assume the interface has a Gaussian PDF. *Answer:* For p'_s being the signal for the Snell's law transmission path at a plane interface, the mean transmitted signal $\langle p' \rangle$ is

$$\langle p' \rangle = p'_s \exp\left(-2\gamma'^2 \sigma^2\right)$$

$$\gamma' \equiv k\left(\cos \theta - \frac{c}{c'}\cos \theta'\right)$$

$$\frac{\sin \theta}{c} = \frac{\sin \theta'}{c'}$$

$$\langle \mathcal{T} \rangle = \mathcal{T} \exp\left(-2\gamma'^2 \sigma^2\right)$$

where
$$\mathcal{T} = 2\rho'c'\frac{\cos \theta}{\rho'c'\cos \theta + \rho c \cos \theta'}$$

10.4.2. A 30 kHz signal is reflected from the sea surface. The surface has a Gaussian PDF and $\sigma = 1$ cm. Over what range of angle is $|\langle \mathcal{R} \rangle|$ greater than 0.7?

10.4.3. Show how the correlation of signals observed at a pair of widely spaced receivers can be used to estimate the roughness of the sea floor. *Answer:* For signals p_a and p_b at the receivers a and b, $\langle \int p_a p_b \, dt \rangle \sim \langle \mathcal{R}_a \rangle \langle \mathcal{R}_b \rangle$.

10.4.4. Show that the PDF of a sinusoidal surface $\zeta = a \sin Kx$, is

$$w(\zeta) = \pi^{-1}(a^2 - \zeta^2)^{-1/2} \quad \text{for} \quad \zeta < a$$
$$= 0 \quad \text{for} \quad \zeta > a$$

10.4.5. Calculate $\langle \mathcal{R} \rangle$ for the reflection at a sinusoidal surface, $\zeta = a \cos Kx$. *Answer:* $\langle \mathcal{R} \rangle / \mathcal{R} = J_0(2ka \cos \theta)$; $J_0(x)$ is a Bessel function. *Hint.* Look up integral expressions of $J_0(x)$.

Problems 10.5

10.5.1. For 100 kHz scattering experiments, determine the maximum size object for which the plane wave approximation is valid. Assume a point source and source object distances of 1 m, 2 m, and R.

10.5.2. Design an experiment in which the plane wave approximation is satisfied over a large area. *Hints.* Melton and Horton, *J. Acoust. Soc. Am.* **47,** 290–298 (1970).

10.5.3. Use the data in Fig. 10.5.5 to calculate the dependence of $\langle \mathcal{R}^2 \rangle$ on beam width.

10.5.4. The reverberation experiment is made by firing a shot and receiving the backscattered signal at nearly the same position. This is similar to Fig. 10.5.3 except that both receiver and source are omnidirectional. The backscattered integral p^2 is measured as a function of time. Use the geometry and definition of $\mathcal{S}$ to write an expression for $\mathcal{S}$ as a function of incident angle θ and integral p^2.

Problem 10.6

10.6.1. A challenging exercise in the interpretation of data. Flower and Hurdle [*J. Acoust. Soc. Am.* **51,** 1109–1111 (1972)] give monostatic scattering data from the ocean bottom. The problem is to calculate a theoretical $\mathcal{S}$ and determine σ, a_1, a_2, b_1, and b_2. Also estimate c' and ρ_1 for the sand. They used a 4° half beam width sonar to measure backscattered sound at 19.5 kHz. The data were taken over the Blake Plateau at about 78°W, 30°N and 900 m depth. The bottom description is sand with patches of ripples. Figure 10.6.4 is a photograph of the bottom. The orientation of patches is random. See also Heezen and Hollister, *The Face of the Deep*, Figs. 9.13 and 9.80. The average values of $\mathscr{S}$ are as follows: incident angle $\theta = 0°$, -3 dB; $10°$, -15 dB; $20°$, -18 dB; $30°$, -24 dB; $40°$, -27 dB, $-50°$, -29 dB; and $60°$, -33 dB. We were able to fit the data from $0°$ to $50°$ by regarding σ as a fitting parameter. We do not give *an* answer because it is part of the exercise to find the range of parameters that yield an $\mathscr{S}$ that fits.

SUGGESTED READING

Appendix A10

C. Eckart, "The scattering of sound from the sea surface," *J. Acoust. Soc. Am.* **25**, 566–570 (1953). This is a fundamental paper.

L. Fortuin, "Survey of the literature on reflection and scattering of sound waves at the sea surface," *J. Acoust. Soc. Am.* **47**, 1209–1228 (1970). A review of the development of the theory, including the scientific argument about the accuracy of the Rayleigh–Marsh method and the literature of the USSR.

Hampton, L., *The Physics of Sound in Marine Sediments*, Plenum Press, New York, 1974.

B. C. Heezen and Charles D. Hollister, *The Face of the Deep*, Oxford University Press, New York, 1971. The bottom photographs and illustrations are superb. The authors describe what is on the bottom and processes that shape the bottom, the materials, and the features. Samples of the sea floor from all over the Earth are described.

C. W. Horton, Sr., "A review of reverberation, scattering, and echo structure," *J. Acoust. Soc. Am.* **51**, 1049–1061 (1972). A review paper that includes the introduction of spherical waves in the theory and scattering of waves from objects.

E. O. La Case and P. Tamarkin, "Underwater sound reflection from a corrugated surface," *J. Appl. Phys.* **27**, 138–148 (1956), Brown University Physics Dept. A comparison of the theories of Rayleigh, Eckart, and Brekhovskikh and laboratory experiments that includes an account of Brekhovskikh's paper (in Russian), *Dokl. Akad. Nauk SSSR*, **79** 585–588 (1951). This work demonstrated the usefulness of scattering theories.

P. F. Lonsdale, R. C. Tyce, and F. N. Spiess, "Near-bottom acoustic observations of abyssal topography and reflectivity," In Hampton, Ed. (1974), pp. 293–317. Observations are made by way of a deep-towed fish having side-scanning sonar (110 kHz), 4 kHz seismic profiling system, cameras, snap-shot TV camera, and magnetometer.

H. Medwin, R. A. Helbig and J. D. Hagyidr, "Spectral characteristics of sound transmission through the rough sea surface,"*J. Acoust. Soc. Am.* **54**, 99-109 (1973). A comparison showing agreement between theory and both laboratory and ocean experiments.

J. M. Proud, R. T. Beyer, and P. Tamarkin, "Reflection of sound from randomly rough surfaces," *J. Appl. Phys.* **31,** 543–552 (1960). Experimental verification of Eckart's theory and evidence that the acoustic measurements can be used to estimate the surface correlation function.

E. Reimnitz, P. W. Barnes, and T. R. Alpha, "Bottom features and processes related to drifting ice," Miscellaneous Field Studies Map MF 532, U.S. Geological Survey (1973); price $0.50. Sonographs, underwater photographs, and sketches of how drifting ice scours the bottom.

J. E. Sanders, K. O. Emery, and E. Uchupi, "Microtopography of five small areas of the continental shelf by side-scanning sonar," *Geol. Soc. Am. Bull.* **80,** 561–572 (1969). A comparison of side-scanning sonar data, geology, and bottom observations from the DSRV *Alvin* are given.

I. Tolstoy and C. S. Clay, *Ocean Acoustics,* McGraw-Hill, New York, 1966, Ch. 6. Eckart's plane wave theory is used as the basis of the derivation of $\langle \Re \rangle$ for the reflection at a randomly rough layered sea floor. The reflection at a rough interface is combined with normal mode expressions for long range transmissions in a fluctuating ocean.

I. Tolstoy, *Wave Propagation,* McGraw-Hill, New York, 1973. An exact, normal coordinate wave theory solution for an impulsive source near a rigid wedge.

FORMULAS FROM MATHEMATICS

Additional formulas are given in the following sources:

Mathematical tables from *Handbook of Chemistry and Physics*, Chemical Rubber Co., Cleveland, Ohio.

L. B. W. Jolley, *Summation of Series*, Dover, New York, 1961.

M. Abramowitz and I. A. Stegun, *Handbook of Mathematical Functions*, U.S. National Bureau of Standards. Applied Mathematical Series, 55. Government Printing Office, Washington, D.C.; price, $6.50. In specialized areas, we refer to specific formulas in Abramowitz and Stegun.

A1.1. ALGEBRA

$$S_N = \sum_0^{N-1} r^n = \frac{1-r^N}{1-r} \qquad (A1.1.1)$$

$$S_\infty = \sum_0^\infty r^n = \frac{1}{1-r} \qquad (A1.1.2)$$

$$(a+b)^n = a^n + na^{(n-1)}b + \frac{n(n-1)}{2!}a^{(n-2)}b^2 + \cdots \qquad (A1.1.3)$$

$$(1+x)^{-1} = 1 - x + x^2 + \cdots \qquad (A1.1.4)$$

$$(1+x)^{1/2} = 1 + \frac{x}{2} - \frac{x^2}{8} + \frac{x^3}{16} - \cdots \qquad (A1.1.5)$$

A1.2. TRIGONOMETRY

Polar and rectangular coordinates. When

$$x = r \cos \phi \qquad \text{and} \qquad y = r \sin \phi \tag{A1.2.1}$$

$$\frac{y}{x} = \tan \phi \qquad \text{and} \qquad \cos^2 \phi + \sin^2 \phi = 1 \tag{A1.2.2}$$

$$\sin \phi = -\sin(-\phi) \tag{A1.2.3}$$

$$\cos \phi = \cos(-\phi) \tag{A1.2.4}$$

$$\sin(\theta + \phi) = \sin \theta \cos \phi + \sin \phi \cos \theta \tag{A1.2.5}$$

$$\cos(\theta + \phi) = \cos \theta \cos \phi - \sin \theta \sin \phi \tag{A1.2.6}$$

$$a \sin(\omega t + \theta) + b \sin(\omega t + \phi) = c \sin(\omega t + \delta) \tag{A1.2.7}$$

where $c^2 = a^2 + b^2 + 2ab \cos(\theta - \phi)$

$$\tan \delta = \frac{a \sin \theta + b \sin \phi}{a \cos \theta + b \cos \phi}$$

$$\sin x = x - \frac{x^3}{3!} + \frac{x^5}{5!} - \frac{x^7}{7!} + \cdots \tag{A1.2.8}$$

$$\cos x = 1 - \frac{x^2}{2!} + \frac{x^4}{4!} - \frac{x^6}{6!} + \cdots \tag{A1.2.9}$$

A1.3. EXPONENTIAL AND LOGARITHMIC FUNCTIONS

$$e = 2.7183$$

when

$$y = e^x = \exp(x) = 1 + x + \frac{x^2}{2!} + \frac{x^3}{3!} + \cdots \tag{A1.3.1}$$

then

$$\ln y = \log_e y = x \tag{A1.3.2}$$

when

$$b = 10^a \tag{A1.3.3}$$

then

$$\log_{10} b = a \tag{A1.3.4}$$

$$\log_{10}(e^x) = x \log_{10} e \tag{A1.3.5}$$

$$\log_{10} e = 0.43429$$

$$\log_e 10 = 2.30258$$

$$\cos \phi = \frac{e^{i\phi} + e^{-i\phi}}{2} \qquad i = \sqrt{-1} \tag{A1.3.6}$$

$$\sin \phi = \frac{e^{i\phi} - e^{-i\phi}}{2i} \tag{A1.3.7}$$

then

$$e^{i\phi} = \cos \phi + i \sin \phi \tag{A1.3.8}$$

$$\log_e (re^{i\phi}) = \log_e r + i\phi \tag{A1.3.9}$$

$$\frac{1}{N} \sum_{m=0}^{N-1} \exp\left(\frac{i2\pi nm}{N}\right) = 0 \qquad \text{for} \quad n = \text{integer} \tag{A1.3.10}$$

$$= 1 \qquad \text{for} \quad n = 0$$

A1.4. IMPLICIT DIFFERENTIATION

$$\frac{d}{dt} f(x, y, z) = \frac{\partial f(x, y, z)}{\partial x}\frac{\partial x}{\partial t} + \frac{\partial f(x, y, t)}{\partial y}\frac{\partial y}{\partial t} + \frac{\partial f(x, y, z)}{\partial z}\frac{\partial z}{\partial t} \tag{A1.4.1}$$

$$\frac{\partial f}{\partial q} = \left(\frac{\partial f}{\partial x}\right)\left(\frac{\partial x}{\partial q}\right) \tag{A1.4.2}$$

A1.5. INTEGRALS

$$\int e^{ax}\,dx = \frac{1}{a}e^{ax} \tag{A1.5.1}$$

$$\int \frac{dx}{x} = \log_e x \tag{A1.5.2}$$

$$\int \frac{x\,dx}{(c + ax^2)^{1/2}} = \frac{(c + ax^2)^{1/2}}{a} \tag{A1.5.3}$$

$$\int \frac{dx}{x(c + ax^2)^{1/2}} = -\frac{1}{\sqrt{c}}\log_e\left[\frac{(c + ax^2)^{1/2} + \sqrt{c}}{x}\right], \qquad c > 0 \tag{A1.5.4}$$

$$\int_0^\infty \exp(-a^2 x^2)\,dx = \frac{\sqrt{\pi}}{2a} \tag{A1.5.5}$$

$$\int_{-\infty}^\infty \exp\left(-\frac{x^2}{a^2} + ibx\right)dx = \sqrt{\pi}a\,\exp\left(-\frac{a^2 b^2}{4}\right) \tag{A1.5.6}$$

$$\int_0^\infty \frac{\sin ax}{x}\,dx = \frac{\pi}{2} \qquad a > 0 \tag{A1.5.7}$$

$$\int_0^\infty \frac{\sin^2 ax}{x^2}\,dx = \frac{\pi a}{2} \tag{A1.5.8}$$

The delta function $\delta(x)$ is zero for $x \neq 0$ and

$$\int_{-\infty}^{\infty} \delta(x)\, dx = 1 \qquad (A1.5.9)$$

$$\frac{1}{2\pi} \int_{-\infty}^{\infty} e^{i\alpha x}\, d\alpha = \delta(x) \qquad (A1.5.10)$$

$$\int_{-\infty}^{\infty} f(x)\, \delta(x - x_0)\, dx = f(x_0) \qquad (A1.5.11)$$

When $x = x(u, v)$ and $y = y(u, v)$, we write

$$\int\int f(x, y)\, dx\, dy = \int\int f(x(u, v), y(u, v)) \left| J\left(\frac{x, y}{u, v}\right) \right| du\, dv$$

$$(A1.5.12)$$

where

$$J\left(\frac{x, y}{u, v}\right) = \begin{vmatrix} \dfrac{\partial x}{\partial u} & \dfrac{\partial y}{\partial u} \\ \dfrac{\partial x}{\partial v} & \dfrac{\partial y}{\partial v} \end{vmatrix} = \frac{\partial x}{\partial u}\frac{\partial y}{\partial v} - \frac{\partial x}{\partial v}\frac{\partial y}{\partial u}$$

Error function for real and complex arguments (from Abramowitz and Stegun, 1964, pp. 297–309). References to their equations are given in brackets. Using $z = x + iy$, for $i = \sqrt{-1}$, we have

$$z^* = x - iy \qquad (A1.5.13)$$

where * indicates complex conjugate.

$$\operatorname{erf} z = \frac{2}{\sqrt{\pi}} \int_0^z e^{-t^2}\, dt \qquad [7.1.1] \qquad (A1.5.14)$$

$$\operatorname{erfc} z = \frac{2}{\sqrt{\pi}} \int_z^\infty e^{-t^2}\, dt = 1 - \operatorname{erf} z \qquad [7.1.2] \qquad (A1.5.15)$$

$$w(z) = e^{-z^2} \operatorname{erfc}(-iz) \qquad [7.1.3] \qquad (A1.5.16)$$

Abramowitz and Stegun give tables of $w(z)$ (pp. 325–328).

$$w(-x + iy) = [w(x + iy)]^* \qquad [\text{p. } 325] \qquad (A1.5.17)$$

INEQUALITY

$$[x + (x^2 + 2)^{1/2}]^{-1} < e^{x^2} \int_x^\infty e^{-t^2}\, dt \leq \left[x + \left(x^2 + \frac{4}{\pi} \right)^{1/2} \right]^{-1} \qquad [7.1.13]$$

$$(A1.5.18)$$

ASYMPTOTIC EXPANSION

$$\sqrt{\pi}\,z e^{z^2}\,\mathrm{erfc}\,z \sim 1 - \frac{1}{2z^2} + \frac{3}{4z^4} - \frac{3 \cdot 5}{8z^6} + \cdots \quad [7.1.23] \quad (\text{A1.5.19})$$

for $|\arctan (y/x)| < 3\pi/4$ and $z \to \infty$.

INTEGRAL EXPRESSIONS OF BESSEL FUNCTIONS OF THE FIRST KIND

$$J_0(z) = \frac{1}{\pi} \int_0^\pi \cos (z \sin \theta)\, d\theta = \frac{1}{\pi} \int_0^\pi \cos (z \cos \theta)\, d\theta \quad [9.1.18]$$
$$(\text{A1.5.20})$$

$$J_n(z) = \frac{i^{-n}}{\pi} \int_0^\pi \exp (iz \cos \theta) \cos (n\theta)\, d\theta \quad [9.1.21] \quad (\text{A1.5.21})$$

$$J_\nu(z) = \frac{(\tfrac{1}{2}z)^\nu}{\pi^{1/2}\Gamma(\nu + \tfrac{1}{2})} \int_0^\pi \cos (z \cos \theta) \sin^{2\nu} \theta\, d\theta \quad [9.1.20]$$
$$(\text{A1.5.22})$$

$$\Gamma\!\left(\frac{3}{2}\right) = \frac{\sqrt{\pi}}{2}$$

A1.6. SERIES EXPANSIONS

Taylor's series expansion about a using $f'(x) = df(x)/dx$:

$$f(x) = f(a) + f'(a)(x - a) + f''(a)\frac{(x - a)^2}{2!} + \cdots \quad (\text{A1.6.1})$$

A1.7. CHANCE EVENTS

A1.7.1. Probability

An event may occur in p ways and fail in q ways. When all ways are equally likely the probability of p occurring $\mathcal{P}(p)$ is

$$\mathcal{P}(p) = \frac{p}{q + p} \quad (\text{A1.7.1})$$

also

$$\mathcal{P}(q) = \frac{q}{q + p} \quad (\text{A1.7.2})$$

$$\mathcal{P}(p) + \mathcal{P}(q) = 1 \quad (\text{A1.7.3})$$

If the probabilities of p_1 and p_2 occurring are $\mathcal{P}(p_1)$ and $\mathcal{P}(p_2)$ respectively, the probability of p_1 *and* p_2 occurring is

$$\mathcal{P}(p_1 \text{ and } p_2) = \mathcal{P}(p_1)\mathcal{P}(p_2) \tag{A1.7.4}$$

The probability of p_1 *or* p_2 occurring is

$$\mathcal{P}(p_1 \text{ or } p_2) = \mathcal{P}(p_1) + \mathcal{P}(p_2) \tag{A1.7.5}$$

Expectation. Event 1 has the "value" E_1 and probability of occurrence $\mathcal{P}(E_1)$, event 2 has E_2 and $\mathcal{P}(E_2)$, and so on. The expectation for all events is

$$\langle E \rangle = \sum_1^N E_n \mathcal{P}(E_n) \tag{A1.7.6}$$

where $\langle \rangle$ indicates the expectation procedure. We also use $\langle \rangle$ for average.

A1.7.2. Probability Density Function (PDF)

The probability density function $w(a)$ is defined as follows:

$$\mathcal{P}(a_2 \geq a \geq a_1) = \int_{a_1}^{a_2} w(a)\, da \tag{A1.7.7}$$

$$\int_{-\infty}^{\infty} w(a)\, da = 1$$

or

$$w(a) = \lim_{(a_2-a_1)\to 0} \frac{\mathcal{P}(a_2 \geq a \geq a_1)}{a_2 - a_1} \tag{A1.7.8}$$

The cumulative distribution $\mathcal{C}(a)$ is

$$\mathcal{C}(a) = \int_{-\infty}^{a} w(a')\, da' \tag{A1.7.9}$$

$$\mathcal{P}(a) = \frac{d\mathcal{C}(a)}{da} \tag{A1.7.10}$$

Expectation of $f(a)$, subject to a PDF $w(a)$ is

$$\langle f \rangle = \int_{-\infty}^{\infty} f(a)w(a)\, da \tag{A1.7.11}$$

A1.7.3. Common PDFs

GAUSSIAN OR NORMAL PDF

$$w_G(a) = \frac{1}{\sigma(2\pi)^{1/2}} \exp\left(\frac{-a^2}{2\sigma^2}\right) \tag{A1.7.12}$$

$$\langle a \rangle = 0$$

$$\langle a^2 \rangle = \sigma^2$$

RAYLEIGH PDF

$$w_R(a) = \frac{|a|}{\sigma^2} \exp\left(\frac{-a^2}{2\sigma^2}\right) \tag{A1.7.13}$$

BIVARIATE (JOINT) PDF

The surface is $z(x, y)$. The mean square roughness of the surface is σ^2. The bivariate PDF depends on the correlation function of the surface $C(\xi, \eta)$

$$C(\xi, \eta) = \frac{1}{4XY\sigma^2} \int_{-X}^{X} \int_{-Y}^{Y} z_1 z_2 \, dx \, dy \tag{A1.7.14}$$

$$\lim X, Y \rightarrow \infty$$

where z_1 is the value at x, y, and z_2 is the value at $x+\xi$, $y+\eta$. The bivariate PDF is

$$w_2 = [2\pi\sigma^2(1 - C^2(\xi, \eta))^{1/2}]^{-1} \times$$

$$\exp\left\{-\frac{z_1^2 + z_2^2 - 2z_1 z_2 C(\xi, \eta)}{2\sigma^2[1 - C^2(\xi, \eta)]}\right\} \tag{A1.7.15}$$

The expectation of f for two variables is

$$\langle f \rangle = \int_{-\infty}^{\infty} \int_{-\infty}^{\infty} f(z_1, z_2) w_2 \, dz_1 \, dz_2 \tag{A1.7.16}$$

When $f = \exp[2i\gamma(z_1 - z_2)]$,

$$\mathcal{W} = \int_{-\infty}^{\infty} w_2 \exp[2i\gamma(z_1 - z_2)] \, dz_1 \, dz_2 = \exp[-4\gamma^2\sigma^2(1 - C(\xi, \eta))] \tag{A1.7.17}$$

WAVE PROPAGATION

The appendix is in two parts. The first, a brief introduction to complex variables, is for students who need the review. The second part, in four sections, presents details for the derivation of the wave equation and its solution. The appendix concludes with an introduction to nonlinear effects.

A2.1. COMPLEX NUMBERS AND VARIABLES

We introduce basic operations for complex numbers and complex variables. Problems are given at the end of the section, and readers should do the appropriate problems as they go through the section.

We start with definitions and basic operations. These are not intended to replace a study of complex variables, but only to fulfill a minimum need or to refresh the reader's memory. A complex number A can be represented as the sum of a real part a and imaginary part b.

$$A = a + ib$$
$$i = (-1)^{1/2}$$

(A2.1.1)

$$i^2 = -1$$
$$i^{-1} = -i$$

(A2.1.2)

Note that the electrical engineering literature uses the letter j to represent the square root of -1. Both a and b are real positive or negative

376

numbers. The sum of complex numbers is the sum of the real and imaginary parts.

$$\text{for } C = c + id$$

$$A + C = a + c + i(b + d) \tag{A2.1.3}$$

$$C + A = a + c + i(b + d)$$

The product of complex numbers is obtained by doing the usual algebraic operation, taking care to let $i^2 = -1$.

$$A \cdot C = (a + ib)(c + id)$$

$$= ac - bd + i(ad + bc) \tag{A2.1.4}$$

and $C \cdot A = A \cdot C$.

The product of a complex number and a pure real or pure imaginary number is

$$cA = ac + ibc$$

$$idA = iad - bd \tag{A2.1.5}$$

Often real numbers are squared to obtain real positive numbers. This has to be altered in complex variables because the operation A^2 yields

$$A^2 = a^2 - b^2 + 2iab$$

a result that is complex. The operation equivalent to squaring is the so-called absolute square or multiplication by the complex conjugate, defined as follows:

$$A = a + ib \tag{A2.1.6}$$

complex conjugate,

$$A^* = a - ib$$

The complex conjugate is obtained by changing all i's in the expression to $-i$. Complex constants correspondingly become A^*. The absolute square is

$$|A|^2 = AA^* = (a + ib)(a - ib) = a^2 + b^2 \tag{A2.1.7}$$

In this context, the term $|A|$ usually means the square root of the absolute square, that is,

$$|A| = (AA^*)^{1/2} \tag{A2.1.8}$$

Any complex number or function plus its complex conjugate is real and the difference is imaginary

$$A + A^* = 2a \quad \text{and} \quad A - A^* = 2ib \tag{A2.1.9}$$

In a sense, an equation involving complex quantities is a system of two equations. The real parts equal the real parts and the imaginary parts equal the imaginary parts as follows:

$$A = B$$

$$a + ib = c + id$$

Thus

$$a = c$$

$$b = d$$

The basic integration formulas for real variables apply to complex variables, but the reader should consult texts on complex variables when the integrand becomes infinite at some point in the region of integration.

The coordinate system (Fig. A2.1.1) is given as

$$z = x + iy$$
$$z^* = x - iy \tag{A2.1.10}$$

From the figure, the components of r are

$$x = r \cos \phi$$
$$y = r \sin \phi$$

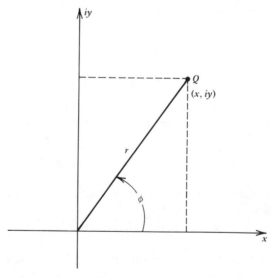

Figure A2.1.1. Complex coordinates: r is distance from origin to point Q, having coordinates $z = x = iy$.

where

$$r = (x^2 + y^2)^{1/2} \quad \text{or} \quad r = (zz^*)^{1/2}$$

$$\tan \phi = \frac{y}{x}$$

(A2.1.11)

Often r is referred to as the modulus and ϕ the angle. Examples of integration formulas are

$$\int az^n \, dz = \frac{az^{n+1}}{n+1}$$

$$\int ae^{bz} \, dz = \frac{ae^{bz}}{b}$$

$$\int a \cos bz \, dz = \frac{a \sin bz}{b}$$

$$\int a \sin bz \, dz = -\frac{a \cos bz}{b}$$

(A2.1.12)

where the constants a and b can be complex.

The function $e^{i\phi}$ has special importance because it expands as follows:

$$e^{i\phi} = \cos \phi + i \sin \phi$$

A way to show this is to use the power series expansions of e^x, $\sin \phi$, and $\cos \phi$. From a standard mathematical handbook, we have the following:

$$e^x = 1 + x + \frac{x^2}{2!} + \frac{x^3}{3!} + \frac{x^4}{4!} + \frac{x^5}{5!} + \cdots$$

$$\sin \phi = \phi - \frac{\phi^3}{3!} + \frac{\phi^5}{5!} + \ldots$$

$$\cos \phi = 1 - \frac{\phi^2}{2!} + \frac{\phi^4}{4!} + \ldots$$

(A2.1.13)

where $4! = 4 \cdot 3 \cdot 2 \cdot 1$, and so on.

Replace x by $i\phi$ in the expansion of e^x. With the aid of the following powers of i.

$$i^2 = -1, \qquad i^3 = -i, \qquad i^4 = 1, \qquad i^5 = i$$

we can separate the real and imaginary parts of $e^{i\phi}$ as follows:

$$e^{i\phi} = \left(1 - \frac{\phi^2}{2!} + \frac{\phi^4}{4!} + \ldots\right)$$

$$+ i\left(\phi - \frac{\phi^3}{3!} + \frac{\phi^5}{5!} + \ldots\right)$$

Comparison of this expansion and the expansions of sin ϕ and cos ϕ shows

$$e^{i\phi} = \cos \phi + i \sin \phi \tag{A2.1.14}$$

Sometimes we would like to go the other way (i.e., cos ϕ to exponentials) Since sin $(-\phi) = -\sin \phi$, we have

$$e^{-i\phi} = \cos \phi - i \sin \phi \tag{A2.1.15}$$

The sum and differences of $e^{i\phi}$ and $e^{-i\phi}$ yield

$$\cos \phi = \frac{e^{i\phi} + e^{-i\phi}}{2}$$
$$\sin \phi = \frac{e^{i\phi} - e^{-i\phi}}{2i} \tag{A2.1.16}$$

To simplify the typography, exponential functions are often written as follows:

$$e^{\phi} = \exp [\phi]$$

Earlier, we let $z = x + iy$. In many manipulations it is convenient to express z in the polar form (i.e., using modulus r and angle ϕ).

$$z = x + iy = r(\cos \phi + i \sin \phi)$$
$$z = re^{i\phi} \tag{A2.1.17}$$

Both forms are used a great deal, and it is possible to convert from one to the other to suit the problem. For example, sums of complex numbers are easily expressed as the components

$$z_1 + z_2 + z_3 = (x_1 + x_2 + x_3) + i(y_1 + y_2 + y_2)$$
$$(z_1 + z_2) + (z_1^* + z_2^*) = 2x_1 + 2x_2$$

Products and powers are convenient to handle in polar form. Let

$$z_1 = x_1 + iy_1 = r_1 \exp (i\phi_1)$$
$$z_2 = x_2 + iy_2 = r_2 \exp (i\phi_2)$$
$$z_1 z_2 = r_1 r_2 \exp [i(\phi_1 + \phi_2)]$$

Then

$$\frac{z_1}{z_2} = \frac{r_1}{r_2} \exp\left[i(\phi_1 - \phi_2)\right]$$

$$z_1^n = r_1^n \exp(in\phi_1)$$

$$z_1^{1/2} = r_1^{1/2} \exp\left[\frac{i(\phi_1)}{2}\right]$$

$$z_1^* = r_1 \exp(-i\phi_1)$$

$$z_1^* z_2 = r_1 r_2 \exp\left[-i(\phi_1 - \phi_2)\right]$$

$$= r_1 r_2 [\cos(\phi_1 - \phi_2) - i \sin(\phi_1 - \phi_2)]$$

The exponent can be complex.

As an example, the direct substitution of z in e^z yields

$$e^{az} = e^{ax + iay} = e^{ax} e^{iay}$$

$$= e^{ax}(\cos ay + i \sin ay)$$

Of course a can be complex, and the real and complex components would have to be sorted out of the exponential.

PROBLEMS

A2.1.1. Perform the indicated operations.

(a) $(4+3i)+(7+2i) =$

(b) $3c + 10b =$, for $c = 2 + i$, $b = 3 - i$.

(c) $(7-5i)(2i) =$

(d) $(7-i)(7+i) =$

(e) $(7-i)^2 =$

(f) $(i-7)^2 =$

(g) $(3-4i)^2 =$

(h) $5(3-4i)^{-1} + 5(3+4i)^{-1} =$

(i) $(7i)(5-12i) - 7i(5+12i) =$

(j) $(7i)(5-12i) + 7i(5+12i) =$

A2.1.2. Perform the indicated operations.

(a) $e^{i\phi} =$ for $\phi = \pi/6$, $30°$.

(b) $e^{i\phi} =$ for $\phi = 0$, $\pi/2$, π, $3\pi/2$, 2π.

(c) Prove $e^{i\phi} = e^{i(\phi + 2n\pi)}$ for $n = 1, 2, 3, \ldots$.

(d) $e^{i\phi} =$ for $\phi = -\pi/3$.

(e) $z = re^{i\phi}$, determine r and ϕ for
$z = 3+4i$, $3-4i$, $4+3i$
$2+2i$, $-4-4i$, $\sqrt{3}+i$
$1-i\sqrt{3}$

(f) Let $z_1 = 1+i$ and $z_2 = \sqrt{3}-i$;

use the polar forms for the operations, determine r and θ.

$$z_1 z_2 = \qquad\qquad z_1^* z_2^* =$$

$$z_1 z_2^{-1} = \qquad\qquad (z_1 z_2)^{-1} =$$

$$z_1^2 = \qquad\qquad z_2^2 =$$

A 2.1.3. Evaluate the following integrals:

(a) $\displaystyle f = \int_a^b dz$ for $a = 1-i,\ b = 3$

(b) $\displaystyle f = \int_0^b z\,dz$ for $b = 2-i$

Ans. $\displaystyle f = \frac{3-4i}{2}$

(c) $\displaystyle f = \int_0^a e^{i\phi}\,d\phi$ for $a = \dfrac{\pi}{4},\ \dfrac{\pi}{2},\ \pi$

(d) $\displaystyle f = \int_{-a}^a e^{i\phi}\,d\phi$ for $a = 0,\ \pi/4,\ \pi/3,\ \pi/2$

A2.2. THREE–DIMENSIONAL WAVE EQUATION

The derivation of the one-dimensional wave equation is given in Section 2.7. We use the results from that section in the extension to three dimensions. The procedure is to combine the results for each of the coordinate directions into one equation. In the derivation we also test two approximations and introduce vector notation.

The force on the volume $\Delta x\,\Delta y\,\Delta z$ in Fig. 2.7.2a along the x direction is $-(\partial \Delta p/\partial x)(\Delta x\,\Delta y\,\Delta z)$. In the three-dimensional problem there are also the forces $-(\partial \Delta p/\partial y)(\Delta x\,\Delta y\,\Delta z)$ and $-(\partial \Delta p/\partial z)(\Delta x\,\Delta y\,\Delta z)$ on the volume. Newton's law for the three directions is the generalization of (2.7.3),

$$\frac{\partial \Delta p}{\partial x} = -\rho_A \frac{du}{dt}$$

$$\frac{\partial \Delta p}{\partial y} = -\rho_A \frac{dv}{dt} \qquad\qquad \textbf{(A2.2.1)}$$

$$\frac{\partial \Delta p}{\partial z} = -\rho_A \frac{dw}{dt}$$

where u, v, and w are the components of velocity along the x, y and z directions. Since u is a function of x and t, the total derivative (du/dt) is found by using implicit differentiation

$$\frac{du}{dt} = \frac{\partial u}{\partial t} + \frac{\partial u}{\partial x}\frac{\partial x}{\partial t} = \frac{\partial u}{\partial t} + u\frac{\partial u}{\partial x} \tag{A2.2.2}$$

with similar expressions for the other components. We wish to drop $u(\partial u/\partial x)$. To show when it is much less than $\partial u/\partial t$, we use the solution obtained after dropping $u(\partial u/\partial x)$. From Section 2.8 we express u as a traveling wave.

$$u = F_+\left(t - \frac{x}{c}\right) = F_+(q)$$

where

$$q = t - \frac{x}{c}$$

Form

$$\frac{\partial u}{\partial t} = \frac{\partial F_+}{\partial q}\frac{\partial q}{\partial t} = \frac{\partial F_+}{\partial q}$$

and

$$u\frac{\partial u}{\partial x} = u\frac{\partial F_+}{\partial q}\frac{\partial q}{\partial x} = -\frac{u}{c}\frac{\partial F_+}{\partial q} = -\frac{u}{c}\frac{\partial u}{\partial t}$$

The ratio $|u(\partial u/\partial x)|/|\partial u/\partial t|$ is u/c. Therefore when u is much less than c we replace du/dt by $\partial u/\partial t$.

Each of the equations in (A2.2.1) represents the force and acceleration along its coordinate axis. These are x, y, and z components of a vector force and a vector acceleration. We use the unit vectors $\mathbf{i}$, $\mathbf{j}$, and $\mathbf{k}$ to indicate the directions along the x, y, and z coordinates as in Fig. A2.2.1, where boldface type designates a vector. Using the unit vector to identify direction, (2.2.1) is summed and written as one equation.

$$\mathbf{i}\frac{\partial \Delta p}{\partial x} + \mathbf{j}\frac{\partial \Delta p}{\partial y} + \mathbf{k}\frac{\partial \Delta p}{\partial z} = -\rho_A\left(\mathbf{i}\frac{\partial u}{\partial t} + \mathbf{j}\frac{\partial v}{\partial t} + \mathbf{k}\frac{\partial w}{\partial t}\right) \tag{A2.2.3}$$

The quantities on the right-hand side of (A2.2.3) are the components of acceleration. We define the velocity of the volume $\mathbf{U}$

$$\mathbf{U} = \mathbf{i}u + \mathbf{j}v + \mathbf{k}w \tag{A2.2.4}$$

and the right-hand side becomes $-\rho_A(\partial\mathbf{U}/\partial t)$. We simplify the notation on

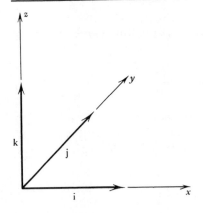

Figure A2.2.1. Rectangular coordinates and unit vectors **i**, **j**, and **k**.

the left by defining ∇ as follows:

$$\nabla \equiv \mathbf{i}\frac{\partial}{\partial x}+\mathbf{j}\frac{\partial}{\partial y}+\mathbf{k}\frac{\partial}{\partial z} \qquad \textbf{(A2.2.5)}$$

Since ∇ operates on Δp to give the left-hand side of (A2.2.3), we have

$$\nabla(\Delta p) = -\rho_A\frac{\partial \mathbf{U}}{\partial t} \qquad \textbf{(A2.2.6)}$$

Here the pressure disturbance does not depend on direction at the point of observation and is called a *scalar*; $\nabla(\Delta p)$ is a vector and is called the *gradient* of (Δp) or grad (Δp); $\partial \mathbf{U}/\partial t$ is the acceleration of the volume and is a vector.

The next step is to calculate the density change and mass flow for the three-dimensional problem. The x component is shown in Fig. 2.7.2b. The net rate of loss of mass for the x component (2.7.4) is $[\partial(\rho u)/\partial x]\, \delta x\, \delta y\, \delta z$. Similarly, the rates of loss along the y and z directions are $[\partial(\rho v)/\partial y)]\, \delta x\, \delta y\, \delta z$ and $[\partial(\rho w)/\partial z]\, \delta x\, \delta y\, \delta z$.

The loss of mass from the volume causes the density to decrease. The rate of mass change is $-[\partial \rho/\partial t]\, \delta x\, \delta y\, \delta z$. The total mass outflow equals the rate of mass change in the cage.

$$\frac{\partial(\rho u)}{\partial x}+\frac{\partial(\rho v)}{\partial y}+\frac{\partial(\rho w)}{\partial z} = -\frac{\partial \rho}{\partial t} \qquad \textbf{(A2.2.7)}$$

Again we use the vector notation for simplication. Here we use the *scalar* product of a pair of vectors. For vectors **A** and **B**, the *scalar product* is $AB \cos (A, B)$, where A and B are the magnitudes and A, B is the angle between them. The scalar products for the unit vectors are

$$\mathbf{i} \cdot \mathbf{i} = 1, \qquad \mathbf{j} \cdot \mathbf{j} = 1, \qquad \mathbf{k} \cdot \mathbf{k} = 1$$
$$\mathbf{i} \cdot \mathbf{j} = 0, \qquad \mathbf{i} \cdot \mathbf{k} = 0, \qquad \mathbf{j} \cdot \mathbf{k} = 0 \qquad \text{(A2.2.8)}$$

By writing the left-hand side of (A2.2.7) as a product of (A2.2.5) and (A2.2.4), we find

$$\nabla \cdot \mathbf{U} = \left(\mathbf{i} \frac{\partial}{\partial x} + \mathbf{j} \frac{\partial}{\partial y} + \mathbf{k} \frac{\partial}{\partial z} \right) \cdot (\mathbf{i}u + \mathbf{j}v + \mathbf{k}w) \qquad \text{(A2.2.9)}$$

Likewise (A2.2.7) simplifies to

$$\nabla \cdot (\rho \mathbf{U}) = -\frac{\partial \rho}{\partial t} \qquad \textbf{(A2.2.10)}$$

The term $\nabla \cdot (\rho \mathbf{U})$ is called the *divergence* of $\rho \mathbf{U}$ or div $(\rho \mathbf{U})$. We wish to simplify the expression by dropping the small terms. Considering the x components, the expansion of the left-hand side gives

$$\frac{\partial}{\partial x} (\rho u) = u \frac{\partial \rho}{\partial x} + \rho \frac{\partial u}{\partial x} \qquad \text{(A2.2.11)}$$

The disturbance causes the density to have a small change $\Delta \rho$ from the very much larger ambient density ρ_A. Letting $\rho = \rho_A + \Delta \rho$, we write

$$\frac{\partial}{\partial x} (\rho u) = \rho_A \frac{\partial u}{\partial x} + \Delta \rho \frac{\partial u}{\partial x} + u \frac{\partial \Delta \rho}{\partial x} \qquad \text{(A2.2.12)}$$

Again we use the solution obtained after dropping the terms to estimate the magnitude of the terms. From Section 2.8, u and $\Delta \rho$ are related by (2.8.17).

$$\frac{u}{c} = \frac{\Delta \rho}{\rho_A}$$

Then the last term on the right-hand side of (A2.2.12) becomes

$$u \frac{\partial \Delta \rho}{\partial x} = \frac{u}{c} \rho_A \frac{\partial u}{\partial x} \qquad \text{(A2.2.13)}$$

We use our earlier assumption that $u/c \ll 1$. Comparison of (A2.2.13) and $\rho_A \, \partial u/\partial x$ shows that $u \, \partial \Delta \rho/\partial x$ is negligible. Since $\Delta \rho/\rho_A$ is also very small $\rho_A (\partial u/\partial x)$ is the remaining term in (A2.2.12). Equation (A2.2.10) therefore becomes

$$\rho_A \nabla \cdot \mathbf{U} = -\frac{\partial \Delta \rho}{\partial t} \qquad \textbf{(A2.2.14)}$$

We have derived a pair of equations, both of which have dependence on **U**. We eliminate **U** by taking the divergence ($\nabla \cdot$) of (A2.2.6) and $\partial/\partial t$ of **(A2.2.14)**. Equating the results gives

$$\nabla \cdot \nabla(\Delta p) = \frac{\partial^2 \Delta \rho}{\partial t^2} \qquad (A2.2.15)$$

The vector product $\nabla \cdot \nabla$ appears very often. When it operates on a *scalar* such as Δp, it is symbolized as ∇^2 and is called the *Laplacian operator*. In rectangular coordinates

$$\nabla^2 \equiv \nabla \cdot \nabla = \frac{\partial^2}{\partial x^2} + \frac{\partial^2}{\partial y^2} + \frac{\partial^2}{\partial z^2} \qquad (A2.2.16)$$

Using the ∇^2 notation, (A2.2.15) is

$$\nabla^2(\Delta p) = \frac{\partial^2 \Delta \rho}{\partial t^2} \qquad (A2.2.17)$$

We use the relation between the disturbances Δp and $\Delta \rho$, (2.7.7), and (2.8.8)

$$\Delta p = \frac{E}{\rho_A} \Delta \rho$$

$$c^2 = \frac{E}{\rho_A} \qquad (A2.2.18)$$

where c is the sound speed.

The substitution of (A2.2.18) into (A2.2.17) gives

$$\nabla^2 p = \frac{1}{c^2} \frac{\partial^2 p}{\partial t^2} \qquad (A2.2.19)$$

and

$$\nabla^2(\Delta \rho) = \frac{1}{c^2} \frac{\partial^2 (\Delta \rho)}{\partial t^2} \qquad (A2.2.20)$$

where we have dropped the Δ from Δp and let p be the sound pressure disturbance. These are the wave equations. Often, we also drop the Δ from $\Delta \rho$ and use ρ as the density fluctuation.

A2.2.1. ∇^2 in Rectangular, Spherical, and Cylindrical Coordinates

The spherical and cylindrical coordinates appear in Fig. A2.2.2. The transformation of the Laplacian from rectangular coordinates

$$\nabla^2 = \frac{\partial^2}{\partial x^2} + \frac{\partial^2}{\partial y^2} + \frac{\partial^2}{\partial z^2} \qquad (A2.2.21)$$

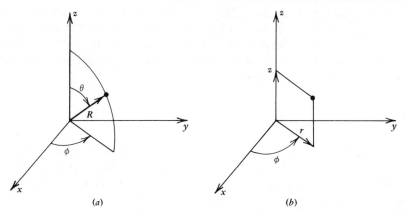

Figure A2.2.2. (a) Spherical coordinates. (b) Cylindrical coordinates. Both systems are superimposed on rectangular coordinates.

to spherical and cylindrical coordinates requires much manipulation. The results are

spherical coordinates R, θ, ϕ:

$$\nabla^2 = \frac{\partial^2}{\partial R^2} + \frac{2}{R}\frac{\partial}{\partial R} + \frac{1}{R^2 \sin^2 \theta}\frac{\partial^2}{\partial \phi^2} + \frac{1}{R^2}\left(\frac{\partial^2}{\partial \theta^2} + \cot \theta \frac{\partial}{\partial \theta}\right)$$

(A2.2.22)

for spherical symmetry (no variation with θ or ϕ):

$$\nabla^2 = \frac{\partial^2}{\partial R^2} + \frac{2}{R}\frac{\partial}{\partial R} \quad \text{or} \quad = \frac{1}{R^2}\frac{\partial}{\partial R}\left(R^2 \frac{\partial}{\partial R}\right) \quad \text{(A2.2.23)}$$

cylindrical coordinates r, z, ϕ:

$$\nabla^2 = \frac{\partial^2}{\partial r^2} + \frac{1}{r}\frac{\partial}{\partial r} + \frac{1}{r^2}\frac{\partial^2}{\partial \phi^2} + \frac{\partial^2}{\partial z^2}$$

(A2.2.24)

The three coordinates in each of these systems are orthogonal, and changes of one variable do not affect the values of the others.

A2.3. SOLUTIONS OF THE WAVE EQUATION AND BOUNDARY CONDITIONS

Our purpose in solving the wave equation is to calculate the sound pressure in the sea. The solution involves three steps:

1. Select the coordinate geometry that fits the problem best.
2. Find solutions that satisfy the sound speed profile in the ocean.
3. Adjust the solutions and fit them together at the boundaries.

As an example of the procedure, we calculate the sound pressure for a plane wave that is incident on an interface between two half spaces. This approximates the reflection of a sound wave at the water sediment interface. The fitting process yields Snell's law, the reflection coefficient, and the transmission coefficient.

A2.3.1. Separation of Variables

We started our derivation of the wave equation by assuming that the motions along the three orthogonal coordinates were independent of one another. Now we use this property and assume that p can be written as a product of functions of the variables

$$p = XYZT \tag{A2.3.1}$$

where X, Y, Z, and T are functions x, y, z, and t, respectively. Substitution into the wave equation for rectangular coordinates (A2.2.19) gives

$$X''YZT + Y''XZT + Z''XYT = \frac{T''}{c^2} XYZ \tag{A2.3.2}$$

where X'' is $\partial X / \partial x^2$, and so on.

We rearrange (A2.3.2) and place all t dependence on one side and all x, y, and z dependence on the other.

$$c^2 \frac{X''}{X} + c^2 \frac{Y''}{Y} + c^2 \frac{Z''}{Z} = \frac{T''}{T} \tag{A2.3.3}$$

Each term is a function of only one variable. That is, T and T'' are only functions of time, and other terms are functions of their spatial coordinates. Since the equation is true for all values of x, y, z, and t, each of the terms must be a constant.

We assume that the time dependence is harmonic and evaluate the constant. For example, assume

$$T = A_1 e^{i\omega t}$$

$$\frac{T''}{T} = -\omega^2 \tag{A2.3.4}$$

where A_1 is a constant. Similarly, try $\exp(-i\omega t)$ and a constant A_2

$$T = A_2 e^{-i\omega t}$$

$$\frac{T''}{T} = -\omega^2 \tag{A2.3.5}$$

and the sum of (A2.3.4) and (A2.3.5) is

$$T = A_1 e^{i\omega t} + A_2 e^{-i\omega t} \quad \text{and} \quad \frac{T''}{T} = -\omega^2 \qquad (A2.3.6)$$

Harmonic time dependence satisfies the right side of (A2.3.3) and T''/T is the constant $-\omega^2$.

If c is a function of x, y, and z, solutions on the left-hand side are difficult and often impossible to obtain. For this illustration, we assume that c is constant. Again we apply logic and say that if X, Y, and Z are independent of each other, then X''/X, Y''/Y, and Z''/Z must equal constants. Using X as an example, we let

$$X = B_1 \exp(ik_x x) + B_2 \exp(-ik_x x) \qquad (A2.3.7)$$

Evaluation of X''/X gives

$$\frac{X''}{X} = -k_x^2 \qquad (A2.3.8)$$

Similar expressions for Y and Z give

$$\frac{Y''}{Y} = -k_y^2 \qquad (A2.3.9)$$

$$\frac{Z''}{Z} = -k_z^2 \qquad (A2.3.10)$$

The substitution of these into (A2.3.3) gives

$$k_x^2 + k_y^2 + k_z^2 = k^2 \qquad \textbf{(A2.3.11)}$$

$$k^2 \equiv \frac{\omega^2}{c^2} \qquad \textbf{(A2.3.12)}$$

where k_x, k_y, and k_z are the x, y, and z components of the wave number **k**, (Fig. A2.3.1). Usually it is clear from the context whether **k** is the vector wave number or the unit vector in the z direction. Both are standard notations.

The equation of a plane wave traveling in the plus x, y, and z direction is

$$p = A \exp[i(\omega t - k_x x - k_y y - k_z z)] \qquad \textbf{(A2.3.13)}$$

where A is a constant. Plane waves for other directions are obtained by changing the signs of k_x, k_y, and k_z. There are eight combinations of signs. The complete solution is the sum of all eight, and each has an arbitrary constant.

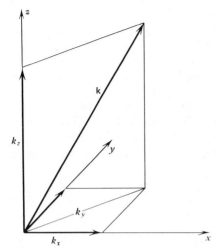

Figure A2.3.1. Components of the wave number **k**.

A2.3.2.   Evaluation at a Boundary Between Two Media

This problem is essentially the same as the calculation in Section 2.9. We give it as an illustration of the method of fitting solutions of the wave equation at a boundary (Fig. A2.3.2).

We simplify the algebra by making the following assumptions. (1) The plane wave is traveling in the x–z plane. It does not depend on y. (2) The *incident* plane wave is traveling in the negative z direction and positive x direction. These two assumptions reduce the eight combinations of signs of k_x, k_y, and k_z to two. In medium 1 we designate k_{1x}, k_{1z}, and $k_1 = \omega/c_1$ for the wave numbers. The constants are A_1 and B_1. The pressure p_1 in medium 1 is

$$p_1 = A_1 \exp\left[i(\omega t - k_{1x}x + k_{1z}z)\right] + B_1 \exp\left[i(\omega t - k_{1x}x - k_{1z}z)\right]$$

$$(A2.3.14)$$

The first term on the right is the wave that is traveling in $+x$ and $-z$ directions. The second term is the reflected wave traveling in $+x$ and $+z$ directions.

In the lower medium, p_2 is

$$p_2 = B_2 \exp\left[i(\omega t - k_{2x}x + k_{2z}z)\right] \qquad (A2.3.15)$$

where k_{2x}, k_{2z}, and $k_2 = \omega/c_2$ are the wave numbers in the lower medium. The wave numbers and their components are shown in Fig. A2.3.2.

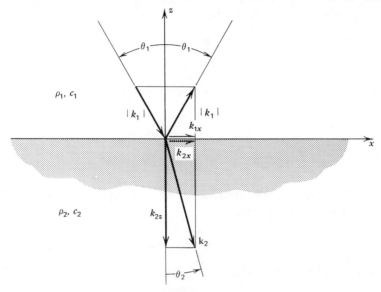

Figure A2.3.2. Geometry for the boundary conditions at a plane interface. Upper medium has density ρ_1 and sound speed c_1. Lower medium has density ρ_2 and sound speed c_2.

The basic assumptions made in fitting solutions at an interface are as follows:

(1) the pressure is continuous across the interface, $p_1 = p_2 |_{z=0}$;

(2) the displacements or velocities in the z direction are equal for each medium, $w_1 = w_2 |_{z=0}$. These are called *boundary conditions*.

First we apply the pressure boundary condition and factor out the time dependence

$$p_1 = p_2 |_{z=0} \tag{A2.3.16}$$

$$(A_1 + B_1) \exp(-ik_{1x}x) = B_2 \exp(-ik_{2x}x) \tag{A2.3.17}$$

This equation holds for all values of x, therefore the exponentials are equal and

$$k_{1x} = k_{2x} \tag{A2.3.18}$$

From the geometry in Fig. A2.3.2, k_{1x} and k_{2x} are found to be

$$k_{1x} = k_1 \sin \theta_1 = \frac{\omega \sin \theta_1}{c_1}$$

$$k_{2x} = k_2 \sin \theta_2 = \frac{\omega \sin \theta_2}{c_2} \tag{A2.3.19}$$

Equating the two gives Snell's law

$$\frac{\sin \theta_1}{c_1} = \frac{\sin \theta_2}{c_2}$$

In general for a plane layered medium, Snell's law applies and all horizontal components of wave number are equal.

$$k_{1x} = k_{2x} = \cdots = k_x \qquad \textbf{(A2.3.20)}$$

Since the exponentials in (A2.3.17) are equal, the equation for the constants is

$$A_1 + B_1 = B_2 \qquad (A2.3.21)$$

To calculate the vertical component of velocity w, we drop Δ and integrate (A2.2.3) over time.

$$w = -\frac{1}{\rho_A} \int \frac{\partial p}{\partial z} \, dt \qquad (A2.3.22)$$

The temporal integration yields

$$w = -\frac{1}{i\rho_A \omega} \frac{\partial p}{\partial z}$$

Now use (A2.3.14) to form

$$\frac{\partial p_1}{\partial z} = ik_{1z}A_1 \exp\left[i(\omega t - k_{1x}x + k_{1z}z)\right] - ik_{1z}B_1 \exp\left[i(\omega t - k_{1x}x - k_{1z}z)\right]$$

$$(A2.3.23)$$

Use $k_{1x} = k_{2x} = k_x$ and obtain (ρ_A becomes ρ_1)

$$w_1 = \frac{-k_{1z}}{\rho_1 \omega}\{A_1 \exp\left[i(\omega t - k_x x + k_{1z}z)\right] - B_1 \exp\left[i(\omega t - k_x x - k_{1z}z)\right]\}$$

$$(A2.3.24)$$

Similarly w_2 is

$$w_2 = \frac{-k_{2z}}{\rho_2 \omega} B_2 \exp\left[i(\omega t - k_x x + k_{2z}z)\right] \qquad (A2.3.25)$$

The vertical components of the velocity are equal at $z = 0$, and $w_1 = w_2$ yields

$$\rho_2 k_{1z}(A_1 - B_1) = \rho_1 k_{2z}B_2 \qquad (A2.3.26)$$

The solution of (A2.3.21) and (A2.3.26) for the ratios B_1/A_1 and B_2/A_1 gives the *pressure reflection coefficient*

$$\mathcal{R}_{12} \equiv \frac{B_1}{A_1} = \frac{\rho_2 k_{1z} - \rho_1 k_{2z}}{\rho_2 k_{1z} + \rho_1 k_{2z}} \qquad \textbf{(A2.3.27)}$$

and the *pressure transmission coefficient*

$$\mathscr{T}_{12} \equiv \frac{B_2}{A_1} = \frac{2\rho_2 k_{1z}}{\rho_2 k_{1z} + \rho_1 k_{2z}}$$ **(A2.3.28)**

These equations take the form of (2.10.1) and (2.10.2) by using

$$k_{1z} = \frac{\omega}{c_1} \cos \theta_1$$

$$k_{2z} = \frac{\omega}{c_2} \cos \theta_2$$ **(A2.3.29)**

This completes our demonstration of the procedure for fitting solutions of the wave equation at boundaries.

A2.3.3. Total Reflection

Beyond the critical angle, the sound pressure is totally reflected. To examine the reflected signal and the sound pressure beneath the interface, we use (A2.3.15) and (A2.3.28) to calculate p_2.

$$p_2 = A_1 \mathscr{T}_{12} \exp\left[i(\omega t - k_x x + k_{2z} z)\right]$$ (A2.3.30)

$$k_{2z} = \frac{\omega \cos \theta_2}{c_2}$$ (A2.3.31)

Since there are no restrictions in the derivation which require θ_1 to be less than critical, we can use the foregoing equations to calculate p_2. By using Snell's law, $\cos \theta_2$ is written

$$\cos \theta_2 = \pm\left[1 - \left(\frac{c_2 \sin \theta_1}{c_1}\right)^2\right]^{1/2}$$ (A2.3.32)

When θ_1 is greater than the critical angle, the square root is imaginary and we write it

$$\cos \theta_2 = -ib_2 \qquad \text{for} \quad \sin \theta_1 > \frac{c_1}{c_2}$$

$$b_2 \equiv \left[\left(\frac{c_2 \sin \theta_1}{c_1}\right)^2 - 1\right]^{1/2}$$ (A2.3.33)

where we have chosen the minus sign of the square root to attenuate p_2 in the negative z direction. The substitution of (A2.3.33) and (A2.3.31) in

(A2.3.30) gives

$$p_2 = A_1 \mathcal{T}_{12} \exp \left[\frac{-b_2 \omega |z|}{c_2} + i(\omega t - k_x x) \right] \qquad (A2.3.34)$$

where we use $|z|$ to avoid confusion with the sign of z and where p_2 is exponentially attenuated in medium 2. The amplitude of p_2 is decreased by the factor e^{-1} at $|z| = c_2/\omega b_2$ and that depth is called the *skin depth*. Layers and objects that are several skin depths below the 1-2 interface have negligible effects on the reflected sound pressures.

Expressions for $\mathcal{R}_{12}$ and the phase angle Φ for reflections beyond the critical angle are given in (2.9.10), (2.9.11), and (2.9.12). Conditions concerning the applicability of these equations to spherically spreading wave fronts near critical angle are given in (2.9.13) and (2.9.14).

A2.4. MEASUREMENTS NEAR AN INTERFACE, STANDING WAVES

Sound pressures near an interface consist of the incident and reflected components of the signal. Since we use short-gated sine wave transmissions in most sonar systems, the incident and reflected waves are separated by their travel times when the receiver is far enough from the interface. Commonly, the arrivals overlap and interfere when the receiver is within a few wavelengths of the interface.

To illustrate the concept of a standing wave, we first assume that the incident wave is traveling vertically downward and reflects upward at a perfectly rigid interface. The sound pressure is the sum of the two waves. Letting Z_i and Z_r be the incident and reflected waves (z is positive upward), we write

$$\begin{align} Z_i &= \exp\left[i(\omega t + k_{1z} z)\right] \\ Z_r &= \exp\left[i(\omega t - k_{1z} z)\right] \end{align} \qquad (A2.4.1)$$

The sum Z is

$$\begin{align} Z &= \exp(i\omega t)\left[\exp(ik_{1z}z) + \exp(-ik_{1z}z)\right] \\ Z_1 &= 2\exp(i\omega t)\cos(k_{1z}z) \end{align} \qquad (A2.4.2)$$

The signal has an envelope $|2\cos(k_{1z}z)|$ and a time dependence $\exp(i\omega t)$. The envelope is stationary in time. For example, the nulls at $k_{1z}z = \pi/2$ are nulls at all times. This is a *standing wave*.

A standing wave occurs when two waves having the same frequency, amplitude, and opposite directions of travel add together. Conversely, a solution that gives a standing wave can be written as the sum of two waves.

Oblique reflection at a partially reflecting interface gives a component that is a standing wave. Using (A2.3.14) and (A2.3.27), the sound pressure in medium 1 is the sum

$$p_1 = A\{\exp[i(\omega t - k_x x + k_{1z}z)] + \Re_{12}\exp[i(\omega t - k_x x - k_{1z}z)]\}$$
$$\text{(A2.4.3)}$$

After factoring out $\exp[i(\omega t - k_x x)]$, (A2.4.3) becomes

$$p_1 = A\exp[i(\omega t - k_x x)]Z_{\Re} \qquad \text{(A2.4.4)}$$

where

$$Z_{\Re} = \exp(ik_{1z}z) + \Re_{12}\exp(-ik_{1z}z) \qquad \text{(A2.4.5)}$$

First we consider the standing waves when the magnitude of $\Re_{12}$ is unity, that is, $|\Re|_{12} = 1$ and $\Re_{12} = \exp(i2\Phi)$: $Z_{\Re}$ becomes

$$Z_{\Re} = \exp(ik_{1z}z) + \exp(-ik_{1z}z + i2\Phi)$$
$$= \exp(i\Phi)[\exp(ik_{1z}z - i\Phi) + \exp(-ik_{1z}z + i\Phi)]$$
$$Z_{\Re} = 2\exp(i\Phi)\cos(k_{1z}z - \Phi) \qquad \text{(A2.4.6)}$$

This shows that total reflection gives standing waves in the z direction as well as a traveling wave along the x direction that moves at the velocity

$$v_x = \frac{\omega}{k_x} \qquad \text{(A2.4.7)}$$

where v_x is the *phase velocity* in (A2.4.4).

Reflections at partially reflecting interfaces have standing waves that are the sum of part of the incident wave and $\Re_{12}$ of the incident wave for the reflected wave. The interference minima are not zero for this case.

A2.5. FINITE AMPLITUDE, NONLINEAR EFFECTS IN SOUND TRANSMISSION

A2.5.1. Harmonic Distortion and the Growth of Shock Waves

When sound pressures become great enough, the approximations of infinitesimal amplitudes made in our derivation of the wave equation become intolerable and we must look at the details of finite amplitude propagation. Although the corrections to the linear theory are small, propagation of intense sounds generates completely new phenomena.

First we show that the speed of sound is a function of ambient pressure. Recall our previous assumption of a Hooke's law linearity between density change $\Delta\rho$ and pressure change Δp.

$$c^2 = \frac{\Delta p}{\Delta \rho} = \frac{E}{\rho_A} \tag{A2.5.1}$$

The physical relation between pressure and density is sketched in Fig. A2.5.1 for air and water. Since the speed of propagation c depends on

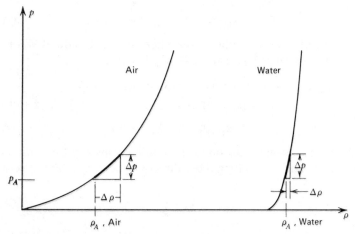

Figure A2.5.1. Equation of state for air and water, $P = P(\rho)$, showing relative incompressibility of water. The curves are diagrammatic; if they were to correct scale, the slope for water would be about 20 times that for air at standard conditions.

the slope, $\Delta p/\Delta\rho$, it is a function of the ambient density

$$c_A = \left[\left(\frac{\Delta p}{\Delta \rho} \right)^{1/2} \right]_{\rho_A} \tag{A2.5.2}$$

The slope, therefore the speed of propagation, is much greater for water than for air. The slope and speed increase as the pressure and density increase.

To calculate the speed we need an analytic expression for the state of the fluid, $p = p(\rho)$. The required relation for sound propagation is the adiabatic (no heat exchange) equation

$$(p\rho^{-\Gamma})_s = K \tag{A2.5.3}$$

where K is a constant. The subscript s identifies the condition of zero heat exchange in a reversible process (constant entropy); water is very close to this state during sound propagation. The adiabatic equation is well known for gases, where Γ is the ratio of the specific heat at constant pressure to the specific heat at constant volume, and K is a constant that depends on the units and mass of gas. When the same form of the equation is used for water, however, the p must be interpreted as including not only external pressure but an internal cohesive pressure of approximately 3000 atm. For water, the constant Γ is not the ratio of specific heats but must be determined empirically. Equation (A2.5.3) may be differentiated to form $(\partial p/\partial \rho)_{s,\rho}$ and thereby to calculate the speed of sound at the ambient density. Using (A2.5.2), we write

$$c^2 = \left(\frac{\partial p}{\partial \rho}\right)_{s,\rho} = \frac{\Gamma p}{\rho} \tag{A2.5.4}$$

$$= K\Gamma\rho^{\Gamma-1}$$

Use the first two terms of the Taylor expansion for ρ near ρ_A

$$\rho = \rho_A\left(1 + \frac{\Delta\rho}{\rho_A}\right)$$

substitute into (A2.5.4) and get

$$c = [K\Gamma\rho_A^{(\Gamma-1)}]^{1/2}\left[1 + \frac{\Delta\rho}{\rho_A}\right]^{(\Gamma-1)/2} \tag{A2.5.5}$$

The first factor is identified as the speed c_A at the reference condition ρ_A and the second bracket is approximated by the first two terms of the binomial expansion. Then

$$c \approx c_A\left[1 + (\Gamma-1)\frac{\Delta\rho/\rho_A}{2}\right] \tag{A2.5.6}$$

The speed at density ρ will reduce to c_A if the increment of density in the sound wave is very small compared to ρ_A. When $\Delta\rho/\rho_A \to 0$, we have infinitesimal amplitude propagation.

An alternative fitting of the curve of Fig. A2.5.1 (Beyer, 1974, p. 98) is in the form of a power series expansion

$$\Delta p \approx A(\Delta\rho/\rho_A) + \frac{B(\Delta\rho/\rho_A)^2}{2} \tag{A2.5.7}$$

Form $c^2 = \dfrac{\partial(\Delta p)}{\partial \rho}$ from (A2.5.7) and compare with the square of (A2.5.6).

The first two terms yield [using (A2.5.4)]

$$A = \rho_A c_A^2 = \Gamma p_A$$
$$B = \Gamma(\Gamma - 1)p_A \tag{A2.5.8}$$

The ratio $B/A = \Gamma - 1$, the "parameter of nonlinearity", has become the preferred description of the medium for high intensity studies. The B/A value for water ranges from 5.0 at temperatures 20°C to 4.6 at 10°C. For air, since $\Gamma = 1.40$, $B/A = 0.40$. In terms of B/A, (A2.5.6) is

$$c \approx c_A\left[1 + \frac{B}{A}\frac{\Delta\rho/\rho_A}{2}\right]$$

Since from Section 2.8 we have (2.8.17)

$$u = \left(\frac{\Delta\rho}{\rho_A}\right)c_A$$

the signal speed $u + c$ at a point of the wave can be written

$$u + c = c_A\left(1 + \beta\frac{\Delta\rho}{\rho_A}\right) \tag{A2.5.9}$$

where

$$\beta \equiv 1 + \frac{B}{2A}.$$

Consider what happens as a *finite* sinusoidal plane wave propagates. Figure A2.5.2a indicates that the crest of the wave is at higher pressure than the undisturbed medium; therefore the speed of sound c is greater than c_A. The particle velocity at the crest is given by (2.8.17). At the axis the particle velocity is zero and the speed of sound c_A. In a rarefaction, on the other hand, $c < c_A$ and u is negative. The net result of these different signal speeds at different parts of the wave is that the crests advance relative to the axial positions and the troughs lag behind. As the effect continues, the wave distorts to a form resembling Fig. A2.5.2b. Finally, if the pressure swing is large enough for a sufficient number of wavelengths, the ultimate finite amplitude wave form (Fig. A2.5.2c) appears. The form is called a "repeated shock" wave or "sawtooth" wave.

Fourier analysis of the various forms of the finite amplitude wave reveals that energy is taken from the fundamental of frequency f (the original sinusoid) and redistributed into second, third, and so on, harmonics as the wave distorts. In the repeated shock form, the pressure amplitude of the second harmonic (frequency $2f$) is half that of the

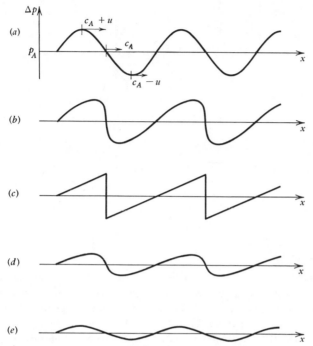

Figure A2.5.2. Various stages of a finite amplitude acoustic wave. The source moves sinusoidally. (*a*) Finite amplitude, close to sinusoidal source. (*b*) Finite amplitude, nonlinear distortion away from source. (*c*) Finite amplitude, fully developed shock wave, farther away from source. (*d*) Finite amplitude, aging shock wave; high frequency dissipation rate is greater than growth rate. (*e*) Infinitesimal amplitude, degenerated shock wave.

fundamental at that point, the third harmonic (3*f*) has one-third the amplitude of the fundamental, and so on. The cascading redistribution of energy from the fundamental to the upper harmonics is accompanied by increasing real loss of energy because the newly generated frequencies, being much higher than the fundamental, dissipate into thermal energy at a much faster rate (see Section 3.3).

We can follow the phenomenon of harmonic distortion into "old age" when the higher frequencies, which define the corners of the shock wave, dissipate too rapidly to be compensated by harmonic growth. At this point the corners of the shock wave erode and become rounded, as the shock wave returns to a form reminiscent of its infancy, but with greatly reduced amplitude (Figs. A2.5.2*d* and *e*).

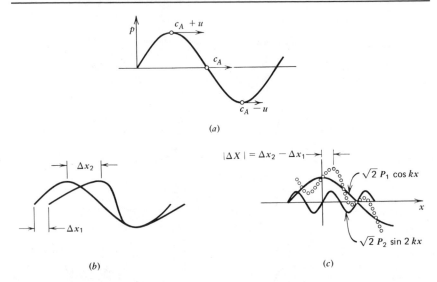

Figure A2.5.3. Growth of a finite amplitude sinusoid. (*a*) Original sine wave form; speed of sound and particle velocity are greater at condensation than for infinitesimal sound (on axis). In the rarefaction sound speed is less than for infinitesimal sound and particle velocity is in the opposite direction. (*b*) Excess propagation Δx_2 at crest, compared to Δx_1 on axis. (*c*) Decomposition of distorted wave into first two harmonics of its Fourier components; P_1 and P_2 are rms pressures of the fundamental and the second harmonic.

A quantitative evaluation of the wave growth effect is obtained by using a geometrical argument (Black, 1940), as in Fig. A2.5.3. Two positions of the wave are shown. The infinitesimal incremental pressure near the axis propagates at speed c_A with negligible particle velocity traveling distance $\Delta x_1 = c_A \, \Delta t$ in time Δt. In the same time interval, using (A2.5.9), the peak of the wave travels

$$\Delta x_2 = c_A \left(1 + \frac{\beta \Delta \rho}{\rho_A}\right) \Delta t$$

The relative advance of the crest compared to an axial region is

$$\Delta X = \Delta x_2 - \Delta x_1 = \beta \frac{\Delta \rho}{\rho_A} c_A \, \Delta t \tag{A2.5.10}$$

Consider the distorted crest of the pressure pulse. To a first approximation, it is made up of a fundamental plus second harmonic with phase relation given in Fig. A2.5.3c. The pressure is

$$p \approx \sqrt{2}\, P_1 \cos kx + \sqrt{2}\, P_2 \sin 2kx$$

To find the relative strengths of the two harmonics notice that the peak, with zero slope, is given by

$$\frac{dp}{dx} = 0$$

therefore

$$P_1 k \sin kx \approx 2\, P_2 k \cos 2kx$$

For small kx (i.e., $k\Delta X$) we approximate $\sin (k\Delta X)$ by $k\Delta X$ and $\cos (2k\Delta X)$ by unity. Using the approximations and letting x become ΔX, P_2 is

$$P_2 \approx P_1 \frac{k\,\Delta X}{2}$$

where ΔX is, again, the relative advance of the peak compared to propagaion at the axis. Putting this together with (A2.5.10) the expression for the second harmonic in terms of the fundamental is

$$P_2 = \frac{\beta(\Delta\rho/\rho_A)kXP_1}{2} \tag{A2.5.11}$$

where $X = c_A \Delta t$. Or, using the approximation from (A2.5.1) with $P_1 \approx \Delta p$

$$\Delta\rho \approx \frac{P_1}{c_A^2}$$

we get

$$P_2 \approx \frac{\beta k X P_1^2}{2\rho_A c_A^2} \tag{A2.5.12}$$

Our derivation is for an unattenuating plane wave. Spherical waves, since they become weaker through divergence, require greater initial amplitudes to achieve the same degree of distortion. For plane waves the second harmonic growth is proportional to the intensity of the fundamental, P_1^2, and the number of wavelengths progressed by the fundamental kX.

Using $B/A = 5.0$ for water, the constant of proportionality β has the value 3.5. The value in air ($B/A = 0.40$) is $\beta = 1.2$. Comparing air and water, the strongest factor is $\rho_A c_A^2$, which is approximately 1.6×10^4 larger in water than in air. Therefore, for the same pressure amplitude of the fundamental, the second harmonic will grow to the same magnitude in a distance only $\frac{1}{5000}$ as far in air as in water. The sensitive listener often hears the popping, crackling sound of shock waves from jet airplane engines during takeoff.

The shock wave may be said to have formed when $P_2 = \frac{1}{2}P_1$. From (A2.5.12), the distance for that condition, for a plane wave is

$$X_0 = \frac{\rho_A c_A^2}{k P_1 \beta} \tag{A2.5.13}$$

or

$$X_0 \simeq (k\mathscr{E}\beta)^{-1} \tag{A2.5.14}$$

where

$$\mathscr{E} \equiv \text{acoustical Mach number} = \frac{u}{c_A} = \frac{P_1}{\rho_A c_A^2} \tag{A2.5.15}$$

For a lossless spherically diverging wave, the mature sawtooth wave form (Fig. A 2.5.2c) would occur at a much greater distance. The calculation of shock conditions at large ranges *must* take into account energy absorption as well as spherical divergence. This difficult problem has been studied by many investigators and is not repeated here (see Beyer 1974, pp. 91–164).

There is a saturation effect that limits the energy that can be put into a sound beam by a source. As the input power increases, more and more power goes into the higher harmonics which, with their larger attenuation rate, rapidly dissipate the new power. Consequently a level is reached where additional expenditure of energy at the source has no effect in producing increased sound pressures at large ranges in the field.

The saturation effect is illustrated in Fig. A2.5.4. On the abscissa are plotted the peak electrical power input to the transducer and the peak source level. There are six ranges at which the field pressure is measured as a function of the source level. If the response were linear, the sound pressure level, SPL, ($=20 \log_{10} P/1\mu$bar, [dB]) at the field position would rise along the 45° line that represents field pressure level proportional to source level.

The saturation effect is increasingly evident as one considers the larger ranges. For the four greatest ranges, the saturation pressure is indicated by the black circles at the right. At 11 yards the SPL is within 2 dB of saturation at a source level of 135 dB; this represents about 8 dB less than what would have been expected for linear behavior.

A subsidiary effect of nonlinear harmonic distortion is the modification of the beam radiation pattern. Since the axial part of the beam is at the greatest intensity, it suffers most relative to the field that is off-axis. As a result, the lesser degradation with increasing angle off-axis causes a

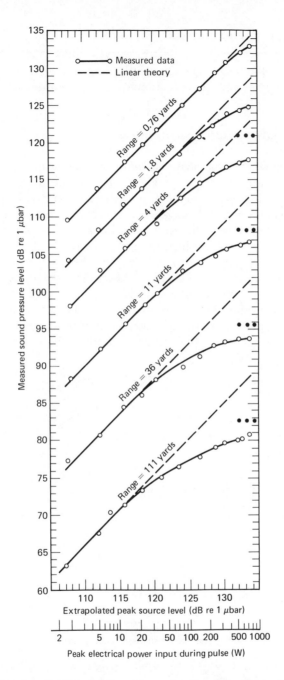

Figure A2.5.4. Amplitude response curves showing the effect of extra losses caused by nonlinear effects. Piston source of 3 in. diameter, $f = 454$ kHz. The black dots are at the saturation pressure levels. (Shooter et al. 1974.)

broadening of the beam. The effect also decreases the number of decibels by which the lobes are suppressed relative to the central part of the beam.

A2.5.2. Acoustic Streaming and Radiation Pressure

Having accepted the nonlinear facts of life (harmonic distortion and shock waves), it is not surprising to learn that intense sound waves cause unidirectional flow in the medium as well. This phenomenon was called the quartz wind in the early days when quartz transducers were used. It is characterized by an outward jetting or drifting of the water in front of any transducer that propagates high intensity sounds. In the case of a plane piston transducer propagating a beam in the open sea, the acoustic streams are strongest on the axis and close to the source, from which they may jet out for distances of the order of meters, gradually circulating back in a large eddy, or vortex, to the side of the transducer (see Fig. A2.5.5). The pattern is three-dimensional, and the sketch should be rotated about the axis of the source to produce the complete picture of the flow.

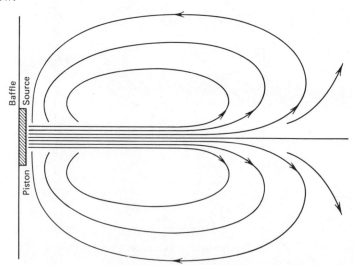

Figure A2.5.5.   Acoustic streaming in front of a piston source in a baffle.

Although the variety of acoustic patterns that occurs is extensive, the basic reason for the flow can be easily comprehended from the point of view of momentum transfer in a sound wave. Following an argument of Fox and Herzfeld (1950), consider a section of a beam (Fig. A2.5.6)

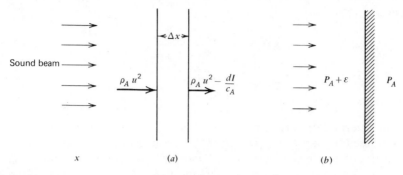

Figure A2.5.6. Acoustic streaming and radiation pressure, (*a*) Sound attenuation across Δx causes a lesser acoustic momentum to the right of Δx which is taken up by streaming momentum. (*b*) Radiation pressure of a sound beam at a wall.

carrying sound in the $+x$ direction. The average momentum carried through unit area in unit time is found by taking the time average of the product of the momentum per unit volume $\rho_A u$ by the particle velocity u.

$$\frac{\text{momentum}}{\text{area} \times \text{time}} = \langle \rho_A u^2 \rangle = \rho_A U^2 \qquad (A2.5.16)$$

where $U = $ rms particle velocity. $\langle\ \rangle$ means average over time.

The quantity we have calculated is the average energy density in the beam $\langle \varepsilon \rangle$. It is also the average intensity $\langle I \rangle = \langle p^2 \rangle / (\rho_A c_A)$ divided by the speed of sound.

$$\langle \varepsilon \rangle = \frac{\langle I \rangle}{c_A} = \frac{P^2}{\rho_A c_A^2} = \rho_A U^2 \qquad (A2.5.17)$$

For a real, dissipative medium such as water, we can calculate the spatial change of average momentum per unit time per unit area, which is equal to the change of average intensity across Δx divided by the speed

$$\frac{\Delta I}{c_A} = \frac{-2\alpha_e I\,\Delta x}{c_A} \qquad (A2.5.18)$$

where we have used the expression for attenuation of intensity,

$$\Delta I = -2\alpha_e I\,\Delta x$$

We have dropped the averaging symbol $\langle\ \rangle$ on I for simplicity.

From Newton's second law, the rate of change of momentum per unit area is the force per unit area or the change of pressure ΔP_A across the slab dx. Therefore

$$\Delta P_A = \frac{\text{change of momentum}}{\text{area} \times \text{time}}$$

$$\Delta P_A = \frac{+2\alpha_e I \, \Delta x}{c_A} \qquad (A2.5.19)$$

The interpretation is that the loss of acoustic momentum in the path Δx has created a change of pressure that if unsupported, will cause the fluid to drift. Otherwise stated, the loss of momentum from the acoustic beam is taken up by a gain of momentum of the fluid mass; thus conservation of momentum is fulfilled.

The detailed flow patterns that result in specific situations depend on the local absorption of energy and conversion from acoustic to drift momentum, the geometry of the region in which flow is driven by these distributed pressure sources, and the sound intensity. For example, Fig. A2.5.7 shows two different flow patterns that may be obtained around a

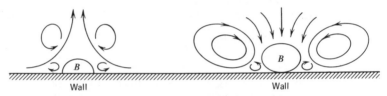

Figure A2.5.7. Acoustic streaming near a bubble attached to a wall. The shape and direction of the flow pattern depend on the sound intensity.

bubble attached to a wall (say, a transducer). The form of the secondary flow depends on the intensity of the sound, the contaminants on the surface, and the viscosity of the fluid (Elder, 1959). Streaming velocities thousands of times greater than the acoustic particle velocity have been observed. If a bubble forms on the face of a transducer, the patterns of Figs. A2.5.5 and A2.5.7 are superimposed. Similar vortex patterns form around particles in a sound field, but the velocities are generally smaller than for bubbles.

As in all nonlinear acoustic phenomena, the magnitude of the effect is proportional to the intensity of the sound; in this case it is the streaming velocity at any point that is proportional to the intensity. For example, if an ideal beam is assumed such that $P(r) = P$ for $r \leq a$ and $P(r) = 0$ for $a_o \geq r \geq a$, where r is the distance from the beam axis, a_o is the radius of an enclosing tube and a is the radius of the nondivergent sound beam, Eckart (1948) predicts that the streaming velocity, u_2, on the axis of the beam, will be

$$u_2(0) = \frac{2\pi^2 f^2 a^2 GIb}{\rho_A c_A^4}$$ (A2.5.20)

where $G = (a^2/a_o^2 - I)/2 - \log(a/a_o)$
$\quad b = \frac{4}{3} + \mu'/\mu$
$\quad I = P^2/(\rho_A c_A)$
$\quad \mu' = $ dynamic bulk viscosity
$\quad \mu = $ dynamic shear viscosity

This equation was used by Liebermann (1949) as a means for calculating the bulk viscosity μ' of water from measurements of the sound beam intensity and streaming velocity.

The streaming pattern is sensitive to the primary acoustic field. When the true primary field is known, it has been shown (Medwin, 1954) that predictions of the streaming velocity can be within 2% of the measured value. This means that a knowledge of both the primary field and the streaming velocity provides an accurate means for the determination of the bulk viscosity of the liquid. It has been so used by a number of researchers. An excellent summary of acoustic streaming has been written by Nyborg (1965).

The average energy density, $\langle \varepsilon \rangle$, is also a measure of the radiation pressure. If we measure the difference in static pressure between the inside and outside of a wall perpendicular to the beam (Fig. A2.5.6b), the inside pressure will be $\langle \varepsilon \rangle$ units larger than the outside static pressure. This greater pressure is called the Langevin radiation pressure. The equivalence of the energy density and the Langevin radiation pressure makes the quantity very suitable for direct determinations of sound beam intensity. The technique is frequently used in ultrasonic measurements where a hydrophone is too large to insert into the field without disturbing it so much that the measurement result is incorrect. Instead, a diaphragm is inserted perpendicular to the beam and the radiation force is measured.

SUGGESTED READING

Robert T. Beyer, *Nonlinear Acoustics*, published by Naval Sea Systems Command, 1974, and the Government Printing Office (0-596-215), 1975, 405 pp.

DIFFRACTION EFFECTS AND LOSSES TO THE MEDIUM

The simple ray theory requires corrections. We need to include the effects of diffraction when rays cross. By our simple theory, the sound pressure would become infinite as the rays cross. Although the pressure does become large, it is limited by diffraction effects.

Losses from the sound wave to the medium require the inclusion of absorption. We describe the physical mechanism and derive a modified wave equation. The solutions have the same form as the observed absorption.

A3.1. SOUND FIELD NEAR A FOCAL POINT

The ray trace intensity has an infinity or singularity where rays cross. Letting $h \sin \theta = L$ in (3.2.26), we have

$$P^2 = P_0^2 R_0^2 \Delta\theta \frac{\rho c}{\rho_0 c_0} \left| \frac{\sin \theta_0}{rL} \right| \tag{A3.1.1}$$

and (A3.1.1) goes to infinity as L goes to zero. This is similar to the focus of light by a lens. In Fig. A3.1.1, a cross section of the sound transmission, F is a focal circle around the z axis because of cylindrical symmetry. Its radius is about 65 km in the deep ocean.

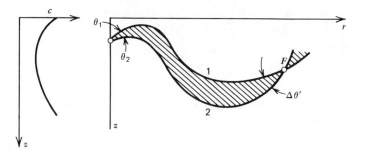

Figure A3.1.1.   Sound rays coming to a focus at F. Travel times along ray paths 1 and 2 are assumed to be the same.

We give a simplified treatment that illustrates a powerful procedure for diffraction problems. We make several approximations and simplifications: F is a straight line, normal to the figure; the sound speed in the region is constant over $\Delta\theta'$; an assumed cylindrical wave front that focuses at F is within $\lambda/8$ of the actual wave front. The theoretical model consists of an incoming cylindrical wave front that passes through F and continues outward.

We use Debye's method to solve the problem. Within the angle $\Delta\theta'$, each incoming ray has an associated plane wave (Fig. A3.1.2). At a given

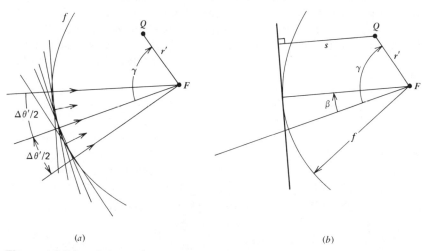

(a)   (b)

Figure A3.1.2.   Converging rays and associated wave fronts. (a) At the same instant of time all plane wave fronts are tangent to the cylinder with axis at F. (b) Distance s from a given plane wave front to Q.

instant all the plane waves are tangent to the cylinder. The field at Q is the integral over $\Delta\theta'$. For a plane wave front with the normal at angle β, the integral is

$$p(Q) \sim e^{i\omega t} \int_{-\Delta\theta'/2}^{\Delta\theta'/2} e^{-iks} \, d\beta \qquad (A3.1.2)$$

We calculate the proportionality constant after evaluation of the integral. From the geometry, we write

$$s = f - r' \cos(\gamma - \beta)$$

The phase kf is the same for all wave fronts and is ignored; $p(Q)$ is expressed

$$p(Q) \sim e^{i\omega t} \int_{-\Delta\theta'/2}^{\Delta\theta'/2} \exp\left[ikr' \cos(\gamma - \beta)\right] d\beta \qquad (A3.1.3)$$

The integral represents the expansion of a cylindrical wave in terms of plane waves and is sometimes called the Debye integral. The integral is similar to an integral expression of $J_0(z)$, the Bessel function of first kind (Equation 9.1.21, Abramowitz and Stegun, 1965), (A1.5.21)

$$J_0(z) = \frac{1}{\pi} \int_0^{\pi} \exp(iz \cos \theta) \, d\theta \qquad (A3.1.4)$$

Cylindrical Bessel functions are solutions of cylindrical coordinate problems. [People use different expressions for the expansion of a spherical wave in terms of plane waves: Brekhovskikh (1960) pp. 238 ff.]

In a large extended medium such as the ocean, we assume that the density of ray paths is a function of β, $D(\beta)$. Since aperture $\Delta\theta'$ is controlled by $D(\beta)$, the new limits are $-\pi/2$ to $\pi/2$. The modified integral is

$$p(Q) \sim e^{i\omega t} \int_{-\pi/2}^{\pi/2} D(\beta) \exp\left[ikr' \cos(\gamma - \beta)\right] d\beta \qquad (A3.1.5)$$

and $D(\beta)$ can be determined numerically by tracing a large number of rays into F.

For example, we specify $D(\beta)$ and give an evaluation of (A3.1.5). We assume that $D(\beta)$ has a broad maximum near $\beta = 0$ and is very small for $|\beta| > \theta'/2$. The error function is convenient, and we let

$$D(\beta) = \exp\left[\frac{-2\beta^2}{(\Delta\theta')^2}\right] \qquad (A3.1.6)$$

The coefficient of β^2 is chosen to match $\langle p(Q)p^*(Q)\rangle$ at large r' to the incoming beam. This choice is verified later in (A3.1.14). The expansion of $\cos (\gamma - \beta)$ for small β gives

$$\cos (\gamma - \beta) = \cos \gamma \cos \beta + \sin \gamma \sin \beta$$

$$\approx \cos \gamma \left(1 - \frac{\beta^2}{2}\right) + \beta \sin \gamma \qquad (A3.1.7)$$

With these substitutions, $p(Q)$ becomes

$$p(Q) \sim [\exp (i\omega t + ikr' \cos \gamma)] \int_{-\pi/2}^{\pi/2} \exp (-a^2\beta^2 + ib\beta) \, d\beta \qquad (A3.1.8)$$

where

$$a^2 \equiv \frac{2}{\Delta\theta'^2} \left(1 + ikr' \Delta\theta'^2 \frac{\cos \gamma}{4}\right) \qquad (A3.1.9)$$

$$b \equiv kr' \sin \gamma$$

We convert this to an integral that is evaluated in closed form by altering the limits. Ordinarily $\Delta\theta' \ll 1$ and $D(\beta)$ is negligible for $\theta > 1$. The limits of integration are unimportant when they are wider than -1 to $+1$. We replace $\pm \pi/2$ by $\pm\infty$ and evaluate (A3.1.8) by using a standard definite integral table (A1.5.6)

$$p(Q) \sim \exp (i\omega t + ikr' \cos \gamma) \frac{\exp (-b^2/4a^2)}{a} \qquad (A3.1.10)$$

We keep the dependence on r' and γ and let other terms go into the constant of proportionality. For $\gamma = 0$, $(\omega t + kr')$ gives an incoming wave and for $\gamma = \pi$, $(\omega t - kr')$ gives an outgoing wave front.

The mean square value of $p(Q)$ is

$$P^2(Q) \approx \langle p(Q)p^*(Q) \rangle$$

$$P^2(Q) = \frac{C}{a_f} \exp \left(-\frac{\Delta\theta'^2 \, k^2 r'^2 \sin^2 \gamma}{4a_f^2}\right) \qquad (A3.1.11)$$

where

$$a_f^2 \equiv 1 + \frac{k^2 r'^2 \, \Delta\theta'^4 \cos^2 \gamma}{16} \qquad (A3.1.12)$$

and C is a constant of proportionality. The focal region has a narrow maximum along $\gamma = 0$.

The last step is to connect $P^2(Q)$ for the focal region calculation to P^2 that is given by (A3.1.1). To do this we move Q along $\gamma = 0$ far enough to be relatively free of diffraction effects, using $kr' \gg 4/\Delta\theta^2$ as a criterion. In this region we have

$$a_f \simeq kr' \, \Delta\theta'^2 \frac{\cos \gamma}{4} \qquad (A3.1.13)$$

The limiting value of $P(Q)$ at very large r' and small γ is

$$P^2(Q) \simeq \frac{C}{kr' \Delta\theta'^2 \cos \gamma} \exp\left[-\frac{\tan^2 \gamma}{(\Delta\theta'/2)^2} \right] \qquad (A3.1.14)$$

The e^{-1} values of $P^2(Q)$ are $\gamma \simeq \Delta\theta'/2$, and this verifies our choice of $D(\beta)$. The width of the beam L is given by $r'\Delta\theta'$. We equate (A3.1.14) and (A3.1.1) to evaluate the constant of proportionality C at $\gamma = 0$. The substitution of the value of C in (A3.1.11), and inclusion of attenuation αR due to absorption, gives

$$P^2(Q) = kP_0^2 R_0^2 \Delta\theta \, \Delta\theta' \frac{\rho c}{\rho_0 c_0} \frac{\sin \theta_0}{r a_f}$$

$$\times \left[\exp\left(-\Delta\theta'^2 k^2 r'^2 \frac{\sin^2 \gamma}{4 a_f^2} \right) \right] 10^{-\alpha R/10} \quad \textbf{(A3.1.15)}$$

where Q is located by the polar coordinates $(r' \cdot \gamma)$; $P^2(Q)$ in a convergence zone is very much larger than the signal for a spherically spreading wave front at the same range.

We have given the solution for a simple symmetric focusing zone. The modified Debye integral (A3.1.5) has the same form of the error function integral when $D(\beta)$ is expressed as $D(\beta) = \exp(-b_1\beta - b_2\beta^2)$ and the limits are θ_1 and θ_2. The sharp edge of a caustic can be represented by letting $D(\beta)$ be zero outside θ_1 or θ_2.

Illustrative Problem. Compute $P(Q)$ at the focus $F(r' = 0)$ for a 1 kW source that transmits a 100 Hz signal in the deep ocean. In Fig. A3.1.3 the source and receiver are at 270 m depth and are separated by approximately 62 km. The receiver is at the focus of the ray paths. The grazing angles for the two limiting ray paths are 3.43° and 9.74° at the source and receiver. Therefore $\gamma = 0$ is along the mean grazing angle 6.58° and $\Delta\theta = \Delta\theta' = 6.31°$ or 0.11 rad.

First we calculate P_0 for a 1 kW source at $R_0 = 1$ m, from

$$P_0^2 = \frac{\rho_0 c_0 \Pi}{4\pi R_0^2}$$

where

$$\rho_0 = 1.04 \times 10^3 \text{ kg/m}^3$$

$$c_0 = 1.52 \times 10^3 \text{ m/s}$$

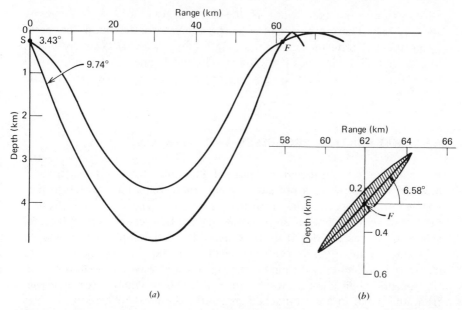

Figure A3.1.3. Limiting ray traces and focal region, drawn with same vertical exaggeration. (*a*) Ray traces taken from Fig. 3.2.7. (*b*) Focus or convergence region. Rays have grazing angles of 12.2° and 15.19° at the axis of the sound channel.

yielding
$$P_0^2 = 1.26 \times 10^8 \text{ Pa}^2$$

Also
$$\alpha = 10^{-3} \text{ dB/km}, \qquad k = 0.413 \text{ m}^{-1},$$
$$\rho c = \rho_0 c_0, \qquad r = 6.2 \times 10^4 \text{ m}$$

At F, $r' = 0$, and $a_f = 1$. The evaluation of (A3.1.15) gives
$$P(F) = 3.0 \text{ Pa}$$

For comparison, we calculate P at the same range for spherical spreading, $R = 6.2 \times 10^4$ m
$$P = \frac{P_0 R_0}{R}$$
$$P = 0.18 \text{ Pa}$$

Comparison shows that the signal in the convergence zone is 24 dB larger than the signal expected for spherical spreading wave fronts.

At $\gamma = \pi/2$, the half width of the focal region, to e^{-1}, is given by $\Delta\theta' kr'/2 = 1$. Therefore $r' = 44$ m perpendicular to the main ray path. Using the e^{-1} criterion again, the length along $\gamma = 0$ is given by $a_f = e$. Therefore (A3.1.12) gives $r' = 2200$ m. The focal zone is long and narrow along the mean ray path of grazing angle 6.58°. It is sketched in Fig. A3.1.3b.

A3.2. RELAXATION PROCESSES AND ATTENUATION

Relaxation processes occur because of the finite time it takes for the medium to respond to a pressure change, or to relax back to the former state after the pressure has returned to normal. The proposed physical models of acoustic relaxation processes have taken several forms: thermal relaxation in gases is described in terms of the activation of vibrational or rotational modes of polyatomic molecules so that the specific heats are changed incrementally during passage of a sound wave; structural relaxation occurs when a sound wave causes a liquid to change, for example, from an icelike to a close-packed structure; chemical relaxation involves ionic dissociation that is alternatively activated and deactivated by sound condensations and rarefactions.

The detailed explanations of the relaxation mechanisms for water are still being actively debated by physical acousticians and physical chemists. This section describes in a general way how relaxation processes can cause absorption and dispersion, and summarizes the quantitative predictions of theory and the empirical formulas of laboratory and ocean attenuation experiments.

A3.2.1. Sound Absorption and Dispersion

To obtain the frequency dependence of the relaxation effect we add a time-dependent term to Hooke's law

$$\Delta p = c^2 \Delta\rho + b\frac{d(\Delta\rho)}{dt} \tag{A3.2.1}$$

where b is a constant. To obtain the "relaxation time," assume that we apply a pressure Δp for all of minus time and release it at $t = 0$. From time $t = 0$, $\Delta p = 0$ and

$$\frac{d(\Delta\rho)}{\Delta\rho} = -\frac{c^2}{b}\, dt \tag{A3.2.2}$$

Integrate, and let $\Delta\rho = \Delta\rho_0$ at $t = 0$; this yields

$$\Delta\rho = \Delta\rho_0 \exp\left(-\frac{t}{\tau_r}\right) \tag{A3.2.3}$$

where

$$\tau_r \equiv \frac{b}{c^2} \tag{A3.2.4}$$

is the relaxation time of the process.

The value of $\Delta\rho$ depends on the fraction of molecules active in the process and this is a function of temperature and pressure. For simplicity we assume at this point that *all* molecules are actively relaxing.

To determine the effect of a relaxation process on the propagating signal, we insert the modified Hooke's law into the derivation of the wave equation. The substitution of (A3.2.1) into (2.7.6), and $\partial/\partial t \simeq d/dt$ gives

$$\frac{\partial^2(\Delta\rho)}{\partial x^2} + \tau_r \frac{\partial^3(\Delta\rho)}{\partial x^2 \partial t} = \frac{1}{c^2}\frac{\partial^2(\Delta\rho)}{\partial t^2} \tag{A3.2.5}$$

Consider harmonic time dependence $\exp(i\omega t)$ and spatial dependence ρ_s

$$\Delta\rho = \rho_s\, e^{i\omega t} \tag{A3.2.6}$$

The wave equation becomes

$$\frac{\partial^2 \rho_s}{\partial x^2}(1 + i\omega\tau_r) + \frac{\omega^2}{c^2}\rho_s = 0 \tag{A3.2.7}$$

$$\frac{\partial^2 \rho_s}{\partial x^2} + \frac{\omega^2}{c^2}(1 - i\omega\tau_r)(1 + \omega^2\tau_r^2)^{-1}\, \rho_s = 0$$

We assume a solution

$$\rho_s \sim \exp[-(ik + \alpha_e)x] \tag{A3.2.8}$$

where the wave number k and the exponential attenuation rate α_e are to be determined. Substitution gives

$$-(k^2 - \alpha_e^2) + \frac{\omega^2}{c^2}(1 + \omega^2\tau_r^2)^{-1} + i\left[2k\alpha_e - \frac{\omega^3\tau_r}{c^2}(1 + \omega^2\tau_r^2)^{-1}\right] = 0 \quad (A3.2.9)$$

In complex variables, reals equal reals and imaginaries equal imaginaries; therefore

$$k^2 - \alpha_e^2 = \frac{\omega^2}{c^2}(1 + \omega^2\tau_r^2)^{-1} \tag{A3.2.10}$$

$$2k\alpha_e = \frac{\omega^3\tau_r}{c^2}(1 + \omega^2\tau_r^2)^{-1} \tag{A3.2.11}$$

We define k in terms of the phase velocity $c_p \equiv \omega/k$, the velocity of the crest of a wave front in the direction of propagation. After some algebraic manipulation, the substitution of k and τ_r and the simultaneous solution of (A3.2.10) and (A3.2.11) gives

$$\alpha_e = \frac{\omega^2 \tau_r c_p}{2c^2(1+\omega^2\tau_r^2)} \tag{A3.2.12}$$

and

$$c_p = (c\sqrt{2})(1+\omega^2\tau_r^2)^{1/2}\,[(1+\omega^2\tau_r^2)^{1/2}+1]^{-1/2} \tag{A3.2.13}$$

The phase velocity depends on the sound frequency and the relaxation frequency; the medium is dispersive. Study of acoustic relaxation mechanisms for water shows that c_p differs from c by less than 1%.

We concentrate on the attenuation, use the approximation $c_p \simeq c$, define the relaxation frequency

$$f_r \equiv \frac{1}{2\pi\tau_r} \tag{A3.2.14}$$

and obtain the characteristic form for relaxation processes in which all molecules participate.

$$\alpha_e = \frac{(\pi f_r/c)f^2}{f_r^2+f^2}\frac{\text{nepers}}{\text{distance}} \tag{A3.2.15}$$

For a fraction F of participating molecules, multiply by F. For attenuation in decibels per unit distance, use the conversion 1 neper $= 8.68$ dB.

The attenuation rate for a relaxation process is therefore

$$\alpha_r = \frac{Af_r f^2}{f^2+f_r^2}\frac{\text{dB}}{\text{distance}} \tag{A3.2.16}$$

where $A = 8.68(\pi/c)F$ and F is the fraction of participating molecules. This reduces to

$$\alpha_r = Af_r \qquad \text{for} \quad f \gg f_r$$

$$\alpha_r = \frac{A}{f_r}f^2 \qquad \text{for} \quad f \ll f_r$$

We note that α_r is proportional to f^2 below f_r and tends to a constant above f_r. This is characteristic of a relaxation process.

Another characteristic appearance of relaxation phenomena is demonstrated by rewriting (A3.2.16) to exhibit the relaxation loss per cycle:

$$\frac{\alpha_r}{f} = A\frac{ff_r}{f^2+f_r^2} \tag{A3.2.17}$$

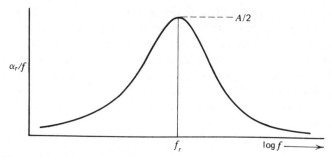

Figure A3.2.1. Attenuation per cycle in a relaxation process.

For $f \ll f_r$, $\alpha_r/f = Af/f_r \ll 1$ and also for $f \gg f_r$, $\alpha_r/f = Af_r/f \ll 1$. The peak occurs at $f = f_r$ where $\alpha_r/f = A/2$. Figure A3.2.1 is the graph of (A3.2.17).

Physically, when the medium is perturbed by a sound wave of period very much less than τ_r, the internal changes cannot respond fast enough, therefore they are not activated. At the other limit, when the sound period is very much greater than the relaxation time, the internal rearrangements can keep step with the impressed sound field. At both extremes there is no significant energy loss *per cycle*. When the sound period is approximately equal to the relaxation time, however, energy can be absorbed internally during the condensation phase of the cycle and dumped into thermal motion during the rarefaction, and this action produces significant energy loss *per cycle*.

A3.2.2. Classical Predictions of Sound Absorption: Attenuation in Fresh Water

Our derivations of the momentum equation (2.7.3) and (A2.2.6) were incomplete, in that the relatively small forces of viscosity were omitted. Their inclusion results in the Navier-Stokes momentum equation, which is derived in any good book on fluid dynamics (e.g. L. D. Landau and E. M. Lifshitz, *Fluid Mechanics*, Addison-Wesley, Reading, Mass, 1959).

$$\rho_A \left(\frac{\partial \mathbf{U}}{\partial t} + \mathbf{U} \cdot \nabla \mathbf{U} \right) = -\nabla p + \left(\frac{4\mu}{3} + \mu' \right) \nabla \nabla \cdot \mathbf{U} - \mu \nabla \times (\nabla \times \mathbf{U})$$

$$\textbf{(A3.2.18)}$$

Comparing with (A2.2.1) and (A2.2.6), the two new terms introduce the dynamic (or absolute) coefficient of shear viscosity μ and the dynamic (or absolute) coefficient of bulk (or compressional) viscosity μ'. Often the adjectives are dropped and μ is called, loosely, the viscosity. The bulk viscosity is less well known; it appears only when the fluid is compressible, so that $\nabla \cdot \mathbf{U}$ exists. These are the problems of acoustics (and high speed compressible flow), and therefore we are interested in the bulk viscosity as well as the shear viscosity. For both μ and μ' the assumption of linearity has been made ("Newtonian" fluids); thus, for example, the relation of the shear stress to the rate of strain is the simple proportionality

$$\frac{F}{A} = \mu\left(\frac{\partial u}{\partial y} + \frac{\partial v}{\partial x}\right)$$

where $\mu = $ constant; F is applied parallel to the area A, and u, v and w are components of the velocity in the x, y, and z directions, respectively.

For sound propagation the viscous terms on the right-hand side of (A3.2.18) are much smaller than Δp. Nevertheless, these terms are all important when we are interested in sound attenuation.

To calculate the attenuation for a plane wave of particle velocity u propagating in the x direction we reduce (A3.2.18) to the one-dimensional form by using $v = w = 0$ and $\partial u/\partial y = \partial u/\partial z = 0$. Also assume $\mathbf{U} \cdot \nabla \mathbf{U} \ll \partial \mathbf{U}/\partial t$, as in Section A2.2 to get

$$\rho_A \frac{\partial u}{\partial t} = -\frac{\partial p}{\partial x} + \left(\frac{4\mu}{3} + \mu'\right)\frac{\partial^2 u}{\partial x^2} \qquad \text{(A3.2.19)}$$

Differentiate with respect to x and use the conservation of mass equation $\rho_A(\partial u/\partial x) = -\partial \rho/\partial t$ and the plane wave relation $p = \rho c^2$ to convert to the wave equation in terms of ρ.

$$\frac{+\partial^2 \rho}{\partial t^2} = \frac{c^2 \partial^2 \rho}{\partial x^2} + \frac{(4\mu/3 + \mu')}{\rho_A}\frac{\partial^3 \rho}{\partial t\,\partial x^2} \qquad \text{(A3.2.20)}$$

If we divide both sides by c^2, direct comparison with the wave equation in terms of relaxation (A3.2.5) allows an interpretation of μ and μ'. The ρ in (A3.2.20) is $\Delta \rho$ in (A3.2.5), and

$$\tau_r = \frac{4\mu/3 + \mu'}{\rho_A c^2} \qquad \textbf{(A3.2.21)}$$

For example, for freshwater at 14°C (subscript F),

$$\mu_F \approx 1.17 \times 10^{-3}\,\text{N} \cdot \text{s/m}^2$$

$$\mu_F' \approx 2.8\,\mu$$

$$\rho_F = 10^3\,\text{kg/m}^3$$

$$c_F \approx 1.48 \times 10^3\,\text{m/s}$$

Therefore

$$\tau_r \simeq 2.1 \times 10^{-12} \text{ s}$$

This very short relaxation time means that for all realistic frequencies in underwater acoustics, we have

$$f \ll f_r = \frac{1}{2\pi\tau_r} \tag{A3.2.22}$$

Therefore from (A3.2.15), using (A3.2.21) and (A3.2.22) for freshwater, we write

$$\alpha_{eF} = \left(\frac{4\mu_F/3 + \mu_F'}{2\rho_F c_F^3} \right) \omega^2 \frac{\text{np}}{\text{distance}} \tag{A3.2.23}$$

It was originally assumed that for all fluids $\mu' = 0$, and publications until the early 1950s implied a discrepancy between the predictions of (A3.2.23) and experiment. Finally, acoustic streaming theory (Section A2.5.2, Eckart, 1948) and experiments (Liebermann, 1949; Medwin, 1954) showed that only in the case of monatomic gases is $\mu' = 0$.

Values of μ_F and μ_F' at atmospheric pressure are plotted as a function of temperature in Fig. A3.2.2. For the temperature range 0°–50°C it turns out that μ_F'/μ_F ranges from 3.11 to 2.68. The attenuation at 14°C and sea level pressure has been calculated from these values and is shown in Fig. 3.3.1.

There is a small change of viscosity coefficients with increasing depth. The pressure effect is less for higher water temperatures (Litovitz and Carnevale, 1955), and in all cases it is small enough for a linear dependence on depth to be a good approximation.

At 0°C this can be approximated by

$$\mu = \mu_{\text{sea level}} \times (1 - 1.2 \times 10^{-4} P_A) \tag{A3.2.24}$$

and

$$\mu' = \mu'_{\text{sea level}} \times (1 - 1.2 \times 10^{-4} P_A) \tag{A3.2.25}$$

where P_A is ambient pressure (atm).

A3.2.3. Attenuation due to Magnesium Sulfate Molecular Relaxation

It became very clear during World War II that the attenuation of sound in seawater was approximately 25 times that in freshwater at military sonar frequencies (20 kHz). Molecular relaxation, first described by Einstein

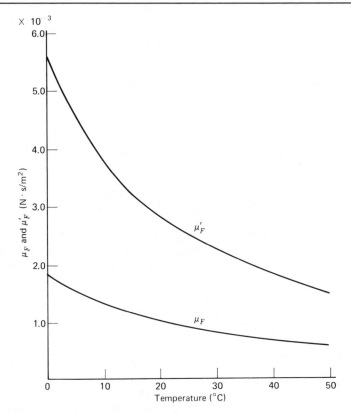

Figure A3.2.2. Dynamic shear and bulk viscosities as a function of temperature at atmospheric pressure for freshwater. Bulk viscosity values derived from Pinkerton (1949).

and later used by H. Kneser to successfully analyze sound absorption in air, was suggested as an explanation for this property in seawater. Sodium chloride was proposed as the culprit.

In 1949, R. W. Leonard, using a water-filled sphere as a resonator, studied the decay times of the sphere as a function of frequency with different salt additives in the distilled water; he identified magnesium sulfate ($MgSO_4$) as the ingredient in seawater that caused the excess attenuation. The sodium chloride, by itself, causes no absorption.

The $MgSO_4$ relaxation is now considered to be a three-step reaction process including ionic dissociation and unimolecular steps (Eigen and Tamm, 1962). The results of numerous laboratory and field experiments have been summarized by an empirical equation (Schulkin and Marsh,

1962) which has the typical form of attenuation due to relaxation. To take account of the depth dependence we multiply by a pressure factor measured by Bezdek (1973). The subscript m indicates magnesium sulfate relaxation.

$$\alpha_m = \frac{SA'f_{rm}f^2}{f^2 + f_{rm}^2}(1 - 1.23 \times 10^{-3}\, P_A)\, dB/m \qquad \textbf{(A3.2.26)}$$

where

$$f_{rm} = 21.9 \times 10^{[6 - 1520/(T+273)]}\, kHz \qquad \textbf{(A3.2.27)}$$
$$= \text{relaxation frequency for MgSO}_4$$
$$T = \text{temperature (°C)}$$
$$A' = 2.03 \times 10^{-5}$$
$$S = \text{salinity (ppt)}$$
$$f = \text{frequency (kHz)}$$
$$P_A = \text{ambient pressure (atm)}$$

The magnesium sulfate effect is shown in Fig. 3.3.1 for $T = 14°C$, $S = 35$ ppt, and $P_A = 1$ atm.

The temperature dependence of the relaxation frequency is presented in Fig. A3.2.3, plotted from (A3.2.27).

For a good summary of this work, see Eigen and Tamm (1962).

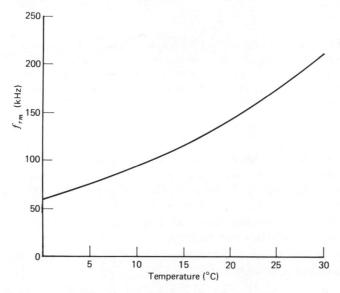

Figure A3.2.3. Variation of relaxation frequency for magnesium sulfate reaction f_{rm} as a function of temperature.

A3.2.4. Attenuation due to Boric Acid Molecular Relaxation

By the mid-1960s substantial evidence had accumulated (Thorp, 1965; Skretting and Leroy, 1971) that at frequencies below 10 kHz the sound attenuation in seawater is up to 10 times the value predicted by the magnesium sulfate relaxation mechanism. The component of seawater that has been indicted for this low frequency relaxation is boric acid (Yeager et. al., 1973; Fisher and Levison, 1973; Simmons, 1975). However carbonates affect the relaxation frequency of the boric acid relaxation as does the $B(OH)_3$ concentration.

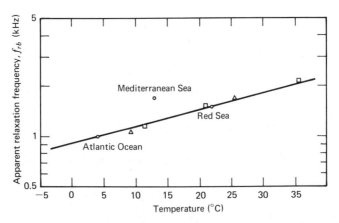

Figure A3.2.4. Relaxation frequency for boric acid reaction f_{rb} as a function of temperature. (Simmons, 1975.) Laboratory data are squares (Simmons) and triangles (Yeager et al., 1973). Ocean data (Mellen and Browning, 1975) are circles.

The relaxation frequency is near 1 kHz, depending on the temperature. The existence of only a weak dependence on pressure is implied by the ability to fit laboratory sea level data on the same curve as deep ocean data (Fig. A3.2.4). However Mediterranean and Pacific Ocean data (not shown) appear to be anomalous, possibly because of a different carbonate content or $B(OH)_3$ concentration.

Adapting the results of Simmons (1975), the attenuation due to boric acid relaxation is written

$$\alpha_{rb} = A'' \frac{f^2 f_{rb}}{f_{rb}^2 + f^2} \text{ dB/m} \qquad \textbf{(A3.2.28)}$$

where

f_{rb} = boric acid relaxation frequency (kHz) = $0.9(1.5)^{T/18}$
T = temperature (°C)   **(A3.2.29)**
f = sound frequency (kHz)
A'' = 1.2×10^{-4} dB/(m·kHz)

The attenuation due to the boric acid relaxation effect is plotted in Fig. 3.3.1 for water temperature 14°C The temperature dependence of the relaxation frequency f_{rb} is plotted in Fig. A3.2.4, from (A3.2.29).

FOURIER TRANSFORMATIONS AND APPLICATIONS

The purpose of the Fourier transformation is to express the signal as a sum of sinusoidal signals. When the Fourier transformation is applied to a time-dependent function, the result is a frequency domain representation of the signal. Application of the transformation to a frequency domain representation gives the time domain. It is a powerful tool because we can use solutions for sinusoidal signals to synthesize complicated signals.

Many digital signal processing procedures are done in the frequency domain because it is easy to express the signals, noise, and system response in the frequency domain. Computational procedures are well known, and the *fast Fourier transformation* (FFT) is a standard subroutine for computers.

In using the FFT for signal analysis the signal is sampled at equally spaced discrete intervals over a finite length of time. This produces a sequence of N numbers. The transformation of these numbers gives N coefficients. Application of the transformation equations to the N coefficients yields exactly the original sequence of N numbers. Since the transformation relates a *finite* number of data points to the *same finite* number of coefficients and the *same finite* number of terms in the expansion, we call it the *finite Fourier transformation* (FFT). Thus FFT stands for both the transformation and an efficient computational algorithm. The transformation is also known as the *discrete Fourier transformation* (DFT).

424

Fourier transformation theory is discussed in many texts in mathematics, physics, and engineering. We give results of the theory without proofs. Texts for further study are suggested at the end of Chapter 4.

A4.1. FINITE FOURIER TRANSFORMATION

The finite Fourier transformation changes N samples of the signal into N amplitude coefficients of sine and cosine functions. We assume that the signal is turned on at $t = 0$ and turned off just before $t = T_1$ (Fig. A4.1.1).

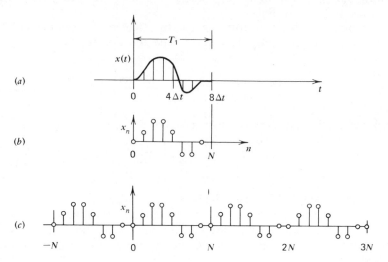

Figure A4.1.1. Representation of sampled data by finite Fourier transformation. The FFT description of (A4.1.4) has generated a periodic series that has the period N. (a) Data. (b) Sampled data. (c) FFT.

The signal is sampled at the times 0, Δt, $2\Delta t$, ..., $(N-1)\,\Delta t$, where $\Delta t = T_1/N$. Since we begin counting at zero, there are N samples. The nature of the signal outside of this region, $0 \le t < T_1$, does not enter into the calculation. The signal can be time dependent $x(t)$, as in this development, or a function of space. We designate the value of the nth sample $x(n\,\Delta t)$ as x_n and the time of the nth sample $n\,\Delta t$ as t_n:

$$x_n \equiv x(t_n)$$
$$t_n \equiv n\,\Delta t$$
$$\Delta t \equiv \frac{T_1}{N}$$

(A4.1.1)

The arguments of the sine and cosine functions ωt are replaced by $2\pi f_m t_n$, where

$$f_m \equiv mf_1$$

$$f_1 \equiv \frac{1}{T_1} \qquad \text{(A4.1.2)}$$

$$2\pi f_m t_n = \frac{2\pi mn}{N}$$

where m represents the frequency mf_1 and n represents the time $n\,\Delta t$. Even though m and n are dimensionless, it is helpful to associate m with frequency and n with time.

The purpose of the transformation is to calculate the coefficients X_m, when the x_n are given, and vice versa. The transformation equation for X_m is

$$X_m = \frac{1}{N} \sum_{n=0}^{N-1} x_n \exp\left(-\frac{i2\pi nm}{N}\right) \qquad \text{(A4.1.3)}$$

Given the X_m, the (inverse) transformation for the x_n is

$$x_n = \sum_{m=0}^{N-1} X_m \exp\left(\frac{i2\pi nm}{N}\right) \qquad \text{(A4.1.4)}$$

The complex amplitude coefficients are capitals (X_m) and lowercase letters are samples (x_n). Proofs of these transformations and most of the other relations in the appendix are left to reader. One step in the proof is important because it leads to the definition of the *discrete delta function* $\delta(n-n')$.

$$\delta(n-n') = 1 \qquad \text{for} \quad n = n'$$

$$= 0 \qquad \text{otherwise} \qquad \text{(A4.1.5)}$$

$$\delta(0) = 1$$

The transformation of $\delta(n-n')$ is

$$\delta(n-n') = N^{-1} \sum_{m=0}^{N-1} \exp\left[\frac{i2\pi(n-n')m}{N}\right] \qquad \text{(A4.1.6)}$$

It can be shown that the summation in (A4.1.6) reduces to $\delta(n-n')$ by recognizing that it is the sum of a geometric series, where the ratio is $r = \exp[2\pi i(n-n')/N]$. The magnitudes of amplitude coefficients of the delta function are equal. The proof is accomplished by substituting (A4.1.4) in (A4.1.3) and using (A4.1.6).

It is a characteristic of the finite Fourier transformation that both x_n and the coefficients X_m are periodic.

$$x_n = x_{n+kN} \quad \text{for} \quad k = 0, \pm 1, \pm 2, \ldots$$

$$X_m = X_{m+kN} \tag{A4.1.7}$$

$$x_{-n} = x_{kN-n}$$

$$X_{-m} = X_{kN-m}$$

When the x_n are real, $x_n = x_n^*$. By taking the complex conjugate of (A4.1.3) and replacing m by $-m$, we obtain for real x_n:

$$X_{-m} = X_m^*$$

$$X_{-m}^* = X_m \tag{A4.1.8}$$

$$X_m = X_{kN-m}^*$$

Even though the initial set of N data points did not have period N, the transformation *creates* a series that is periodic, as illustrated Fig. A4.1.1. Given $x(t)$, a short segment of data (Fig. A4.1.1a). The sample data points (Fig. A4.1.1b) are substituted in (A4.1.3), whereupon the X_m are determined. In turn, the X_m can be substituted in (A4.1.4) to express the sequence x_n. The result is a series having the period N. The effect of doubling N for the same signal and sampling rate is represented in Fig. A4.1.2. The signal is relatively isolated from the periodic repetition in Fig. A4.1.2, whereas in Fig. A4.1.1c it takes on the character of a sine wave.

Because of the periodicity of the coefficients, the summation is complete when it is over any sequence of N terms. In the interpretation of

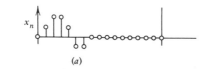

(a)

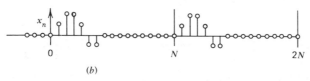

(b)

Figure A4.1.2. Representation of sampled data (a) by finite Fourier transformation (b). The value of N has been doubled relative to that in Fig. A4.1.1, but the data signal is the same.

experimental data, we write the summation

$$x_n = \sum_{m=1-N/2}^{N/2} X_m \exp\left(\frac{i2\pi nm}{N}\right) \tag{A4.1.9}$$

to put the frequency range in its smallest absolute range; that is, $|m| \leq N/2$. This corresponds to the frequency coefficients for less than half the sampling frequency.

The information in the range of $mf_1 > Nf_1/2$ does not represent information at higher frequencies, but rather, information in the range of $-Nf_1/2$, which is shifted up. We call $Nf_1/2$ the folding frequency. Obviously there is a problem if the original signal had frequency components above the folding frequency. The essential result is that our knowledge of the existence of frequency components for $|f| > |Nf_1/2|$ is lost, but the effect is retained because the higher frequency signal components are shifted to the region $|f| < |Nf_1/2|$.

The sampling theorem states that the signal can be reconstructed if it is sampled at a frequency greater than twice the bandwidth of the signal. When we sample at a rate of at least 4 or 5 times the bandwidth—that is, $|f| < |Nf_1/4|$—smoothing between data points is a *simple* reconstruction procedure.

A4.1.1. Numerical Examples of FFT Operations

We use the discrete delta function $\delta(n)$ to demonstrate the operations (a) computation of the FFT coefficients, (b) time shifting, and (c) addition. These numerical calculations are the first check on our FFT computer programs. The δ function (Fig. A4.1.3a) has the value unity at $n = 0$ and is zero for other values of n.

$$x_n = \delta(n)$$

COMPUTATION OF THE FFT COEFFICIENTS

The values of n go from zero to $N-1 = 7$,

$$N = 8$$

Values of $\exp(-i2\pi mn/N)$ (for all the transformations) are given in Table A4.1.1.

Using (A4.1.3), we write

$$X_m = \frac{1}{8} \sum_{n=0}^{7} x_n \exp\left(-\frac{i\pi nm}{4}\right)$$

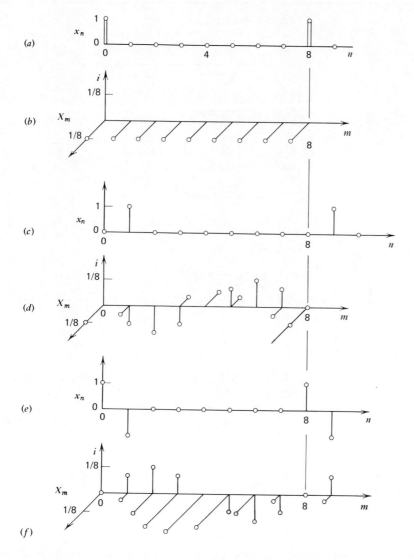

Figure A4.1.3. Examples of the finite Fourier transform, giving function x_n and its corresponding spectrum X_m for three functions.

TABLE A4.1.1.1

mn	$\pi nm/4$	$\exp(-i\pi mn/4)$
0	0	1
1	$\pi/4$	$(1-i)/\sqrt{2}$
2	$\pi/2$	$-i$
3	$3\pi/4$	$-(1+i)/\sqrt{2}$
4	π	-1
5	$5\pi/4$	$-(1-i)/\sqrt{2}$
6	$3\pi/2$	$+i$
7	$7\pi/4$	$(1+i)/\sqrt{2}$

For $x_0 = 1$, $x_1 = \cdots = x_7 = 0$,

$$X_m = \tfrac{1}{8}\exp(0) = \tfrac{1}{8}$$

The spectral coefficients appear in Fig. A4.1.3b. All the X_m are equal and real.

TIME SHIFTING

The time shifted signal is $N = 8$, $x_n = 0, 1, 0, 0, 0, 0, 0, 0$; X_m is given in Fig. A4.1.3c. The summation (A4.1.3) reduces to

$$X_m = \tfrac{1}{8}\exp\left(-\frac{i\pi m}{4}\right) \tag{A4.1.10}$$

The spectrum of the time-shifted signal is shown in Fig. A4.1.3d. Real and imaginary components are plotted separately and are not summed in the figure.

ADDITION

The FFT of the linear combination of two signals is the sum of the transformation of each series. Let the signal be as shown in Fig. A4.1.3e,

$$x_n = \delta(n) - \delta(n-1), \qquad N = 8 \tag{A4.1.11}$$

$$x_n = 1, -1, 0, 0, 0, 0, 0, 0 \tag{A4.1.12}$$

This is equivalent to the linear combination,

$$x_n = ax_n' + bx_n'' \tag{A4.1.13}$$

where

$$x'_n = \delta(n)$$
$$x''_n = \delta(n-1)$$
$$a = 1, \qquad b = -1 \tag{A4.1.14}$$

The spectral coefficients are

$$X_m = aX'_m + bX''_m \tag{A4.1.15}$$

Using the spectra involving computation of the FFT coefficients and time shifting, we find

$$X_m = \frac{1}{8}\left[1 - \exp\left(-\frac{i\pi m}{4}\right)\right] \tag{A4.1.16}$$

The spectrum is plotted in Fig. A4.1.3f.

In these examples, the x_n are real. The spectra have the conjugate relation $X_m = X^*_{N-m}$.

A4.2. THE FFT AND THE INFINITE SERIES AND INTEGRAL

Many oceanographers report the results of data processing as continuous functions of frequency or time rather than index numbers as obtained from the FFT. To do this, quantities such as time t, sampling interval Δt, record length T_1, and frequency f must be related to n, m, and N of the FFT. Three steps are required. First, the periodic properties of x_n and X_m are used to alter the summation interval. Second, n, m, and N are expressed in terms of t, f, Δt, and T_1. Third, passing to the limits yields the usual Fourier series and integral transformations.

A4.2.1. Infinite Series

For periodic $x(t)$, $x(t) = x(t + kT_1)$, where $k = 0, \pm 1, \ldots$. The usual infinite series expressions for Fourier series are

$$x(t) = \sum_{m=-\infty}^{\infty} X_m \exp(i2\pi mf_1 t) \tag{A4.2.1}$$

$$X_m = \frac{1}{T_1} \int_{T_0}^{T_1+T_0} x(t) \exp(-i2\pi mf_1 t)\, dt \tag{A4.2.2}$$

where $f_1 = 1/T_1$
      T_0 = an arbitrary (convenient) time

We use $x(t)$ instead of x_n in (A4.2.1) because the infinite series represents the unsampled $x(t)$. If $x(t)$ has discontinuous steps, there are difficulties, known as Gibbs' effect, in using (A4.2.1) to calculate $x(t)$ in the vicinity of the discontinuity.

A4.2.2. Fourier Integral

Nonperiodic functions are expressed by means of the Fourier integral. Such functions include solitary waves, signals from explosions, a single sonar ping, a click, and many other phenomena. In all these cases the receiver is quiet until the signal comes in; it has an output, and finally is quiet for a long time afterward (Fig. A4.2.1). This portion of the signal

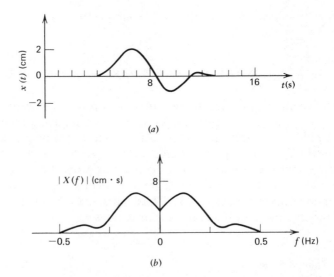

Figure A4.2.1.   Signal and its spectrum, giving absolute value of spectrum $|X(f)|$. Signal is sketch of a solitary water wave.

can be isolated and processed to determine its frequency content. Actually, most digital measurements are made by sampling the signal and applying the FFT. Our purpose is to show the way to translate the X_m obtained by means of the FFT to the equivalent function for an aperiodic signal. When the following integral is finite, we write

$$\int_{-\infty}^{\infty} |x(t)|\, dt = \text{finite value}$$

The Fourier integral expressions of $x(t)$ and $X(f)$ are

$$x(t) = \int_{-\infty}^{\infty} X(f) \exp(i2\pi ft) \, df \qquad \textbf{(A4.2.3)}$$

$$X(f) = \int_{-\infty}^{\infty} x(t) \exp(-i2\pi ft) \, dt \qquad \textbf{(A4.2.4)}$$

On comparing (A4.2.2) and (A4.2.4) we see that the latter follows from (A4.2.2) by forming $T_1 X_m$ and letting T_1 become infinite. The relations are

$$X(mf_1) \equiv T_1 X_m, \qquad T_1 = \frac{1}{f_1} \qquad \textbf{(A4.2.5)}$$

$$X(mf_1) \rightarrow X(f) \qquad \textbf{(A4.2.6)}$$

Here $X(f)$ is referred to as the *complex amplitude spectrum, amplitude spectrum, spectrum,* and *amplitude density* in the literature; $X(f)$ is a function of frequency from $-\infty$ to ∞. Usually, in practice, the frequencies at which the contributions are important have a relatively small range. The absolute value of $X(f)$ is given in Fig. 4.2.1b. The units of X_m and $X(f)$ follow from (A4.1.3), (A4.2.4), and (A4.2.5). When $x(t)$ is in centimeters, x_n and X_m are in centimeters; $X(f)$ has the units of centimeter-seconds or centimeters per hertz.

A4.2.3. Effect of Sampling and FFT in Computing $X(f)$

The relation of the FFT and the Fourier integral is demonstrated by examples. If the signal in Fig. A4.2.1 is sampled at 1 s intervals, the time series x_n in Fig. A4.2.2 results. The choice of N is arbitrary, but it does affect the FFT representation. We have used $N = 8$ and $N = 16$ to compare the results. Equation (A4.2.5) was used to calculate $X(f)$, where $f = m/T_1$. For one case, $T_1 = 16$ s, for the other $T_1 = 8$ s. The estimates of $X(f)$ are at the discrete frequencies given by m/T_1. The moduli of the FFT coefficients $|X_m| = (X_m X_m^*)^{1/2}$ are shown. The magnitude of the peak for $N = 16$ is half the magnitude for $N = 8$. Translating these to $X(f)$, the magnitudes of $X(f)$ are about the same for both $N = 16$ and $N = 8$. The $X(f)$ is computed every 1/16 Hz for $N = 16$ and every 1/8 Hz for $N = 8$. As a rule of thumb, for a signal having time duration T, enough zeros should be added to have $T_1 > 2T$.

Frequently the signal $x(t)$ and its spectrum $X(f)$ are known, and it is desired to construct the set of FFT coefficients. The process is reversed.

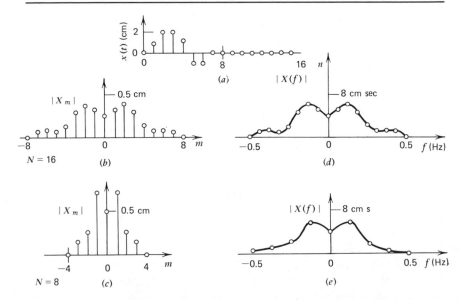

Figure A4.2.2. x_n and transforms. (a) x_n; sample interval is $\Delta t = 1$ s. (b), (c) FFT transforms for $N = 16$ and $N = 8$. (d), (e) Points are FFT approximations to the integral transform $X(f)$ for $N = 16$, $f_1 = \frac{1}{16}$ Hz, $N = 8$, $f_1 = \frac{1}{8}$ Hz.

Choose T_1 and the sampling rate and calculate N, m, and n. Equation (A4.2.5) gives the transfer from $X(f)$ to X_m; $X(f)$ has the range of $-f$ to f and this corresponds to X_{-m} to X_m. If $x(t)$ is real, then $X_{-m} = X_m^*$. You can show this by using (A4.1.8) and requiring that the imaginary components of x_n are zero.

A4.3. IMPULSIVE SOURCE AND $\delta(T)$

Many geophysical measurements use impulsive sources. The signal from an explosion has a large initial pressure and decays rapidly. Ignoring the impulsive source strength for the present, an approximation to the explosive source function is

$$s(t) = 0 \qquad \text{for} \quad t < 0$$

$$s(t) = \frac{1}{T_e} \exp\left(-\frac{t}{T_e}\right) \quad \text{for} \quad t > 0 \tag{A4.3.1}$$

We "normalize" $s(t)$

$$\int_{-\infty}^{\infty} s(t)\, dt = 1 \tag{A4.3.2}$$

Using (A4.2.4), the spectrum of $s(t)$ is found to be

$$S(f) = \frac{1}{T_e} \int_0^\infty \exp\left[-\left(\frac{1}{T_e} + i2\pi f\right)t\right] dt \tag{A4.3.3}$$

$$S(f) = \frac{1}{1 + i2\pi f T_e}$$

To decrease the duration of the explosion, we decrease T_e. Correspondingly, the bandwidth over which $S(f)$ is nearly constant increases. As T_e tends to zero, in the limit we obtain a "delta function":

$$\delta(t) = \lim_{T_e \to 0} \int_{-\infty}^\infty S(f) \exp(i2\pi ft) \, df$$

$$\delta(t) = 0 \qquad \text{for} \quad t \neq 0 \tag{A4.3.4}$$

$$\int_{-\infty}^\infty \delta(t) \, dt = 1$$

$$S_\delta(f) = 1$$

Since $S_\delta(f) = 1$ is the Fourier transformation of $\delta(t)$, $\delta(t)$ is

$$\delta(t) = \int_{-\infty}^\infty \exp(i2\pi ft) \, df \tag{A4.3.5}$$

Usually $\delta(t)$ appears under the integral sign as

$$\int_{-\infty}^\infty f(t)\,\delta(t)\,dt = f(0) \tag{A4.3.6}$$

or more generally as a convolution (described in Section A4.4).

$$\int_{-\infty}^\infty f(t')\,\delta(t - t')\,dt' = f(t) \tag{A4.3.7}$$

Since we follow the custom of defining the time integral of the $\delta(t)$ as being unity (A4.3.4), $\delta(t)$ has the units of time^{-1}. Correspondingly, its spectrum is dimensionless.

By interchanging f and t in (A4.3.5) we obtain the frequency-dependent delta function $\delta(f - f')$. It appears as a result of a time integration in multiple integrals. For example, the integral I is

$$I = \int\!\!\int\!\!\int_{-\infty}^\infty F(f)G(f') \exp[i2\pi(f - f')t] \, df \, df' \, dt \tag{A4.3.8}$$

The time-dependent integral yields $\delta(f-f')$. The f' integration follows, and the expression reduces to

$$I = \int F(f)G(f)\, df \qquad \text{(A4.3.9)}$$

This is an example of the identification of a "delta function integral" inside a multiple integral. This kind of reduction appears quite often.

A4.4. SIGNALS AND FILTERS

A filter is a device that may alter the phase and amplitude of the signal as a function of frequency. The frequency response of the filter for the frequency mf_1 is

$$F_m = \frac{\text{output } (mf_1)}{\text{input } (mf_1)} \qquad \text{(A4.4.1)}$$

and F_m may be complex. On applying this relation to each frequency coefficient X_m, each output coefficient is found to have the amplitude $X_m F_m$. We calculate the output h_n by using (A4.1.9) and replacing X_m by $X_m F_m$

$$h_n = \sum_{m=1-N/2}^{N/2} X_m F_m \exp\left(\frac{i2\pi mn}{N}\right) \qquad \text{(A4.4.2)}$$

The digital operation is illustrated in Fig. A4.4.1. By comparing (A4.4.2) and (A4.1.9), we are able to see that the frequency domain represented in the output signal H_m is

$$H_m = X_m F_m \qquad \text{(A4.4.3)}$$

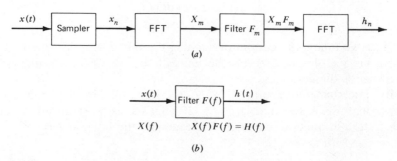

Figure A4.4.1. Filters in frequency domain. (a) Digital filter. (b) Analogue filter: $X(f)$ is spectrum of $x(t)$, $H(f)$ is spectrum of output $h(t)$.

Ordinarily we use x_n for the input and require the output h_n to be real. Recalling (A4.1.8), real h_n and x_n require that

$$H_m^* = H_{-m} \tag{A4.4.4}$$

$$X_m^* F_m^* = X_{-m} F_{-m}$$

$$F_m^* = F_{-m} \tag{A4.4.5}$$

The phase shift of the output signal relative to the input signal depends on the amplitude dependence of $|F_m|$. The choices of the ratios of real to imaginary parts of F_m affects the amount of the phase shift. The design of filters for minimum phase and maximum phase conditions is a sophisticated art.

When the signal is an aperiodic function and the Fourier integral is applicable, we have

$$X(mf_1) = T_1 X_m$$

$$H(f) = X(f)F(f) \tag{A4.4.6}$$

$$h(t) = \int_{-\infty}^{\infty} X(f)F(f) \exp{(i2\pi ft)} \, df \tag{A4.4.7}$$

These relations are useful because the spectrum of the output is the product of the spectrum of the input and the filter response. We adjust $F(f)$ to pass a signal by choosing $F(f)$ to be large when $X(f)$ is large. To reject a signal, we choose $F(f)$ to be zero when $X(f)$ is large. The relations apply to both analogue and digital filters (Fig. 4.4.1b).

An alternative way to express the input-output relations of a filter is the *convolution*. The convolution of the signals $x(t)$ and $f(t)$ is

$$\text{convol}(x, f) = \int_{-\infty}^{\infty} f(\tau)x(t-\tau) \, d\tau$$

$$= \int_{-\infty}^{\infty} f(t-\tau)x(\tau) \, d\tau \tag{A4.4.8}$$

The spectrum of $f(t)$ is $F(f)$ and the spectrum of $x(t)$ is $X(f)$. By using the Fourier integral expressions for $f(\tau)$ and $x(t-\tau)$ and substituting these into (A4.4.8), we obtain (after much manipulation)

$$\text{convol}(x, f) = \int_{-\infty}^{\infty} X(f)F(f) \exp{(i2\pi ft)} \, df \tag{A4.4.9}$$

$$\text{convol}(x, f) = h(t)$$

Usually calculations of the convolution are made by using the Fourier transformations.

The integral of the absolute square of $h(t)$ appears often in sonar signal processing systems. Designating it as I_{hh}, and using (A4.4.7), we write

$$I_{hh} \equiv \int_{-\infty}^{\infty} h(t)h^*(t)\, dt \tag{A4.4.10}$$

$$= \int_{-\infty}^{\infty} \int_{-\infty}^{\infty} \int_{-\infty}^{\infty} X^*(f')F^*(f')X(f)F(f)$$

$$\exp\left[i2\pi(f-f')t\right] df\, df'\, dt \tag{A4.4.11}$$

We do the integral on t and recognize that it has the same form as the example (A4.3.8) and A(4.3.9). We perform the δ function integration by replacing f' by f and obtain

$$I_{hh} = \int_{-\infty}^{\infty} |X(f)|^2 |F(f)|^2\, df \tag{A4.4.12}$$

If $x(t)$ and $h(t)$ have the units of volts, I_{hh} has the units of volts squared-seconds or volts squared per hertz.

A4.5. COVARIANCE AND CORRELATION OF SIGNALS

The detection of a signal requires a method of determining the presence of a *particular wave form* $x(t)$. We also want to known when $x(t)$ is sensed by our hydrophone. Simple wave forms are easy to identify visually when interferences and noise are small. Complicated wave forms and large interferences make the task difficult.

Our purpose is to develop computational procedures for measuring the similarity of the reference signal $x(t)$, and $y(t)$, which may be the sum of $x(t)$ and other wave forms. In sonar $x(t)$ is delayed by the travel time. Thus we also need a method of determining the arrival time of $x(t)$. Since most quantitative measurements are made with digital computers, we formulate the problem for sampled data.

For a quantitative measure, we have chosen to use the mean square of the differences of the signals. This is intuitively attractive, it leads naturally to the *correlation* and *covariance functions*, and it shows that the correlation operation is an optimum way to detect a known signal in the presence of noise.

Although the correlation function can be defined for finite deterministic functions, the original concepts came out of statistics and studies of random functions.

A4.5.1. Correlation and Least Squares

As a preliminary exercise, let us test the assertion that y *is a multiplicative constant times* x, *and that it is shifted relative to* x. This procedure is appropriate when we want to detect the presence of signals that can occur at any time.

The analysis is simpler if x and y have mean values of zero; if not, the means can be subtracted to give x and y having means of zero. We assume that the mean values are

$$\langle x \rangle = \frac{1}{N}\sum_{n=0}^{N-1} x_n = 0 \quad \text{and} \quad \langle y \rangle = \frac{1}{N}\sum_{n=0}^{N-1} y_n = 0 \quad \textbf{(A4.5.1)}$$

The mean square values of x and y are

$$\sigma_x^2 = \frac{1}{N}\sum_{n=0}^{N-1} x_n^2 \quad \text{and} \quad \sigma_y^2 = \frac{1}{N}\sum_{n=0}^{N-1} y_n^2 \quad \textbf{(A4.5.2)}$$

where σ_x and σ_y are the rms values of sequences x_n and y_n. The operation of taking the mean as in (A4.5.1) is used often and is symbolized by $\langle \ \rangle$; for example, $\sigma_x^2 = \langle x_b \rangle^2$. We assume that when the sequence y_n is shifted by k steps relative to x_n, it becomes y_{n+k}. The mean and σ_y are assumed to be unchanged.

For our test, we define the mean squared value $e^2(k)$

$$e^2(k) = \frac{1}{N}\sum_{n=0}^{N-1}\left[\left(C_{xy}(k)\frac{x_n}{\sigma_x}\right) - \frac{y_{n+k}}{\sigma_y}\right]^2 \quad \textbf{(A4.5.3)}$$

We use (x_n/σ_x) and (y_{n+k}/σ_y) to make $C_{xy}(k)$ dimensionless and to give those sequences unity rms values. The object is to choose $C_{xy}(k)$ such that $e^2(k)$ is a minimum. We regard k as a parameter. We can shift y relative to x after determining $C_{xy}(k)$ for the minimum. From calculus, we take $\partial/\partial C_{xy}(k)$ of (A4.5.2) and set the result equal to zero. After manipulation, the value of $C_{xy}(k)$ for minimum $e^2(k)$ is

$$C_{xy}(k) = \frac{1}{\sigma_y\sigma_y N}\sum_{n=0}^{N-1} x_n y_{n+k} \quad \textbf{(A4.5.4)}$$

The multiplicative constant for minimum error is proportional to the mean value of the products $x_n y_{n+k}$. Students of statistics know $C_{xy}(k)$ as the *correlation function*. It has limits of 1 and -1. A value of zero means that the functions are not alike; that is, they are uncorrelated. If the functions are alike, but one is shifted with respect to the other, the value of k for which $C_{xy}(k)$ is a maximum gives the amount of the shift. We use the notation $C_{xy}(k)$ to indicate the correlation of x and y when y is shifted by k.

Equation (A4.5.4) gives the result for real x_n and y_n. When x_n and y_n are complex, the square in (A4.5.3) is replaced by the absolute square. Taking $\partial/C_{xy}^*(k)$ of $|e(k)|^2$ gives

$$C_{xy}(k) = \frac{1}{\sigma_x \sigma_y N} \sum_{n=0}^{N-1} x_n^* y_{n+k} \qquad \textbf{(A4.5.4a)}$$

We use the complex form in the balance of the discussion. By the *least squares criterion, $C_{xy}(k)$ is an optimum test of the similarity of x and y.* If $x \sim y$, then $|C_{xy}(k)|$ has a maximum of 1.

The covariance function is the correlation function after multiplication by the rms values of x_n and y_n. The covariance is

$$\text{cov}\,[x, y(k)] \equiv \sigma_x \sigma_y C_{xy}(k) = \langle x^* y(k) \rangle \qquad (A4.5.5)$$

An example of the calculation of the correlation and the covariance of x and y for $k = 0$ is given in Fig. A4.5.1. The mean value of the product is zero, the correlation is zero, and the dissimilarity of x and y is confirmed by inspection. It is evident that shifting y by k will not alter the null correlation of x and y. The student might recalculate the correlation after shifting y by one or two increments, to prove this point.

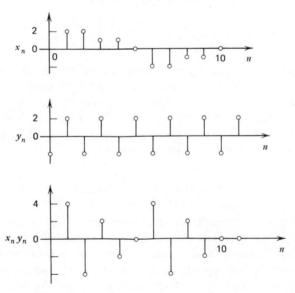

Figure A4.5.1. Correlation and covariance of x and y.

Next we give an example of the dependence of the covariance and the correlation on k when $x = y$. The notation for correlation of x with itself is $C_{xx}(k)$.

$$\sigma_x^2 C_{xx}(k) = \frac{1}{N} \sum_{n=0}^{N-1} x_n^* x_{n+k} \qquad (\text{A4.5.6})$$

where $C_{xx}(k)$ is often called the "autocorrelation." The dependence on k is shown in Fig. A4.5.2. The reader can prove that $C_{xx}(k)$ is symmetric;

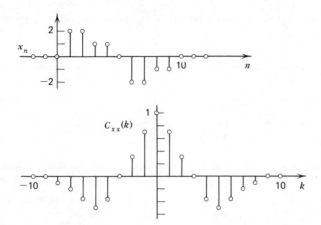

Figure A4.5.2. Correlation of x_n and x_{n+k}: x is the same function as in Fig. A4.5.1.

that is, that $C_{xx}(k) = C_{xx}(-k)$. (*Hint.* Let $n = n' - k$. After the change, the summation is still over the N terms of x_n.)

A4.5.2. Integral Expressions

The summations of the finite covariance function can be passed to the integral form by letting N tend to infinity as follows. Define

$T \equiv$ duration of the study
$t \equiv$ time of the nth sample
$\Delta t \equiv$ time between samples = reciprocal of
    sampling frequency
$N \equiv T/\Delta t$
$\tau \equiv k\Delta t \equiv kT/N \equiv$ time lag between x_n and y_{n+k}
$x_n \rightarrow x(t), \qquad y_{n+k} \rightarrow y(t + \tau)$

Then (A4.5.4) becomes

$$\text{cov}\,[x.\ y(\tau)] = \frac{1}{T}\sum_0^T x^*(t)y(t+\tau)\,\Delta t \qquad (A4.5.7)$$

Taking the limit, Δt becomes dt and

$$\text{cov}\,[x,\ y(\tau)] = \frac{1}{T}\int_0^T x^*(t)y(t+\tau)\,dt \qquad \textbf{(A4.5.8)}$$
$$= \langle x^*(t)y(t+\tau)\rangle$$

where $\langle\ \rangle$ again means the averaging operation. Correspondingly (A4.5.2) becomes

$$\sigma_x^2 = \frac{1}{T}\int_0^T |x(t)|^2\,dt \qquad \text{and} \qquad \sigma_y^2 = \frac{1}{T}\int_0^T |y(t)|^2\,dt \qquad \textbf{(A4.5.9)}$$

The correlation function is

$$C_{xy}(\tau) = \frac{\text{cov}\,[x,\ y(\tau)]}{\sigma_x\sigma_y} \qquad (A4.5.10)$$

The region of integration is rather arbitrary, and the limits can be chosen as T_0 and $T_0 + T$. The limits $-T/2$ to $+T/2$ are convenient for many functions. If σ_x^2 and σ_y^2 are finite and different from zero as T tends to infinity in (A4.5.9), we can write the limiting operation

$$\sigma_x\sigma_y C_{xy}(\tau) = \lim_{T\to\infty}\frac{1}{T}\int_{-T/2}^{T/2} x^*(t)y(t+\tau)\,dt \qquad (A4.5.11)$$

Acoustic signals are often aperiodic, and (A4.5.9) tends to zero as T becomes infinite. The definition of $C_{xy}(\tau)$ for this case is

$$C_{xy}(\tau) = \frac{1}{(I_{xx}I_{yy})^{1/2}}\int_{-\infty}^{\infty} x^*(t)y(t+\tau)\,dt$$
$$\hspace{6cm} \textbf{(A4.5.12)}$$
$$C_{xx}(\tau) = \frac{1}{I_{xx}}\int_{-\infty}^{\infty} x^*(t)x(t+\tau)\,dt$$

$$I_{xx} \equiv \int_{-\infty}^{\infty} |x(t)|^2\,dt$$
$$\hspace{6cm} (A4.5.13)$$
$$I_{yy} \equiv \int_{-\infty}^{\infty} |y(t)|^2\,dt$$

The concepts of covariance and correlation that we have developed in these past two sections have been discussed both for continuous functions and time series. Either of these functions can be used to represent a varying voltage. And the varying voltage, in turn, can be generated by transducers that may be used to sense temperatures, sound pressures, ocean surface heights, and so on. The continuous functions and the time series therefore might represent ocean variables. The correlation functions of these variables are one of the important ways to characterize the quantity being measured.

The reader can find various correlation terminology in the literature. *Autocorrelation* means $C_{xx}(\tau)$ and *autocovariance* means $\sigma_x^2 C_{xx}(\tau)$. *Cross-correlation* refers to $C_{xy}(\tau)$. Sometimes the terms "correlation" and "covariance" are used interchangeably because σ_x and σ_y are suppressed and the signals are referred to arbitrary levels. When the correlation is calculated for quantities that are functions of time, as previously, it is sometimes called the *temporal correlation*. When the quantities are functions of position, for example, the height of the sea floor, the parameters are distances rather than time and the similarity between adjacent points is called the *spatial correlation* function. Surfaces that have both time and space dependence have a temporal and spatial correlation function.

Illustrative Example. Calculate the correlation, $C_{xx}(\tau)$, of $x = a \cos 2\pi f t$.

$$(a) \quad \sigma_x^2 = \frac{1}{T} \int_{-T/2}^{T/2} a^2 \cos^2 2\pi f t \, dt$$

$$= \frac{a^2}{2} \qquad T = \frac{1}{f}$$

$$(b) \quad \sigma_x^2 C_{xx}(\tau) = \frac{1}{T} \int_{-T/2}^{T/2} a^2 \cos(2\pi f t) \cos[2\pi f(t+\tau)] \, dt$$

$$= \frac{a^2}{2} \cos(2\pi f \tau)$$

$$C_{xx}(\tau) = \cos 2\pi f \tau$$

The correlation of a pair of sinusoidal wave forms having the same frequency is a cosine function. If they have different frequencies, the correlation is zero. The latter can be shown by computing the correlation of $x = \exp(2i\pi f_1 t)$ and $y = \exp(2i\pi f_2 t)$. Let T be very large.

A4.5.3. Signals and Noise

The correlation procedure is an optimum way to detect the presence of a desired signal when undesired signals from other sources may be present.

The transmitter sends a wave form $x(t)$, and after time T the backscattered signal $x(t-T)$ is received. Signals from other sources are different from $x(t)$ and unwanted. We call the sum of those unwanted quantities the noise n. The input to the receiver $a(t)$ consists of signal $x(t-T)$ and noise $n(t)$

$$a(t) = x(t-T) + n(t) \qquad (A4.5.14)$$

or if the continuously time-varying signal is sampled periodically (increment, Δt) to produce a time series

$$a_n = x_{n-j} + n_n \qquad (A4.5.15)$$

where $t = n\Delta t$ and $T = j\Delta t$.

Using (A4.5.4a), the covariance of x and a is found to be

$$\sigma_x \sigma_a C_{xa}(k) = \frac{1}{N} \sum_{n=0}^{N-1} x_n^*(x_{n-j+k} + n_{n+k}) \qquad (A4.5.16)$$

$$\sigma_a^2 \approx \sigma_x^2 + \sigma_n^2 \qquad \text{for} \quad |C_{xn}(k)| \ll 1$$

and the correlation is

$$C_{xa}(k) = \frac{\sigma_x}{\sigma_a} C_{xx}(k-j) + \frac{\sigma_n}{\sigma_a} C_{xn}(k) \qquad (A4.5.17)$$

where $C_{xx}(k-j)$ has its maximum value at $k-j = 0$. When $C_{xn}(k)$ is zero, $C_{xa}(k)$ has a peak at $k = j$ and $T = j\Delta t$. Even when x and n are uncorrelated, particular samples of n often give nonzero values of $C_{xn}(k)$. These cause fluctuations in the measured values of $C_{xa}(k)$. When the fluctuations are much less than σ_x/σ_a, $C_{xa}(k)$ has a peak at T and the signal is recognized in the presence of the noise.

A4.6. CORRELATION FUNCTION AND SPECTRAL DENSITY OF SIGNALS

Our purpose is to show the relation of the *spectral density* of a signal to the covariance of the signal. (Some of the older literature refers to the spectral density as being the "power spectrum" or "power spectral density.")

For complex x_n and y_n, the covariance (A4.5.4a) is

$$\sigma_x \sigma_y C_{xy}(k) = \frac{1}{N} \sum_{n=0}^{N-1} x_n^* y_{n+k} \qquad (A4.6.1)$$

The Fourier transformation of both sides gives

$$\frac{\sigma_x\sigma_y}{N}\sum_{k=0}^{N-1}C_{xy}(k)\exp\left(-\frac{i2\pi mk}{N}\right)=\frac{1}{N^2}\sum_{n=0}^{N-1}\sum_{k=0}^{N-1}x_n^*y_{n+k}$$

$$\exp\left(-\frac{i2\pi mk}{N}\right) \tag{A4.6.2}$$

Change of summation index from k to n', using $k=n'-n$, gives

$$=\frac{1}{N}\sum_{n=0}^{N-1}x_n^*\exp\left(\frac{i2\pi nm}{N}\right)\frac{1}{N}\sum_{n'=0}^{N-1}y_{n'}\exp\left(\frac{-i2\pi n'm}{N}\right)$$

$$=X_m^*Y_m \tag{A4.6.3}$$

where X_m^* and Y_m are the complex amplitude coefficients of x_n^* and y_n. We have the transformation pair

$$\sigma_x\sigma_y\sum_{k=0}^{N-1}C_{xy}(k)\exp\left(\frac{-i2\pi mk}{N}\right)=NX_m^*Y_m \tag{A4.6.4}$$

$$\sigma_x\sigma_yC_{xy}(k)=\sum_{m=0}^{N-1}X_m^*Y_m\exp\left(\frac{i2\pi mk}{N}\right) \tag{A4.6.5}$$

The last two equations are the finite Fourier transformations of $C_{xy}(k)$ and $X_m^*Y_m$. We can change these summations to the infinite Fourier integrals by following the same procedure as in Sections A4.2.1 and A4.2.2. In the change, τ replaces t; $\sigma_x\sigma_yC_{xy}(\tau)$ replaces $x(t)$ and $(X_m{}^*Y_m)$ replaces X_m. The result is

$$C_{xy}(k)\equiv C_{xy}(\tau) \tag{A4.6.6}$$

$$S_{xy}(f)\equiv T_1X_m^*Y_m \tag{A4.6.7}$$

$$S_{xy}(f)=\sigma_x\sigma_y\int_{-\infty}^{\infty}C_{xy}(\tau)\exp(-i2\pi f\tau)\,d\tau \tag{A4.6.8}$$

$$\sigma_x\sigma_yC_{xy}(\tau)=\int_{-\infty}^{\infty}S_{xy}(f)\exp(i2\pi f\tau)\,df \tag{A4.6.9}$$

where $S_{xy}(f)$, *cross spectral density*, has the units of volts squared multiplied by time. Because it is a density function in frequency coordinates, it is customary to express the units as volts squared per hertz.

The corresponding transformation pair for $x(t)$ and $x(t+\tau)$ are obtained by letting $X_m=Y_m$ in (A4.6.6) and (A4.6.7); $S_{xx}(f)$ is the absolute square of X_m and is symmetric. Expansion of $\exp(ix)=\cos x+i\sin x$ and change of the limits of integration gives

$$\sigma_x^2 C_{xx}(\tau) = 2 \int_0^\infty S_{xx}(f) \cos (2\pi f \tau) \, df \qquad \textbf{(A4.6.10)}$$

The cosine is also an even function which means that $C_{xx}(\tau)$ is even or symmetric; thus

$$S_{xx}(f) = 2\sigma_x^2 \int_0^\infty C_{xx}(\tau) \cos (2\pi f \tau) \, d\tau \qquad \textbf{(A4.6.11)}$$

and $S_{xx}(f)$ is the *spectral density*. Equations (A4.6.8) to (A4.6.11) were derived by Weiner and Khinchine and are known as the Wiener–Khinchine theorem. For $\tau = 0$, we have Parseval's theorem

$$\sigma_x^2 = 2 \int_0^\infty S_{xx}(f) \, df \qquad (A4.6.12)$$

Although the theory has been given for finite signals and signals that are known over a time duration T, the Wiener–Khinchine theorem also applies to random functions or signals. Random functions are unpredictable in the sense that knowing a sample of the waveform of the signal between zero and T does not enable us to predict the waveforms between T and $2T$, $2T$ and $3T$, and so on. If, within fluctuations, values of σ_x are the same for each of the samples and if the $C_{xx}(\tau)$ are the same, the random function is said to be stationary. If the random signal is due to a stationary process, the spectral density of the noise is a convenient measure of the process. Incidentally, in its common usage, *noise usually means a stationary random signal.*

There is a problem with the definition and reporting of spectral density measurements. What we have given is consistent with most of the recent texts in communication and signal theory. It is also the same as the definition used by Blackman and Tukey in their spectral analysis procedure. In (A4.6.9) the frequency ranges from $-\infty$ to ∞, whereas in (A4.6.10) it ranges from zero to ∞ and a factor 2 is in front of the integral. *Some people have absorbed the 2 into the definition of spectral density;* that is, they use

$$P_{xx}(f) \equiv 2S_{xx}(f)$$

$$P_{xx}(f) = 4\sigma_x^2 \int_0^\infty C_{xx}(\tau) \cos 2\pi f \tau \, d\tau \qquad \textbf{(A4.6.13)}$$

$$\sigma_x^2 C_{xx}(\tau) = \int_0^\infty P_{xx}(f) \cos 2\pi f \tau \, df \qquad \textbf{(A4.6.14)}$$

There is a rationale and a bit of history for doing this. Analogue measurements of signals and filter responses give the sinusoidal components as functions of positive frequency and phase. Hence users of analogue instruments to measure spectral densities have tended to report $P_{xx}(f)$. [Sometimes they clarify the measurements as being $P_{xx}(f)$ by adding the phrase "*for positive frequencies.*"] However users of the FFT and Blackman and Tukey procedures have tended to report $S_{xx}(f)$. We call $S_{xx}(f)$ and $S_{xy}(f)$ the spectral densities because they occur in that form.

A4.7. MATCHED FILTER AND CORRELATION RECEIVER

The matched filter is one member of a more general class of optimum filters often called Wiener filters. Because these filters can be designed to produce a desired output, they have become extremely important in exploration geophysics. The general topic of Wiener filters is beyond our scope. The matched filter has a signal output $h(t)$ and noise output σ_{nF} for signal plus noise input. The filter response $F(f)$ is chosen to maximize $|h(t)|^2/\sigma_{nF}^2$. We derive a simplified expression for $F(f)$ by combining results from Sections 4.4 and A4.6. The filter input is $g(t)$ and $h(t)$ is the output.

The correlator is an optimum way to detect the signal $g(t)$.

$$C_{gg}(\tau) \sim \int_{-\infty}^{\infty} g^*(t)g(t+\tau)\, dt \qquad (A4.7.1)$$

Using the Wiener–Khinchine theorem,

$$S_{gg}(f) \sim \int_{0}^{\infty} C_{gg}(\tau) \cos(2\pi f\tau)\, d\tau \qquad (A4.7.2)$$

$$S_{gg}(f) \sim G(f)G^*(f) \qquad (A4.7.3)$$

because $g(t)$ has the spectrum $G(f)$. If $g(t)$ is the input and we require $h(t) \sim C_{gg}(t)$ to be the output of the filter, the spectrum of the signal $G(f)$ times the spectrum of the filter $F(f)$ is $S_{gg}(f)$.

$$H(f) \sim G(f)G^*(f) \sim G(f)F(f) \sim S_{gg}(f) \qquad (A4.7.4)$$

$$F(f) \sim G^*(f) \qquad (A4.7.5)$$

The matched filter output is

$$h(t) \sim \int_{-\infty}^{\infty} G^*(f)G(f) \exp(i2\pi ft)\, df \qquad (A4.7.6)$$

$$h(t) \sim \int_{-\infty}^{\infty} g^*(t')g(t'+t)\, dt' \qquad (A4.7.7)$$

Proportionality signs are used because the amplitudes are arbitrary. The peak of $h(t)$ at $t=0$ is proportional to the integral of $|g(t)|^2$ over the whole duration of the signal, and this is an important reason for the signal to noise improvement of the matched filter. For a bandwidth Δf, the temporal resolution Δt is approximately $\Delta t = 1/\Delta f$.

To show the relation between the matched filter and the tapped delay line, we impulse the matched filter; that is, let $g(t)$ be $\delta(t)$ and $G(f) = 1$. The output $h_I(t)$ is

$$h_I(t) = \int_{-\infty}^{\infty} G^*(f) \exp(i2\pi ft)\, df \qquad (A4.7.8)$$

Taking the conjugate of both sides and replacing t by $-t'$, we write

$$h_I(-t') = \int_{-\infty}^{\infty} G(f) \exp(i2\pi ft')\, df \qquad (A4.7.9)$$

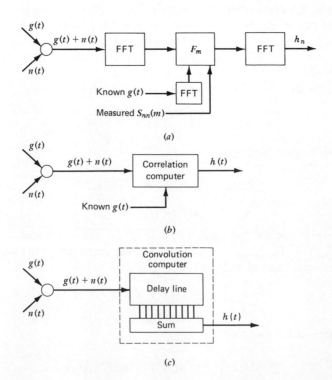

Figure A4.7.1. Matched filters. (a) Digital. (b) Correlation. (c) Convolution. Here $g(t)$ is signal, $n(t)$ is noise, $h(t)$ is filtered output of signal and noise, and $S_{nn}(m)$ is spectral density of the noise.

where $h_I(-t')$ is the temporal reversal of the signal. Excluding a fixed time delay, the output of a tapped delay line is also $h_I(-t')$ or $g(-t)$.

In matched filter applications to sonars, often the bandwidth of the signal is small compared to the carrier frequency, and the noise spectrum is nearly constant over the band. The signal band for seismic profiling may run from about 10 to several hundred hertz. The noise spectrum (Fig. 4.2.4) changes a great deal over the seismic signal band and should be included in the design of the matched filter for this band. Developments which include $S_{nn}(f)$ give the matched filter

$$F(f) \sim \frac{G^*(f)}{S_{nn}(f)} \tag{A4.7.10}$$

Three designs, or constructions, of matched filters are illustrated in Fig. A4.7.1. When the filter is a digital operation (Fig. A4.7.1a), Fourier transformations are made of the incoming signals and noise. The estimate of the noise spectrum can be made between transmissions so that $S_{nn}(f)$ can be "updated." In Fig. A4.7.1b a replica of the transmitted signal is stored in a correlation computer. The correlations of the incoming signal and noise are computed. Presumably the noise spectrum is constant over the signal frequency band and the correlation of the replica signal and the noise tends to zero. A time domain or convolution filter (Fig. A4.7.1a) can be contructed from a computer program or a hard-wired device using a delay line.

SOURCES AND ARRAYS

A5.1. MONOPOLE AND DIPOLE SOURCES

A5.1.1. The Monopole

The simplest of all sound sources is the small pulsating sphere. Since its surface "breathes" uniformly in all directions, the radiation is independent of direction. Bodies that radiate isotropically are called monopoles.

To calculate the dependence of the radiated pressure on the sphere motion, assume that the instantaneous radial velocity at the surface $R = a$ is

$$u_r = U_a e^{i\omega t}]_{R=a} \qquad (A5.1.1)$$

where U_a is the rms radial velocity.

The isotropic radiated pressure at range R is

$$p = \frac{P_0 R_0 \exp[i(\omega t - kR)]}{R} \qquad (A5.1.2)$$

where $P_0 = $ rms pressure at reference distance $R = R_0$.

We can evaluate P_0 in terms of U by calculating the radial particle velocity of the acoustic wave and equating it to the radial velocity at the surface of the pulsating sphere.

Using (3.1.7) with (A5.1.2) we get at $R = a$

$$u_r = \frac{P_0 R_0 \exp[i(\omega t - ka)]}{a \rho_A c} \left(1 + \frac{1}{ika}\right) \qquad (A5.1.3)$$

Equating u_r from (A5.1.1) to u_r from (A5.1.3) at $R = a$, we have

$$U_a e^{i\omega t} = \frac{-i}{\rho_A \omega a^2} \{P_0 R_0 \exp[i(\omega t - ka)](1 + ika)\} \qquad \text{(A5.1.4)}$$

Solve for

$$P_0 R_0 = i\rho_A \omega a^2 U_a [\exp(ika)]/(1 + ika)$$

and (A5.1.2) becomes

$$p = \frac{ik(\rho_A c) U_a S_a \exp[i(\omega t - kR + ka)]}{4\pi R(1 + ika)} \qquad \textbf{(A5.1.5)}$$

where $S_a = 4\pi a^2 = $ surface area of source.

The result of this little exercise tells us a few things. First, although the source motion (A5.1.1) has radial velocity proportional to $\exp(i\omega t)$, the radiated pressure for $ka \ll 1$ is not in phase with the velocity U_a; it is proportional to $i \exp(i\omega t)$ and therefore leads by approximately 90° (depending on ka). The effect is readily identified in the case of an explosive charge; an outward flow, the "afterflow," of the fluid always follows the radiated pressure.

Second, for a constant velocity amplitude in a given medium, the radiated pressure is proportional to the sound frequency $k = \omega/c$. This means, for example, that a sound source for which the electrical voltage produces a velocity amplitude inversely proportional to frequency will produce an acoustic pressure that is independent of frequency. The electrodynamic source such as the household conical loudspeaker is designed to operate in this manner.

Third, the pressure is proportional to the volume flow $U_a S_a$. In fact when $ka \ll 1$, regardless of the detailed shape and motion of the source the radiated pressure for a "monopole," by definition, depends only on the volume flow V (m³/s) normal to the moving surface. Let the velocity at the surface be $\mathbf{U}$,

$$V = \int_{\text{surface}} \mathbf{U} \cdot d\mathbf{S} \qquad \text{(A5.1.6)}$$

In the case of a pulsating sphere, since $\mathbf{U}$ is in the direction of the normal, the integration gives $V = 4\pi a^2 U_a$.

Finally, for the same volume flow and sound frequency, the radiated sound pressure is proportional to the density of the medium. For example, a transducer that has a surface velocity independent of the medium will radiate about 10^3 greater pressure in water than in air because $\rho_{\text{water}} \simeq 10^3 \rho_{\text{air}}$.

A5.1.2. The Dipole

The combination of two out-of-phase monopole sources separated by a distance small compared to the radiated wavelength is called a dipole. Examples of the dipole are not difficult to find: an oscillating membrane (Fig. A5.1.1) simultaneously creates a condensation in one direction and a rarefaction in the other. When it reverses, the condensation is replaced by a rarefaction and the rarefaction by a condensation, and so on. When

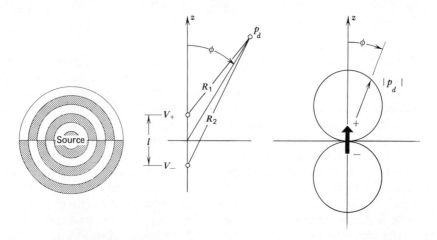

Figure A5.1.1. The dipole source. Two point sources of equal strength but opposite phase, $\pm V$ are separated by the distance l, where $kl \ll 1$. The far field directionality in polar coordinates is a figure-eight pattern.

the membrane is in a small baffle, $kl \ll 1$, where l is the effective separation between the two sides, a dipole exists. Physically, it can be seen that the two opposing radiations will completely cancel each other along a plane perpendicular to the line joining the two poles, and partially cancel at other points, as well.

Call the summation pressure $p_d = p_+ + p_-$ and define $V_+ = -V_- = US$.

$$p_d = \frac{i \rho_A \omega V_+}{4\pi} \left\{ \frac{\exp[i(\omega t - kR_1)]}{R_1} - \frac{\exp[i(\omega t - kR_2)]}{R_2} \right\}$$

The geometry shows that at large ranges, $R \gg l$, R_1 and R_2 differ only slightly from R.

The small corrections to R in the denominator are negligible. However, in the numerator, as corrections to the phase, they are everything. Doing the algebra, under the restriction $l/R \ll 1$, we get

$$p_d = \frac{i\rho_A \omega V_+ \exp\left[i(\omega l - kR)\right]}{4\pi R} \left[2i \sin\left(\frac{kl}{2} \cos\phi\right)\right] \quad \text{(A5.1.7)}$$

The coefficient in front of the brackets is identified from (A5.1.5) as the pressure due to the monopole $ka \ll 1$ at $R \gg a$.

When, also, $kl \ll 1$ we have the ideal dipole

$$|p_d| = |p_m|\,(kl \cos\phi) \qquad \textbf{(A5.1.8)}$$

where p_m = pressure due to a monopole of strength V. Comparing the far field dipole behavior with that of the monopole, there are two significant effects; the radiated pressure is (1) reduced by the factor $kl \ll 1$ and (2) modified by a directionality $\cos\phi$, resulting in a maximum radiation along the line of the dipole and zero sound pressure in the central plane perpendicular to the dipole line. The radiation pattern is sometimes called a figure-eight pattern because of the appearance of the cosine when the magnitude is plotted in polar coordinates (Fig. A5.1.1).

A5.2. CIRCULAR PISTON TRANSDUCER

The geometry for a circular piston transducer is given in Fig. A5.2.1. Starting with (2.5.23) we suppress $\exp\left[i(\omega t - kR)\right]$ and the absolute pressure magnitude and write the real part of the pressure as proportional to

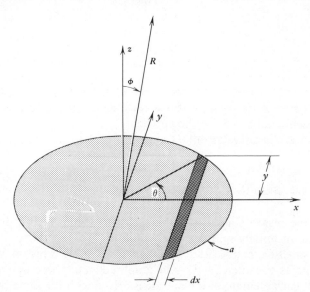

Figure A5.2.1. Geometry for pressure due to a circular piston.

the integral of the areal elements $2y\,dx$ of the circle.

$$p \sim 2 \int_{-a}^{a} y \cos{(kx \sin{\phi})}\,dx \tag{A5.2.1}$$

We change to polar coordinates

$$x = a \cos{\theta}$$
$$dx = -a \sin{\theta}\,d\theta$$
$$y = a \sin{\theta} \tag{A5.2.2}$$

and define $z \equiv ka \sin{\phi}$

$$P \sim 2a^2 \int_0^{\pi} \cos{(z \cos{\theta})} \sin^2{\theta}\,d\theta \tag{A5.2.3}$$

The integral has the form of the integral expression of the Bessel Function $J_1(z)$ and, using (A1.5.22) for $\nu = 1$, is

$$J_1(z) = \frac{z}{\pi} \int_0^{\pi} \cos{(z \cos{\theta})} \sin^2{\theta}\,d\theta \tag{A5.2.4}$$

Substitution for the integral and z gives

$$P \sim \pi a^2 \frac{2J_1(ka \sin{\phi})}{ka \sin{\phi}} \tag{A5.2.5}$$

The directionality is

$$D = \frac{2[J_1(ka \sin{\phi})]}{ka \sin{\phi}} \tag{A5.2.6}$$

The complete expression for the pressure radiated by a circular piston in a large baffle is the same as for a rectangular piston (5.2.2) with the proper directionality (A5.2.6) being used.

Table A5.2.1 gives values of D and D^2.

A5.3. BEAM SHAPING FOR ARRAYS

Although the piston source is perhaps the best known transducer, it is only one of an infinite number of possibilities. To generalize, consider a mosaic of small elements, fitted together to resemble a piston, with each of the elements capable of having its amplitude and phase independently adjusted. If each element of the mosaic is driven with the same frequency, phase, and velocity amplitude, and if the elements are a very small

TABLE A5.2.1. $2J_1(z)/z$

z	$\dfrac{2J_1(z)}{z}$	$\left[\dfrac{2J_1(z)}{z}\right]^2$	z	$\dfrac{2J_1(z)}{z}$	$\left[\dfrac{2J_1(z)}{z}\right]^2$
0.0	1.0000	1.0000	7.0	−0.0013	0.00000
0.2	0.9950	0.9900	7.016	0	0
0.4	0.9802	0.9608	7.5	+0.0361	0.0013
0.6	0.9557	0.9134	8.0	0.0587	0.0034
0.8	0.9221	0.8503	8.5	0.0643	0.0041
			9.0	0.0545	0.0030
1.0	0.8801	0.7746	9.5	0.0339	0.0011
1.2	0.8305	0.6897			
1.4	0.7743	0.5995	10.0	+0.0087	0.00008
1.6	0.7124	0.5075	10.173	0	0
1.8	0.6461	0.4174	10.5	−0.0150	0.0002
			11.0	−0.0321	0.0010
2.0	0.5767	0.3326	11.5	−0.0397	0.0016
2.2	0.5054	0.2554	12.0	−0.0372	0.0014
2.4	0.4335	0.1879	12.5	−0.0265	0.0007
2.6	0.3622	0.1326			
2.8	0.2927	0.0857	13.0	−0.0108	0.0001
			13.324	0	0
3.0	0.2260	0.0511	13.5	+0.0056	0.00003
3.2	0.1633	0.0267	14.0	0.0191	0.0004
3.4	0.1054	0.0111	14.5	0.0267	0.0007
3.6	0.0530	0.0028	15.0	0.0273	0.0007
3.8	+0.0068	0.00005	15.5	0.0216	0.0005
3.832	0	0			
			16.0	0.0113	0.0001
4.0	−0.0330	0.0011	16.471	0	0
4.5	−0.1027	0.0104	16.5	−0.0007	0.00000
5.0	−0.1310	0.0172			
5.5	−0.1242	0.0154	17.0	−0.0115	0.00013
6.0	−0.0922	0.0085	17.5	−0.01868	0.00035
6.5	−0.0473	0.0022			

For a circular piston, $z = ka \sin \phi$.

fraction of a wavelength apart, the mosaic will be equivalent to a piston. Now, if the amplitudes are adjusted to cause the greatest velocities to occur for the central elements, and the particle velocity decreases as one goes out toward the edge of the mosaic, a completely new beam pattern can be generated. The near field will also be altered, of course. This treatment is called shading.

This discussion is essentially a continuation of Sections 2.5 and 2.6. On reviewing the derivation of the directional response of equally spaced multiple sources, it is evident that these expressions are Fourier transformations in space coordinates where k and y play the roles of $2\pi f$ and t, respectively. In general, the directional response D for a line source of extent W is

$$D = \frac{1}{W_0} \int_{-W/2}^{W/2} A(y) \exp[iky \sin \phi] \, dy \qquad (A5.3.1)$$

where

$$W_0 = \int_{-W/2}^{W/2} A(y) \, dy \qquad (A5.3.2)$$

is the normalizing constant and $A(y)$ is the amplitude of the drive of each point element.

Thus far we have dealt with either continuous transducers or arrays of point transducers. Sonar arrays are usually constructed by combining many identical elements that cannot be assumed to be points. Suppose the linear array consists of M equally spaced line elements, each having an extent w. If A_m is the amplitude of the mth line transducer and D is the sum of the integrals over the individual transducers, we have

$$D = \sum_{m=0}^{M-1} A_m \int_{mW/(M-1)-w/2}^{mW/(M-1)+w/2} \exp(iky \sin \phi) \, dy \qquad (A5.3.3)$$

The change of variable

$$y' = y - \frac{mW}{M-1}$$

makes the limits of integration $-w/2$ to $w/2$. Factor $\exp[ikmW \sin \phi/(M-1)]$ from the integral, yielding

$$D = \left\{ \sum_{m=0}^{M-1} A_m \exp\left[\frac{ikm(\sin \phi)W}{M-1} \right] \right\} \left[\int_{-w/2}^{w/2} \exp(iky' \sin \phi) \, dy' \right] \qquad (A5.3.4)$$

We identify the term in braces as the array response D_A and the term in brackets as the transducer element response D_T. The response of an array of transducers is therefore the product of the response of the directional pattern of the individual transducers D_T and the array D_A.

$$D = D_T D_A \qquad \textbf{(A5.3.5)}$$

Equation (A5.3.5) is called the product theorem for transducers.

It is simpler to show the effects of shading the response of an array by assuming that the array elements are very close together and replacing the summations by an integral. To illustrate shading design, we assume that the reader has decided what beam width he desires and is about to select an array design to give that beam width. Side lobe reduction depends on $A(y)$ and will require arrays of different lengths. We compare the responses of three shaded arrays with the uniform array and then briefly discuss arrays which have equal side lobes. The D is subscripted to indicate the shading being used (e.g., D_u = uniform array). The radiation patterns are compared in Fig. A5.3.1.

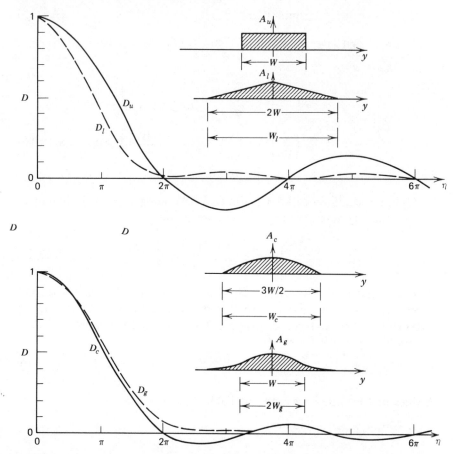

Figure A5.3.1. Response of shaded line transducers. For all examples, $\eta = kW \sin \phi$. The widths W_l and W_c were chosen to give the first null at $\eta = 2\pi$ as for the uniform line source of extent W.

1. *Uniform line array* (line piston).

$$A_u(y) = \frac{1}{W} \qquad \frac{-W}{2} \leq y \cdot \leq \frac{W}{2} \qquad\qquad \text{(A5.3.6)}$$

$$= 0 \qquad \text{otherwise}$$

$$D_u = \frac{\sin(\eta/2)}{\eta/2} \qquad\qquad \text{(A5.3.7)}$$

where

$$\eta \equiv kW \sin\phi \qquad\qquad \text{(A5.3.8)}$$

$$\text{1st null} = 2\pi \qquad\qquad \text{(A5.3.9)}$$

Values of $(\sin x)/x$ are given in Table A5.3.1.

2. *Linear tapered line array.*

$$
\begin{aligned}
A_l(y) &= \left(\frac{2}{W_l}\right)\left(1 - \frac{2y}{W_l}\right) \qquad \text{for} \quad 0 \leq y \leq \frac{W_l}{2} \\
&= \left(\frac{2}{W_l}\right)\left(1 + \frac{2y}{W_l}\right) \qquad \text{for} \quad \frac{-W_l}{2} \leq y \leq 0 \qquad \text{(A5.3.10)} \\
&= 0 \qquad \text{otherwise}
\end{aligned}
$$

After substitution of (A5.3.10) in (A5.3.1), integration, and manipulation of the trigonometric expressions, we obtain

$$D_l = \frac{\sin^2\left[(kW_l/4)\sin\phi\right]}{\left[(kW_l/4)\sin\phi\right]^2} \qquad\qquad \text{(A5.3.11)}$$

The first null is at $kW_l \sin\phi = 4\pi$. For comparison with the piston line array, we adjust the first null so that it occurs as $\eta = 2\pi$ by letting the extent of the linear tapered array be $W_l = 2W$, and we express D_l as a function of $\eta = (kW_l/2)\sin\phi$.

$$D_l = \frac{\sin^2(\eta/2)}{(\eta/2)^2} \qquad\qquad \text{(A5.3.12)}$$

Values of $(\sin^2 x)/x^2$ are listed in Table A5.3.1.

3. *Cosine tapered line array.*

$$
\begin{aligned}
A_c(y) &= \frac{\pi}{2W_c} \cos\frac{\pi y}{W_c} \qquad \text{for} \quad \frac{-W_c}{2} \leq y \leq \frac{W_c}{2} \\
&= 0 \qquad \text{otherwise}
\end{aligned}
\qquad \text{(A5.3.13)}
$$

TABLE A5.3.1. $(\sin x)/x$

x	$\dfrac{\sin x}{x}$	$\left(\dfrac{\sin x}{x}\right)^2$	x	$\dfrac{\sin x}{x}$	$\left(\dfrac{\sin x}{x}\right)^2$
0	1	1	6.0	-0.0466	0.0022
0.2	0.9933	0.9867	6.2	-0.0134	0.0002
0.4	0.9735	0.9478	2π	0	0
0.6	0.9411	0.8856	6.4	$+0.0182$	0.0003
0.8	0.8967	0.8041	6.6	0.0472	0.0022
			6.8	0.0727	0.0053
1.0	0.8415	0.7081			
1.2	0.7767	0.6033	7.0	0.0939	0.0088
1.4	0.7039	0.4955	7.2	0.1102	0.0122
1.6	0.6247	0.3903	7.4	0.1214	0.0147
1.8	0.5410	0.2927	7.6	0.1274	0.0162
			7.7252	0.1284	0.0165
2.0	0.4546	0.2067	7.8	0.1280	0.0164
2.2	0.3675	0.1351			
2.4	0.2814	0.0792	8.0	0.1237	0.0153
2.6	0.1983	0.0393	8.2	0.1147	0.0132
2.8	0.1196	0.0143	8.4	0.1017	0.0104
			8.6	0.0854	0.0073
3.0	$+0.0470$	0.0022	8.8	0.0665	0.0044
π	0	0			
3.2	-0.0182	0.0003	9.0	0.0458	0.0021
3.4	-0.0752	0.0056	9.2	0.0242	0.0006
3.6	-0.1229	0.0151	9.4	$+0.0026$	0.0000
3.8	-0.1610	0.0259	3π	0	0
			9.6	-0.0182	0.0003
4.0	-0.1892	0.0358	9.8	-0.0374	0.0014
4.2	-0.2075	0.0431			
4.4	-0.2163	0.0468	10.904	-0.0913	0.0083
4.4934	-0.2172	0.0472			
4.6	-0.2160	0.0467	4π	0	0
4.8	-0.2075	0.0431			
			14.066	$+0.0709$	0.0050
5.0	-0.1918	0.0368			
5.2	-0.1699	0.0289	5π	0	0
5.4	-0.1431	0.0205			
5.6	-0.1127	0.0127	17.221	-0.0580	0.0034
5.8	-0.0801	0.0064			
			6π	0	0
			20.371	$+0.0482$	0.0023

For a line, $x = k(W/2)\sin\phi$.

Integration and manipulation yields

$$D_c = \frac{\cos\left[(kW_c/2)\sin\phi\right]}{1 - \left[(kW_c/\pi)\sin\phi\right]^2} \tag{A5.3.14}$$

The first null occurs at $kW_c \sin\phi = 3\pi$. On letting the array length be $W_c = 3W/2$ and expressing D_c as a function of η, we obtain

$$D_c = \frac{\cos(3\eta/4)}{1 - (3\eta/2\pi)^2} \tag{A5.3.15}$$

4. *Gaussian tapered line array.*

$$A_g(y) = W_g^{-1}(\pi)^{-1/2}\exp\left(\frac{-y^2}{W_g^2}\right) \tag{A5.3.16}$$

With the aid of an integral table, evaluation of the integral over infinite limits yields.

$$D_g = \exp\left[-\left(\frac{kW_g}{2}\right)^2\sin^2\phi\right] \tag{A5.3.17}$$

The "ideal" Gaussian shaded transducer has an infinite extent; however the contributions of elements at distances greater than $kW_g \sin\phi \simeq 1.5$ is small. Since D_g does not have a null for comparison with the uniform array, we choose the beam width by letting $W_g = W/2$. Then

$$D_g \simeq \exp\left(\frac{-\eta^2}{16}\right) \tag{A5.3.18}$$

We use the Gaussian shaded transducer in our discussion of sound signals scattered at a rough surface.

Comparative responses of each of the types of transducer shading are presented in Fig. A5.3.1. The widths of the transducers were adjusted to keep the width of the main lobe about the same.

5. *Equal side lobes (Dolph-Chebyshev).* Array shading procedures that yield beam patterns having equal side lobes were introduced by Dolph (1946) when he showed that the Chebyshev polynomials could be used to determine the shading and phases of the elements of an array. The arrays are optimum in the sense that for a given number of elements and separations between elements, the choice of beam width sets the optimum side lobe level, or vice versa. Shading allows one to obtain lower side lobes, but at the cost of a greater beam width in the central lobe. Since Dolph's paper there have been many studies of transducer shading including effects of unequally spaced arrays and three-dimensional volume arrays.

SCATTERING OF SOUND BY BUBBLES

Bubbles are compressible objects in a much less compressible medium, water; thus their effect on the transmission of sound is far larger than their number or volume might indicate.

A6.1. BUBBLE RESONANCE FREQUENCIES, DAMPING CONSTANTS, AND ACOUSTICAL CROSS SECTIONS: DEPENDENCE ON PHYSICAL PARAMETERS

This section derives the expressions for the acoustical cross sections and the resonance frequency of a gas bubble. We show the dependence of these quantities on the physical parameters of the bubble gas, the liquid that surrounds it, and the interface between the two.

Assume that the bubble is irradiated by a sound field of wavelength much greater than the bubble radius ($ka \ll 1$). The incident plane wave is therefore uniform at all points of the bubble.

$$p_p = P_p \, e^{i\omega t} \tag{A6.1.1}$$

where P_p is the rms value.

The incident plane wave intensity at the bubble is

$$I_p = \frac{|P_p|^2}{\rho_A c} \tag{A6.1.2}$$

Assuming that the bubble is small compared to the wavelength and that

461

it is not fixed leads to a spherically symmetrical scattered pressure wave

$$p_s = P_s \frac{R_1}{R} \exp\left[i(\omega t - kR)\right] \tag{A6.1.3}$$

where R_1 is the reference distance from the scatterer ($= 1$ m). The scattered intensity is

$$I_s = \frac{|P_s|^2 (R_1/R)^2}{\rho_A c} \tag{A6.1.4}$$

The scattering cross section is calculated from (A6.1.2) and (A6.1.4).

$$\sigma_s = \frac{\displaystyle\int_A I_s \, dA}{I_p} = 4\pi R_1^2 \frac{|P_s|^2}{|P_p|^2} \tag{A6.1.5}$$

To determine the pressure ratio in (A6.1.5) we express the pressure and particle velocity boundary conditions at the bubble surface. Denoting the interior acoustic pressure by

$$p_i = P_i \, e^{i\omega t} \tag{A6.1.6}$$

we first state the pressure condition

$$[\text{pressure inside} = \text{pressure outside}]_{R=a}$$

To the pressures given by (A6.1.1), (A6.1.3), and (A6.1.6), we must add a shear viscous stress at the surface. This stress is proportional to the radial rate of strain u_r/R. The proportionality constant $C_1\mu$ includes the dynamic coefficient of shear viscosity for water μ and the dimensionless constant C_1, which depends on the geometry.

$$p_i = p_p + p_s + C_1\mu \left(\frac{u_r}{R}\right)_{R=a} \tag{A6.1.7}$$

To obtain u_r in terms of the scattered pressure, use the radial component of the acoustic force equation at the surface

$$\rho_A \left(\frac{\partial u_r}{\partial t}\right)_{R=a} = -\left(\frac{\partial p_s}{\partial R}\right)_{R=a}$$

Using (A5.1.3) and assuming harmonic time dependence for u_r, the radial component of the particle velocity of the radiated wave for $ka \ll 1$ is

$$u_r = -iP_s R_1 \frac{e^{i\omega t}}{\rho_A c k a^2} \tag{A6.1.8}$$

To simplify p_s, expand the exponential in (A6.1.3) and apply the condition $ka \ll 1$ at the surface to obtain

$$p_s = P_s\left(\frac{R_1}{a}\right)(1 - ika)\, e^{i\omega t} \qquad (A6.1.9)$$

Using (A6.1.6), (A6.1.1), (A6.1.9), and (A6.1.8) in (A6.1.7), we find the rms pressure condition at $R = a$,

$$P_i = P_p + P_s\left(\frac{R_1}{a}\right)(1 - ika) - \frac{i\mu C_1 P_s R_1}{\rho_A c k a^3} \qquad (A6.1.10)$$

Next we turn to the particle velocity. The interior radial particle velocity at the surface can be evaluated in terms of the interior pressure p_i. Consider the relation between the bubble volume and the interior pressure. If the bubble is large, the reversible adiabatic relation $p_{iT} V^\gamma = $ constant will hold, where p_{iT} is the total interior pressure. However for small bubbles the action becomes isothermal, and the effective value of γ approaches unity. Equally important, the bubble pressure and temperature do not instantly follow the volume variation. For example, as the bubble volume is decreasing during a cycle, the temperature will be increasing. But at the minimum volume some heat may still be escaping, which means that the gas temperature and pressure is thereby decreasing. Therefore we rewrite the adiabatic law replacing γ by $\gamma(b + id)$ to allow the phase between pressure and volume and the effective γ to be a function of driving frequency and bubble size

$$p_{iT} V^{\gamma(b+id)} = \text{constant} \qquad (A6.1.11)$$

where

$$p_{iT} = P_{iA} + p_i = \text{total interior pressure} \qquad (A6.1.12)$$

and P_{iA} is the average value of the interior pressure including the surface tension effect, b and d are real dimensionless numbers that account for the change of magnitude and phase of γ.

Differentiate with respect to time and rearrange to ($P_{iA} \gg |p_i|$)

$$\frac{dV}{dt} = -\frac{i\omega p_i V}{\gamma(b+id)P_{iA}} \qquad (A6.1.13)$$

Use $dV/dt = 4\pi a^2 u_r]_{R=a}$ and (A6.1.6) and (A6.1.13) to obtain the interior radial particle velocity at $R = a$.

$$u_r]_{R=a} = -i\omega a P_i \frac{e^{i\omega t}}{[3\gamma(b+id)P_{iA}]} \qquad (A6.1.14)$$

Equate to the particle velocity of the radiated wave at $R = a$, given by (A6.1.8).

Rearrange to form

$$\frac{P_i}{P_s} = \frac{3\gamma(b+id)P_{iA}R_1}{\rho_A c^2 k^2 a^3} \tag{A6.1.15}$$

Following the derivation that led to (6.3.9), define the natural frequency of the bubble f_R in terms of the average interior pressure $P_{iA} = \beta P_A$, where P_A is the exterior hydrostatic pressure, and γb the effective ratio of the specific heats (dropping the id).

$$f_R \equiv \frac{1}{2\pi a}\left(\frac{3\gamma b\beta P_A}{\rho_A}\right)^{1/2} \tag{A6.1.16}$$

Using (6.3.9). the resonance frequency is written

$$f_R = f_R'(\beta b)^{1/2} \tag{A6.1.17}$$

The detailed expressions for b and β are given in (A6.1.26) and (A6.1.27). Fig. A6.1.1 shows their behavior for sea level air bubbles.

Therefore (A6.1.15) becomes

$$\frac{P_i}{P_s} = \left(\frac{R_1}{a}\right)\left(\frac{f_R}{f}\right)^2 (1 + id/b) \tag{A6.1.18}$$

The fluctuating interior pressure can be eliminated by using (A6.1.18) in (A6.1.10). Also, since (A6.1.8) and (A6.1.14) show that u_r lags p_s by 90°

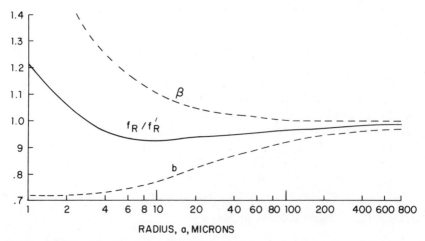

Figure A6.1.1. Frequency constants for resonance of a clean, free air bubble at sea level. See (A6.1.17), (A6.1.26), and (A6.1.27).

and leads p_p by 90°, when evaluated at $R = a$ and $f \ll f_R$, we replace complex p_p by real $-p_p$. This effectively fixes the phase of complex P_s in terms of the incident wave as reference. Then

$$\frac{P_s}{P_p} = \frac{-a/R_1}{[(f_R/f)^2 - 1] + i[ka + (d/b)(f_R/f)^2 + C_1\mu/(\rho_A\omega a^2)]} \quad \text{(A6.1.19)}$$

We now define the damping constants

$$\delta = ka + \frac{d}{b}\left(\frac{f}{f_R}\right)^2 + \frac{4\mu}{\rho_A\omega a^2}$$

$$= \delta_r + \delta_t + \delta_v \quad \textbf{(A6.1.20)}$$

where $\delta_r = ka = $ damping constant due to reradiation
   $\delta_t = (d/b)(f_R/f)^2 = $ damping constant due to thermal conductivity
   $\delta_v = 4\mu/(\rho_A\omega a^2) = $ damping constant due to shear viscosity

Our proportionality constant C_1 has been given the value 4, as found in other studies (Devin, 1959).

A6.1.1. Resonance and the Total Scattering Cross Section

The scattering cross section is obtained by using (A6.1.19) in (A6.1.5)

$$\sigma_s = 4\pi R_1^2 \frac{|P_s|^2}{|P_p|^2} = \frac{4\pi a^2}{[(f_R/f)^2 - 1]^2 + \delta^2} \quad \textbf{(A6.1.21)}$$

When $f = f_R$, the values of δ are crucial to the size of the scattering cross section. The damping constants have been evaluated in terms of the physical constants of the gas bubble and water. [See Devin, 1959, Medwin, 1977.]

To calculate δ_t, the thermal damping constant, we must evaluate

$$\frac{d}{b} = 3(\gamma - 1)\left[\frac{X(\sinh X + \sin X) - 2(\cosh X - \cos X)}{X^2(\cosh X - \cos X) + 3(\gamma - 1)X(\sinh X - \sin X)}\right]$$

$$\text{(A6.1.22)}$$

where

$$X = a\left(\frac{2\omega\rho_g C_{pg}}{K_g}\right)^{1/2} \quad \text{(A6.1.23)}$$

Most of the constants are conveniently given in cgs units:

$K_g = $ thermal conductivity of gas ($\approx 5.6 \times 10^{-5}$ cal/(cm)(s)(°C) for air)
$\rho_g = $ density of gas $= \rho_{gA}[1 + 2\tau/(P_A a)](1 + 0.1z)$ (A6.1.24)

ρ_{gA} = density of free gas at sea level ($\approx 1.29 \times 10^{-3}$ g/cm^3 for air)

τ = surface tension ($\approx 75.$ dynes/cm for air-water surface)

$P_A = 1.013 \times 10^6 (1 + 0.1z)$ dynes/cm^2 (A6.1.25)

z = bubble depth (m)

C_{pg} = specific heat at constant pressure of gas (≈ 0.24 cal/g for air)

To calculate δ_v, we need the value

$$\mu = \text{shear viscosity of water} \qquad [\approx 0.01 \text{ g/(cm)(s)}]$$

While we are about it, let us give the constants needed for the resonance frequency equation (A6.1.16)

$$\beta = \frac{P_{iA}}{P_A} = 1 + \frac{2\tau}{P_A a}\left(1 - \frac{1}{3\gamma b}\right) \tag{A6.1.26}$$

$$b = \left[1 + \left(\frac{d}{b}\right)^2\right]^{-1}\left[1 + \frac{3\gamma - 1}{X}\frac{\sinh X - \sin x}{\cosh X - \cos X}\right]^{-1} \tag{A6.1.27}$$

The resonance frequency and the damping constants can be calculated by the following procedure. Use the foregoing values for air, or suitable tables, to obtain the physical constants of the bubble at the depth z. This will include the constants of the gas, ρ_{gA}, K_g, C_{pg}, γ, the constants of the water, ρ_A, μ, and the surface tension between them τ. For a given sound frequency and bubble radius calculate X, and then d/b, b, and β in that order. The resonance frequency for that bubble radius can then be obtained from (A6.1.16) and the damping constants readily fall out of (A6.1.20).

Figure A6.1.2 gives the damping constants as a function of radius for air bubbles at sea level. The resonance radii a_R are indicated.

A6.1.2 The Mechanical Damping Constant

An alternative description of bubble oscillation is sometimes useful. In addition to the inertial force and the stiffness force, (6.3.8), there must be a damping force, $R_M\dot{\xi}$, which we assume to be proportional to velocity. The mechanical damping constant R_M represents energy losses caused by sound reradiation, shear viscosity, and thermal conductivity. The equation describing the motion of the bubble

$$m\ddot{\xi} + R_M\dot{\xi} + s\xi = 4\pi a^2 P_p\, e^{i\omega t} \tag{A6.1.28}$$

where the right-hand side is the external force on the bubble.

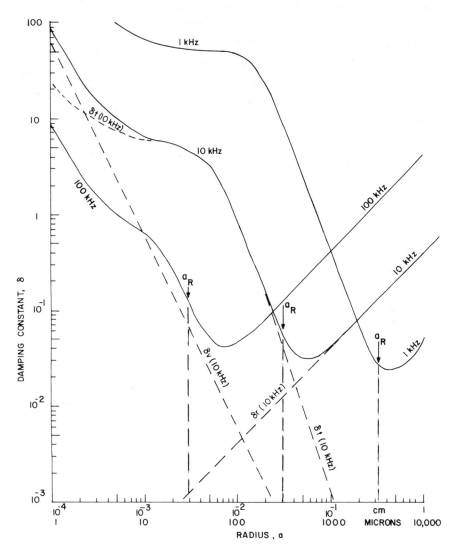

Figure A6.1.2 Damping constants of air bubbles at sea level as a function of bubble radius for three sound frequencies: contributing parts δ_v, δ_t, δ_r are shown for 10 kHz curve; resonance radii a_R are given for three frequencies. The damping constants at resonance δ_R are not the minimum values of δ.

Substitution into (A6.1.28) shows that the radial displacement is

$$\xi = \frac{-i4\pi a^2 (P_p/\omega) e^{i\omega t}}{R_M + i(\omega m - s/\omega)} \tag{A6.1.29}$$

We show that R_M is simply related to δ. Comparing $u_r = \dot{\xi}$ from (A6.1.29) with (A6.1.14), and using (A6.1.15) and (A6.1.19)

$$R_M + i\left(\omega m - \frac{s}{\omega}\right) = -i4\pi a^3 \rho_A \omega \left\{ \left[\left(\frac{f_R}{f} \right)^2 - 1 \right] + i\delta \right\}$$

Equating the reals shows that

$$R_M = 4\pi a^3 \rho_A \omega \delta$$

and with (6.3.7) we have

$$R_M = \omega m \delta \tag{A6.1.30}$$

The damping constant *at resonance* δ_R is defined in terms of the resonance frequency so that

$$\delta_R = \frac{R_M}{\omega_R m} \tag{A6.1.31}$$

Similarly, we define the component damping constants at resonance so that

$$\delta_R = \delta_{Rr} + \delta_{Rv} + \delta_{Rt} \tag{A6.1.32}$$

where, from (A6.1.20) with $f = f_R$, we find

$$\delta_{Rr} = k_R a_R$$

$$\delta_{RT} = \left. \left(\frac{d}{b} \right) \right]_{f=f_R}$$

$$\delta_{Rv} = \frac{4\mu}{\rho_A \omega_R a_R^2} \tag{A6.1.33}$$

The damping constants at resonance for an air bubble at sea level are plotted in Fig. 6.3.1.

A6.1.3.   The Extinction Cross Section

The extinction cross section of a bubble can be calculated from

$$\sigma_e = \frac{\Pi_e}{I_p}$$

The extinguished power (scattered and absorbed) is obtained from the rate at which work is done on the bubble by the incident pressure. The calculation requires integration of the product of the force by the velocity, at $R = a$, over a period $T = 1/f$.

$$\Pi_e = \frac{1}{T}\int_T (p_p)(4\pi a^2)(\dot{\xi})]_{R=a}\, dt \qquad (A6.1.34)$$

The difficulties associated with the product of complex quantities can be avoided by using the real values of p_p and $\dot{\xi}$ from (A6.1.1) and (A6.1.29) before multiplying and integrating in (A6.1.34). Substitution of $\delta = R_M/\omega m$ from (A6.1.30) allows simplification and we get

$$\sigma_e = \frac{\Pi_e}{P_p^2/\rho c} = \frac{4\pi a^2(\delta/ka)}{[(f_R/f)^2 - 1]^2 + \delta^2} \qquad \textbf{(A6.1.35)}$$

A6.2. DISPERSION IN BUBBLY WATER

The dependence of the sound speed on the compressibility and density is given by

$$c^2 = \frac{1}{\rho_A K} \qquad \textbf{(A6.2.1)}$$

where the compressibility K is the reciprocal of the bulk elasticity E. The compressibility is made up of a part due to the bubble-free water K_0 and a complex part due to the bubbles themselves K_1

$$K = K_0 + K_1 \qquad (A6.2.2)$$

where K_0 is expressed in terms of the speed of sound through bubble-free water which we call c_0 given by (3.2.20) and the ambient density ρ_A.

$$K_0 = \frac{1}{\rho_A c_0^2} \qquad (A6.2.3)$$

The complex compressibility due to the bubbles is found by using the displacement from (A6.1.29)

$$K_1 = \frac{(\Delta V/V)}{\Delta P} = \frac{N\Delta v}{\Delta P} = \frac{NS\xi}{P_p\, e^{i\omega t}} = \frac{NS^2}{m\omega^2[(-1 + \omega_R^2/\omega^2) + iR_M/(\omega m)]}$$

$$(A6.2.4)$$

where N is the number of bubbles per unit volume, Δv is the change in volume of each bubble, $S = 4\pi a^2$ is the surface area of each bubble, and ξ is the radial displacement of the bubble surface. To simplify, we use

$\delta = R_M/(\omega m)$ from (A6.1.30) and define the frequency ratio

$$Y = \frac{f_R}{f} = \frac{\omega_R}{\omega} \tag{A6.2.5}$$

Then

$$K_1 = \frac{N4\pi a[Y^2 - 1 - i\delta]}{\rho_A \omega^2[(Y^2 - 1)^2 + \delta^2]} \tag{A6.2.6}$$

The speed of sound in the bubbly medium can now be written in the form

$$c = \left(\frac{1}{\rho_A K}\right)^{1/2} = \frac{c_0}{(1 + A - iB)^{1/2}} \tag{A6.2.7}$$

where

$$A = \frac{Y^2 - 1}{(Y^2 - 1)^2 + \delta^2} \frac{4\pi a N c_0^2}{\omega^2} \quad \text{and} \quad B = \frac{\delta}{(Y^2 - 1)^2 + \delta^2} \frac{4\pi a N c_0^2}{\omega^2}$$

The interpretation of the complex speed is clearer if we consider the complex propagation constant

$$k = \frac{\omega}{c} = \frac{\omega(1 + A - iB)^{1/2}}{c_0} \tag{A6.2.8}$$

Since A and B are extremely small in the ambient ocean, the expression whose square root is extracted is of the form $(1 + \text{small quantity})$. Therefore the first terms of the Taylor's expansion for k give

$$k \simeq k_0\left(1 + \frac{A}{2} - \frac{iB}{2}\right) \tag{A6.2.9}$$

where $k_0 = \omega/c_0$.

The equation for a plane wave propagating through a bubbly medium is

$$p_p = P_p \exp[i(\omega t - kx)] \tag{A6.2.10}$$
$$= P_p \exp(-k_{im}x) \exp[i(\omega t - k_{re}x)]$$

where $k_{im} = k_0 \dfrac{B}{2}$

$$k_{re} = k_0\left(1 + \frac{A}{2}\right)$$

In this form it is clear that the imaginary part of the complex propagation constant represents the attenuation of the wave through the bubbly region; this was called α_b in (6.4.1). The real part k_{re} is the wave number for the propagation of constant phase surfaces. The ratio ω/k_{re} is the phase velocity, which we call $\text{Re}\{c\}$. It is a function of the sound frequency, and the medium is said to be dispersive.

$$\text{Re}\,\{c\} = \frac{\omega}{k_{re}} = c_0\left[1 - \frac{2\pi a N c_0^2}{\omega^2}\frac{Y^2-1}{(Y^2-1)^2+\delta^2}\right] \qquad \text{(A6.2.11)}$$

It is useful to write the speed in terms of the fraction of gas in bubble form U

$$U = N(\tfrac{4}{3}\pi a^3) \qquad \text{(A6.2.12)}$$

Then

$$\text{Re}\,\{c\} = c_0\left[1 - \frac{3UY^2}{2a^2k_R^2}\frac{Y^2-1}{(Y^2-1)^2+\delta^2}\right] \qquad \textbf{(A6.2.13)}$$

where $k_R = \omega_R/c_0$ is the value of k_0 at resonance.

Consider two special cases. At low frequency, $f \ll f_R$, the speed is

$$c_{lf} = c_0\left(1 - \frac{3U}{2a^2k_R^2}\right) \qquad \textbf{(A6.2.14)}$$

Since ak_R is a constant for a given gas at a given depth, the low frequency asymptotic speed depends only on the total gas volume U.

At the high frequency extreme for $f \gg f_R$, we have

$$c_{hf} = c_0\left[1 + \frac{3UY^2}{2a^2k_R^2(1+\delta^2)}\right] \to c_0 \qquad \textbf{(A6.2.15)}$$

Therefore bubbles do not affect the sound phase speed if the frequency is high enough. Sound velocimeters, which operate in the megahertz range, yield values of $c = c_0$ even in bubbly water, because $f \gg f_R$ for the significant fractions U in the sea.

Figure A6.2.1 illustrates the fractional change in phase velocity for two different bubble densities N resonant at the same frequency.

The generalization to a medium with bubbles of random radii is accomplished by replacing N by $n(a)\,da$ and U by $u(a)\,da$ in (A6.2.12). Because all contributions to the compressibility are very small quantities, they add linearly and the speed of sound in the bubbly region can be written in terms of the integral over all radii.

$$\text{Re}\,\{c\} = c_0\left[1 - \frac{3}{2}\int_a \frac{u(a)\,Y^2(Y^2-1)\,da}{a^2k_R^2[(Y^2-1)^2+\delta_R^2]}\right] \qquad \text{(A6.2.16)}$$

The effect of a mixture of bubble sizes is to smear the dispersion curve so that although the magnitudes of the deviation from the bubble-free value are increased, the frequency range between the peak and the trough is also increased.

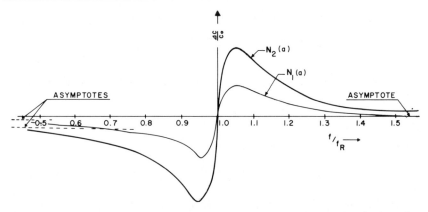

Figure A6.2.1.   Dispersion of sound phase velocity for bubbles of a single radius. The fractional change in speed is proportional to N, the number of bubbles per unit volume. The high frequency asymptote is zero. The low frequency asymptote is proportional to the volume fraction U. $N_2(a) > N_1(a)$.

A6.3.   BUBBLE MEASUREMENTS AT SEA

In situ bubble research has been done with equipment sketched in Fig. A6.3.1. The devices were 2 to 7 m in extent and were constructed so that the water medium freely entered the insonified space.

A6.3.1.   Pulse Echo Technique

The pulse echo technique has featured a sinusoidal wave train lasting approximately 0.5 ms, with the transducer electronically switched to

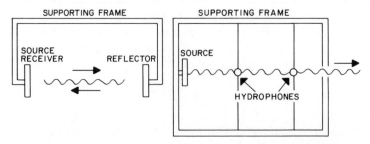

Figure A6.3.1.   Two experimental schemes for measuring bubble effects at sea. *Left*: pulse-echo technique (Medwin, 1970); *right*: CW technique (Medwin, et al. 1975).

receive the reflected echoes. The oscilloscope screen was photographed to record the echo patterns (Fig. A6.3.2).

In principle a single oscillogram can be analyzed in three different ways to obtain the variables of the medium: (1) comparison of the exponential attenuation shown by the echo pattern at sea with that in clean water provides the excess absorption and scattering by objects (assumed to be

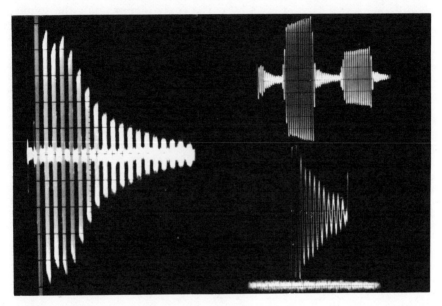

Figure A6.3.2. *Left*: oscillogram showing 19 echoes of a pulse echo pattern at 200 kHz. *Right* (top): two echoes and backscatter between them at 30 kHz; (*middle*) amplified backscatter between two echoes; (*bottom*) system noise level on same scale as backscatter.

bubbles) at sea; (2) the relative reverberation between echoes, compared to the preceding and following echo levels, measures the scatter due to bubbles along the path; (3) the time between echoes permits a calculation of the local speed of propagation. In practice, because of phase shift at the reflectors, the dispersion of sound velocity at sea is often too small to measure accurately with this pulse echo system. However the extinction cross sections and the scattering cross sections have been determined as a function of frequency at sea. By using both the extinction and the scatter data, the absorption cross section can be found and the number of bubbles in a given radius increment per unit volume of water is directly calculable. It is assumed that the scattering and absorption cross sections of dirty gas bubbles are the same as for clean bubbles (Fig. 5.4.4).

A6.3.2. CW Technique

It is also possible to obtain *in situ* bubble information by CW acoustic measurements (Medwin et al., 1975). The study requires only the sound source and two very small hydrophones separated by a fixed distance (Fig. A6.3.1).

The speed is easily found as a function of frequency by measuring the number of wavelengths between the two hydrophones. Then

$$c = f\lambda = f\frac{x}{M - \phi/360}. \qquad (A6.3.1)$$

where x is the distance between the two hydrophones and $M + \phi/360$ is the integral plus fractional number of wavelengths. The number M is obtained by using a rough value of the bubble-free speed, c_0, or data from a sound velocimeter. A good phasemeter can give ϕ to the nearest tenth of a degree, f is known to 1 part in 10^6 by using a frequency synthesizer or stable oscillator and frequency counter. In fact it is the

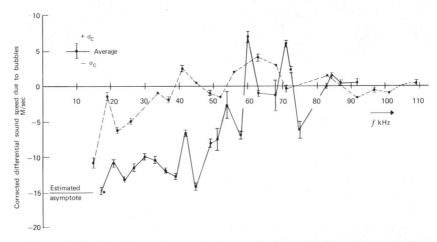

Figure A6.3.3. Variation of sound phase velocity with frequency including standard deviation bars. Dashed line data taken at 3.3 m below surface in water 15 m deep, 1 mile from shore off San Diego during 6 knot winds. Solid line data obtained 5 m below surface adjacent to an ocean tower in Bass Strait, Australia, during 8 to 10 knot winds.

measurement of distance x that is the crudest part of the experiment and limits the absolute accuracy. The distance can be calibrated by using a velocimeter as the reference $c = c_0$ for high frequencies. Some results for the near surface ocean are presented in Fig. A6.3.3.

The same CW equipment can be used to measure the attenuation between the two hydrophones and, thereby, to assess the bubble density or volume fraction in particular bands of bubble radii. The attenuation can be found by either analogue or digital measurements. In digital analysis high rate sampling at the two hydrophones can provide two digital time series from which both magnitude and phase of the frequency components of the signal at each of the two hydrophones can be obtained by FFT. The ratio of the magnitudes then gives the attenuation. The difference of the phases, used in (A6.3.1), yields the speed. When a harmonic-rich signal serves as the sound source, both the attenuation and phase shift can be determined, from the same data, for a large number of frequencies up to half the sampling frequency.

SUGGESTED READING

P. M. Morse and K. U. Ingard, *Theoretical Acoustics*, McGraw-Hill, New York, 1968. Chapter 8 (pp. 400–466) is an advanced treatment of scattering from cylinders, spheres, bubbles and surfaces. Multiple scatter is considered.
Herman Medwin, "In Situ Acoustic Measurements of Microbubbles at Sea" (Paper 6C0696), J. Geoph. Res., Oceans and Atmospheres, **82,** 971–976 (1977).

FISH ECHOES AND RESONANCE

In this simplified description of the random character of the echoes from fish, the average number of fish per unit volume is assumed to be constant within the insonified volume. Applications of the bubble theory to the swim bladders of fish are in Section A7.2.

A7.1. PDF OF ECHOES

Our derivation is for a single species of fish. We choose to receive the echoes on the source transducer. For simplicity, we assume that the transducer is a baffled circular piston.

We have two variabilities to consider and each is random. We assume that the scattered pressure depends on the position of the fish in the sonar beam D^2 and is independent of how it scatters l, where l is the *effective* length, which is a function also of fish aspect and sound frequency.

On the basis of studies of reverberation measurements, we choose the Rayleigh PDF to describe the fluctuations of l. This general form of the Rayleigh PDF (Fig. A7.1.1) is

$$w_R(x) = |x|\, \sigma^{-2} \exp\left(\frac{-x^2}{2\sigma^2}\right) \qquad (A7.1.1)$$

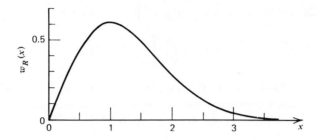

Figure A7.1.1. Rayleigh PDF.

It was derived by Rayleigh in his calculation of the probability of occurrence of an amplitude x caused by the sum of many sine waves having the same frequency but random phases. The Rayleigh PDF is consistent with our qualitative description of the scattering process as due to the superposition of a large number of randomly phased Huygens wavelets.

Assuming the Rayleigh PDF for l, we write

$$w_R(l) = \left(\frac{|l|}{\langle l^2 \rangle}\right) \exp\left(-\frac{l^2}{2\langle l^2 \rangle}\right) \tag{A7.1.2}$$

where

$$\langle l^2 \rangle = \langle \mathcal{S}_{bs}A \rangle = \sigma_{bs}$$

and $[w_R(l)][\langle l^2 \rangle^{1/2}]$ is dimensionless. In the absence of the directional effects due to the transducer $(D = 1)$ we would expect the amplitude of the envelopes of the echoes to have a Rayleigh PDF.

To calculate the effect of the position of the fish in the sonar beam, we assume that the fish is within $\Delta\phi\,\Delta R$ in Fig. 7.2.1 and the incremental gated volume ΔV_G is

$$\Delta V_G = 2\pi R \sin\phi\, d\phi R\,\Delta R$$

with cylindrical symmetry about the z axis. Assuming that the *back radiation is negligible*, the gated volume V_G is

$$V_G = \frac{2\pi}{3}\left[\left(R + \frac{\Delta R}{2}\right)^3 - \left(R - \frac{\Delta R}{2}\right)^3\right]$$

$$V_G \simeq 2\pi R^2\,\Delta R \tag{A7.1.3}$$

The ratio $\Delta V_G/V_G$ is the probability of the fish being within ΔV_G

$$\mathcal{P}(\Delta V_G) = \sin\phi\, d\phi \tag{A7.1.4}$$

For each angle ϕ, D^2 has the response b.

The PDF for the response being within $b \pm \Delta b/2$ is

$$w_I(b) = \frac{1}{\Delta b} \int_{\Delta \phi} \sin \phi \, d\phi \qquad (A7.1.5)$$

where $\Delta \phi$ is evaluated for all angles for which b is within $b \pm \Delta b/2$. Small values of b have multiple values of ϕ because of the side lobes. This is illustrated for several values of b in Fig. A7.1.2.

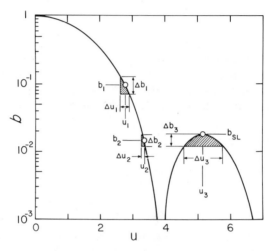

Figure A7.1.2.   Insonification function b, where b is insonification and reception for a circular piston transducer $u = ka \sin \phi$. From Peterson et al, 1976.

We use the variable

$$u \equiv ka \sin \phi$$

where a is the radius of the circular piston transducer. This change of variable alters (A7.1.5) to

$$(ka)^2 \, w_I(b) = \frac{1}{\Delta b} \int_{\Delta u} \frac{u \, du}{[1 - u^2/(ka)^2]^{1/2}} \qquad (A7.1.6)$$

where Δu is for all values of b within the range. An example of $w_I(b)$ for $ka = 2\pi$ is given in Fig. A7.1.3. The high peak at $b = 0.0175$ is due to the maximum of the first side lobe; $w_I(b)$ is complicated and large for $b < 0.0175$.

Using the PDFs of l and b, the cumulative probability of observing a

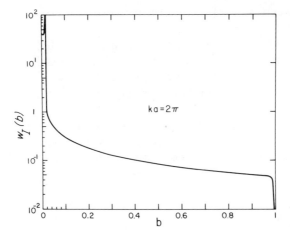

Figure A7.1.3. PDF $w_I(b)$ for insonification and reception at amplitude b. High peak at 0.0175 is due to maximum of first side lobe of b. From Peterson et al. 1976.

value of an echo peak equal to or less than e is the product

$$\mathscr{C}(e) = \int_0^1 w_I(b)\, db \int_0^{e/b} w_R(l)\, dl \tag{A7.1.7}$$

where the upper limit of l is the maximum value of l for a particular b. The PDF of e, $w_E(e)$, is the derivative of $\mathscr{C}(e)$ with respect to e and is

$$w_E(e) = \int_0^1 \frac{w_I(b)w_R(e/b)}{b}\, db \tag{A7.1.8}$$

We evaluate $w_I(b)$ and $w_E(e)$ numerically.

Numerical procedures. We give a three-step outline of our numerical method for evaluating (A7.1.8), discussing transducer directional response, $w_I(b)$, and $w_E(e)$ in the following three sections.

A7.1.1. Transducer Directional Response

Ordinarily, D and D^2 are given as functions of u. For the circular transducer, see Table A5.2.1.

$$D = \frac{2J_1(u)}{u} \tag{A7.1.9}$$

$$b = D^2$$

b ranges from 0 to 1. It is convenient to express Δb as $1/M$ for M levels and b or b_m as m/M.

The calculation requires u as a function of b. We use Newton's method to calculate the values of u_m for a set of b_m. This can be done once and stored as a "look-up table" in the computer. We also calculate $(du/db)_m$ for the set of b_m and store them. We suggest using at least 100 amplitude steps.

We find that we can simplify the calculation considerably by limiting this part of the calculation to the main lobe or $0 \le u \le 3.83$. This eliminates multiple values of u. The simplification is dealt with in connection with $w_I(b)$.

A7.1.2. $w_I(b)$

The conversion of the integral for $w_I(b)$ to a summation gives

$$(ka)^2 w_I(b_m) = \sum_m u_m \left(\frac{\Delta u}{\Delta b}\right)_m \left[1 - \frac{u_m^2}{(ka)^2}\right]^{-1/2} \qquad (A7.1.10)$$

where the summation is over all values of u_m for which b_m is within $b_m \pm \Delta b/2$. For b_m greater than the maximum of the first side lobe b_{SL}, the expression is single valued and there is one term for each level. Then (A7.1.10) reduces to

$$(ka)^2 w_I(m\Delta b) = u_m \left(\frac{\Delta u}{\Delta b}\right)_m \left[1 - \frac{u_m^2}{(ka)^2}\right]^{-1/2} \qquad \text{for} \quad m > \frac{b_{\mathrm{SL}}}{\Delta b}$$

$$\textbf{(A7.1.11)}$$

where u_m and $(\Delta u/\Delta b)_m$ are the values of u and the derivative for the level b_m: these come from the look-up tables calculated in Section A7.1.1.

The dependence of $w_I(b)$ on b is very complicated for b less than b_{SL}. Since the region is very narrow, we suggest using its average value. The integral of $w_I(b)$ over 0 to 1 is unity, and we let the average value w_{SL} be calculated as follows:

$$w_{\mathrm{SL}} b_{\mathrm{SL}} = 1 - \int_{b_{\mathrm{SL}}}^{1} w_I(b) \, db \qquad (A7.1.12)$$

The equivalent summation is

$$w_{\mathrm{SL}} = (b_{\mathrm{SL}})^{-1} \left[1 - \frac{1}{M} \sum_m^M w_I\left(\frac{m}{M}\right)\right] \qquad \text{for} \quad m > \frac{b_{\mathrm{SL}}}{\Delta b} \quad \textbf{(A7.1.13)}$$

where $\Delta b = 1/M$. The set of $w_I(m/M)$ and w_{SL} are stored as a look-up table for Section A7.1.3.

A7.1.3. $w_E(e)$

The equivalent summation for (A7.1.8) follows directly after we express b, db, and e in terms of incremental quanties. We let

$$b = \frac{m}{M}$$

$$db \rightarrow \Delta b = \frac{1}{M} \tag{A7.1.14}$$

$$e \rightarrow e_n = ne_0 \langle l^2 \rangle^{1/2}$$

where e_0 is the dimensionless amplitude increment. The change of the integral to the summation gives

$$w_E(e_n) = \sum_{m=0}^{M} m^{-1} w_I \left(\frac{m}{M}\right) w_R \left(\frac{e_n}{m/M}\right) \tag{A7.1.15}$$

where db/b reduces to $1/m$. The $m = 0$ term is zero as $m \rightarrow 0$ because $w_R(e_n M/m)$ goes to zero more rapidly than $1/m$ becomes infinite. We drop $m = 0$, express $w_R(\ )$, and write

$$w_E(e_n)(ka)^2 \langle l^2 \rangle^{1/2} = (ka)^2 \sum_{m=1}^{M} m^{-1} w_I \left(\frac{m}{M}\right) g_{mn} \exp \left(\frac{-g_{mn}^2}{2}\right) \tag{A7.1.16}$$

$$g_{mn} \equiv ne_0 \frac{M}{m}$$

where we use the approximate independence of $(ka)^2 w_I(m/M)$ of ka for tabulation. The summation is over the terms in the "look-up table" of (A7.1.11) and (A7.1.12). Table A7.1.1 is for $ka = 4\pi$.

For very small $e(e_n < \simeq b_{SL})$, the values of $w_E(e)$ are approximate because of the simplification in the calculations of $w_I(b)$. The main sources for these small values are fish in the side lobes. Usually the inaccuracy is negligible because echoes having very small amplitudes are lost in the noise.

TABLE A7.1.1. Probability Density of Echoes $w_E(e)$, $ka = 4\pi$

$e/\langle l^2 \rangle^{1/2}$	$w_E(e)(ka)^2 \langle l^2 \rangle^{1/2}$	$e/\langle l^2 \rangle^{1/2}$	$w_E(e)(ka)^2 \langle l^2 \rangle^{1/2}$
0.01	4600	0.60	2.07
0.02	2067	0.65	1.87
0.03	273	0.70	1.70
0.04	29.3	0.75	1.54
0.05	16.7	0.80	1.40
0.06	14.5	0.85	1.28
0.07	13.0	0.90	1.16
0.08	11.8	0.95	1.06
0.09	10.8	1.00	0.961
0.1	10.0	1.2	0.652
0.15	7.39	1.4	0.436
0.20	5.89	1.6	0.284
0.25	4.90	1.8	0.181
0.30	4.19	2.0	0.111
0.35	3.64	2.2	0.0666
0.40	3.21	2.4	0.0385
0.45	2.85	2.6	0.0215
0.50	5.55	2.8	0.0116
0.55	2.29	3.0	0.00605

A7.2. SWIM BLADDER RESONANCE CALCULATIONS

A7.2.1. Resonance Frequency

The theoretical treatment of bubbles in Chapter 6 forms the framework for the calculation of bubble effects of fish with swim bladders. The resonance frequency (6.3.9) and (6.3.11) is changed in two ways.

1. The simplest bladders of fish are closer to prolate ellipsoids than to spheres. For an axis ratio of $2:1$, the calculation of the resonance frequency of the bladder can proceed with only 2% discrepancy when the equivalent radius a' is calculated from the volume equation $a' = (3V/4\pi)^{1/3}$. Then a' replaces a in (6.3.9). If the ratio of the major to the minor axis is very much greater than unity, Fig. 7.1.1 is used to calculate the increase in the resonance frequency due to shape.

2. The frequency correction for surface tension must be replaced by the more relevant effects of shear elasticity of the tissues surrounding the gas bubble. The correction is calculated from the physics of vibration of a

rubberlike spherical shell (McCarthney and Stubbs, 1970). It is written as an additive term to the numerator rather than the factor β in (6.3.11)

$$f_{RF} = \frac{[(3\gamma P_A + 4\mu_1 \times 3t/a')/\rho_A]^{1/2}}{2\pi a'} \tag{A7.2.1}$$

where t = thickness of bladder tissue $\simeq 0.2a'$
μ_1 = real part of complex elastic shear modulus of bladder membrane

The μ_1 term takes account of the tendency of the elastic forces of the fish bladder tissue to resist deformation as the bladder is vibrated. The force required to extend the bladder creates an effective pressure within the cavity which increases the resonance frequency of the bubble. Correction for thermal conductivity is ignored for simplicity. Measured values of μ_1 are of the order of 1 or 2 atm (see Table A7.2.1) and increase approximately as the square of the frequency. Because this effect is added to $3\gamma P_A$, it is significant only at shallow depths. The complex dynamic elastic shear modulus is defined by

$$\mu = \mu_1(1 + id) = \frac{\text{shear stress}}{\text{shear strain}} \tag{A7.2.2}$$

TABLE A7.2.1. Components of Complex Shear Modulus for Fish Tissue at Arbitrary Orientation of Fibers*

Type of Fish	Frequency (kHz)	$\mu = \mu_1(1 + id)$	
		μ_1 (atm)	d (dimensionless)
Plaice, muscle	11.8	3.9	0.23
tissue from skin	11.4	3.6	0.23
	8	1.1	0.24
	1.58	0.12	0.30
Stauridia, muscle	11.4	1.3	0.17
tissue from skin	9.8	0.80	0.17
	9.68	0.76	0.18
	3.70	0.09	0.25
Corvina nigra,	11.6	1.5	0.26
muscle tissue	9.72	0.65	0.27
	3.68	0.12	0.31
	3.53	0.10	0.30

* *Source.* Lebedeva (1965).

where μ_1 = real part of shear modulus

$\qquad d$ = loss factor for shear

The loss factor d enters into the damping constant, as we discuss later.

Since the resonance frequency is inversely proportional to the bubble radius, we consider what happens to the effective radius of a fish swim bladder as the animal changes depth. There are two limiting possibilities:

1. If the volume and shape of the fish bladder are kept constant by physiological actions (active control), then assuming that the bladder elasticity is much less than $3\gamma P_A$ in (A7.2.1), we have

$$f_{RF} \sim P_A^{1/2} \sim (1+0.1z)^{1/2} \qquad \text{(A7.2.3)}$$

where z = water depth (m)

2. If the fish bladder maintains constant mass and constant temperature, its volume changes according to Boyle's law PV = constant (passive behavior),

$$P_A \frac{4\pi a'^3}{3} = \text{constant}$$

Then a' will be proportional to $(P_A)^{-1/3}$ and

$$f_{RF} \sim P_A^{+5/6} \sim (1+0.1z)^{5/6} \qquad \text{(A7.2.4)}$$

Hersey and Backus (1962) have provided evidence that there are fauna that behave actively and those that behave passively when water pressure changes.

A7.2.2. Damping Constant

Let us now consider the damping constant of a fish bladder.

If we assume that thermal and reradiation effects are the same as for the free bubble (6.3.14), the total damping constant for a fish bubble at resonance is

$$\delta_{RF} = \delta_{Rr} + \delta_{Rt} + \delta_{RvF} \qquad \textbf{(A7.2.5)}$$

where δ_{RvF} is the damping constant at resonance due to the viscous effects of the fish swim bladder.

Andreeva (1964) has proposed that

$$\delta_{RvF} = \frac{4\mu_1 d}{3\gamma P_A + 4\mu_1} \qquad \textbf{(A7.2.6)}$$

Illustrative Problem. Calculate the approximate damping constant and the backscattering cross section of an unknown fish resonant at 10 kHz at sea level.

Assume the bubble gas is air ($\gamma = 1.4$).

From Table A7.2.1 assume $\mu_1 \approx 1$ atm $= 10^5$ Pa

$$d \approx 0.25$$

Then (A7.2.6) gives

$$\delta_{RvF} = 0.12$$

From Fig. 6.3.1 find $\delta_{R_r} = 0.014$ and $\delta_{R_t} = 0.04$.
Therefore, using (A7.2.5),

$$\delta_{RF} = 0.014 + 0.04 + 0.12$$
$$= 0.17$$

The bubble radius is calculated from (A7.2.1) with $t/a' \approx 0.2$,

$$a' = \frac{\{[(3)(1.4)(1.0 \times 10^5) + (4)(10^5)(0.2)]/1035\}^{1/2}}{2\pi \times 10^4}$$
$$= 4 \times 10^{-4} \ m = 400 \text{ microns}$$

Then from (7.1.5) at resonance

$$\sigma_{bs} = \frac{a'^2}{\delta_{RF}^2} = \frac{(4 \times 10^{-4})^2}{0.17^2} = 5.5 \times 10^{-6} \ m^2$$

MARINE SEDIMENTS

The reflection of sound pressures at the sea floor depends on the physical properties of the sediments, namely, density, compressional wave velocity, shear wave velocity, and absorption loss. The presence of shear waves and absorption losses alters the reflection.

A8.1. ELASTIC CONSTANTS AND THE SOUND VELOCITY IN MARINE SEDIMENTS

When sound velocity measurements are not available, we use the description of the sediments to estimate sound velocity. The data include the density, mineralogy, porosity, and particle size distribution. To use these quantities to estimate *in situ* sound velocities of the sediments, we derive the dependence of the sediment sound velocity on the frame, the particles, and the water.

The derivation uses the bulk modulus of elasticity E and modulus of rigidity G. From the theory of elasticity, the compressional wave velocity c and shear wave velocity c_s are

$$c^2 = \frac{E + 4G/3}{\rho_0}$$

$$c_s^2 = \frac{G}{\rho_0}$$

(A8.1.1)

Usually the dynamic values of E and G are much larger than the static values. Sound velocity measurements yield the dynamic values which we use. From elasticity, a pressure increment ΔP on the volume V causes a volume change $-\Delta V$,

$$\frac{-\Delta V}{V} = \frac{\Delta P}{E} \qquad \textbf{(A8.1.2)}$$

The volume of water in the sediment V_w is

$$V_w = nV \qquad \text{(A8.1.3)}$$

where n is the fraction of the volume occupied by water, or the porosity. The volume of the solids V_s is

$$V_s = (1-n)V \qquad \text{(A8.1.4)}$$

The bulk moduli of the water and the solids are E_w and E_s.

We assume the frame and the water move together and have an effective bulk modulus E. For an increment of pressure ΔP, the volume of the frame and its water change by $-\Delta V$. Following Gassmann (1951) and Hamilton (1969), we separate ΔP and ΔV into their components inside the volumes: ΔP_h is the hydrostatic pressure that is effective on all surfaces of the volume and determines the additional hydrostatic pressure in the pores; ΔP_e is the effective pressure on the frame. The sum is

$$\Delta P = \Delta P_h + \Delta P_e \qquad \text{(A8.1.5)}$$

The change of volume ΔV is the sum of ΔV_w and ΔV_s

$$\Delta V = \Delta V_w + \Delta V_s \qquad \text{(A8.1.6)}$$

where ΔV_s has two components: the reduction due to ΔP_e on the volume of the frame V and the compression of the solids V_s due to ΔP_h. Using (A8.1.4), ΔV_s is found to be

$$\frac{-\Delta V_s}{V} = (1-n)\Delta P_h E_s^{-1} + \Delta P_e E_s^{-1} \qquad \text{(A8.1.7)}$$

The change of volume of the water in the pores ΔV_w is calculated using (A8.1.3)

$$\frac{-\Delta V_w}{V} = n\,\Delta P_h E_w^{-1} \qquad \text{(A8.1.8)}$$

and

$$\frac{-\Delta V}{V} = n\,\Delta P_h E_w^{-1} + (1-n)\Delta P_h E_s^{-1} + \Delta P_e E_s^{-1} \qquad \text{(A8.1.9)}$$

Gassmann defines a frame bulk modulus E_f in the following equation:

$$\frac{-\Delta V}{V} = \Delta P_h E_s^{-1} + \Delta P_e E_f^{-1} \qquad \textbf{(A8.1.10)}$$

We calculate E as a function of E_w, E_s, and E_f. The algebraic manipulation eliminates ΔP_h, ΔP_e, and reduces the expressions. We first substitute $\Delta P - \Delta P_h$ for ΔP_e in (A8.1.9) and (A8.1.10), then equate the two. After rearrangement, the equation is

$$\Delta P(E_s^{-1} - E_f^{-1}) = \Delta P_h[n(E_s^{-1} - E_w^{-1}) + (E_s^{-1} - E_f^{-1})] \quad (A8.1.11)$$

For another expression, we equate (A8.1.2) and (A8.1.10). The result is

$$\Delta P(E^{-1} - E_f^{-1}) = \Delta P_h(E_s^{-1} - E_f^{-1}) \quad\quad (A8.1.12)$$

On taking the ratio of (A8.2.11) and (A8.2.12) and solving for E^{-1}, we obtain

$$E^{-1} = E_s^{-1} \frac{E_s^{-1} - E_f^{-1} + nE_s E_f^{-1}(E_s^{-1} - E_w^{-1})}{(E_s^{-1} - E_f^{-1}) + n(E_s^{-1} - E_w^{-1})} \quad\quad (A8.1.13)$$

Equation (A8.1.13) has the form of the compressibility where the E^{-1} s are the compressibilities. This is the same as obtained from Biot's theory (by Stoll, 1974, for the frame and water moving together). Hamilton (Gassmann) gives a different form. To obtain their form, we factor $nE_f^{-1}(E_s^{-1} - E_w^{-1})$ from the top and bottom of (A8.1.13). After rearrangement, E is

$$E = E_s \frac{E_f + F}{E_s + F} \quad\quad \textbf{(A8.1.14)}$$

where

$$F \equiv \frac{E_f(E_s^{-1} - E_f^{-1})}{n(E_s^{-1} - E_w^{-1})} \quad \text{or} \quad F \equiv \frac{E_w(E_s - E_f)}{n(E_s - E_w)} \quad\quad (A8.1.15)$$

The bulk modulus of the frame E_f depends on the particles, their size, and the structure. We now consider the calculation of E from measurements of E_s, E_w, n, and E_f.

Static measurements of the frame bulk modulus can be made by compressing drained samples. Static measurements are difficult to make on fragile samples such as cores of unconsolidated deep sea cores. Laughton (1957) reports large margins of error for his static measurements. An alternative procedure is to use measurements of E, E_s, E_w, and n to calculate E_f. From the ratio of (A8.1.11) and (A8.1.12), the expression for E_f is

$$E_f = E_s \frac{nE(E_s - E_w) - E_w(E_s - E)}{nE_s(E_s - E_w) - E_w(E_s - E)} \quad\quad \textbf{(A8.1.16)}$$

Hamilton (1971a) used computations of E_f to develop a relationship between porosity n and E_f (Fig. A8.1.1). To compute E using (A8.1.14),

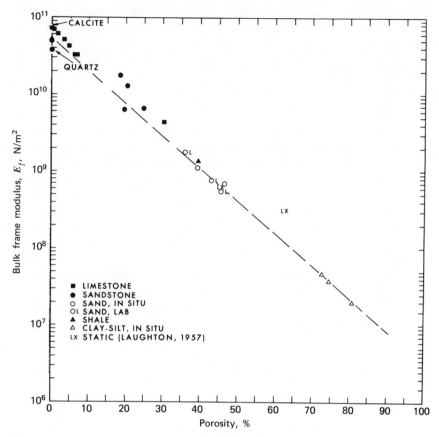

Figure A8.1.1.   Dynamic bulk frame modulus E_f. (Hamilton, 1971a).

the porosity and Fig. A8.1.1 are used to estimate E_f, and E_w for the water is tabulated or calculated from $E_w = \rho c_w^2$. Presumably temperature and salinity of the pore water and seawater just above the interface are the same.

For E_s, Hamilton suggests using the Reuss–Voigt–Hill average for the bulk modulus of the aggregate of mineral grains. On designating E_R for the Reuss average and R_V for the Voigt average, the R–V–H average is

$$E_R^{-1} = n_1 E_1^{-1} + n_2 E_2^{-1} + \cdots$$

$$E_V = n_1 E_1 + n_2 E_2 + \cdots$$

$$E_{\text{RVH}} = \frac{E_R + E_V}{2} \qquad (A8.1.17)$$

where n_1, n_2, ..., are the fractions of the volume for each mineral or phase and E_1, E_2, ..., are the bulk moduli. Hamilton (1971a) gives values of E_s:

medium sand: $E_s = 5.1 \times 10^9 \text{ N/m}^2$

silt-clays: $E_s = 5.0 \times 10^9 \text{ N/m}^2$

clayey-silt ("Clay"): $E_s = 5.4 \times 10^9 \text{ N/m}^2$

We use the description, density, and any other two constants to compute the elastic constants. For example, velocity of compressional waves, density, and porosity are routinely measured for sediment cores. Values of E_f and E_s are estimated from porosity and composition, respectively, and E_w is the value for the pore water. Equations (A8.1.14) and (A8.1.15) give E. Calculated values of G and c_s are obtained from (A8.1.1). Hamilton's (1971a) measurements and computations are given in Table 8.2.1. The dependence of computed G on porosity is illustrated in Fig. A8.1.2.

Hamilton (1971b) gives an extensive discussion of the prediction of the elastic properties of surficial sediments on the sea floor. He assumes that E_f, E_s, and n are the same in the laboratory and on the sea floor. The density is a little larger because of the increase of pore water density.

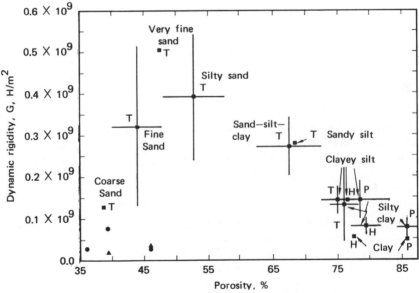

Figure A8.1.2. Dynamic rigidity G versus porosity. (Hamilton, 1971a.)

A8.2.   ATTENUATION COEFFICIENT AND Q

Often the damping factor Q^{-1} is used to describe the attenuating properties of materials. To show the relationship of Q to the attenuation coefficient, we begin with particle displacements ξ of the plane wave.

$$\xi = A \exp\left[-\alpha_e x + i(\omega t - kx)\right] \tag{A8.2.1}$$

Letting W be the mean energy in the wave field and ΔW be the loss of energy in one cycle, Q^{-1} is defined as

$$Q^{-1} \equiv \frac{\Delta W}{2\pi W} \tag{A8.2.2}$$

The mean energy in the wave is

$$W = \rho \left\langle \left|\frac{d\xi}{dt}\right|^2 \right\rangle \tag{A8.2.3}$$

$$W = \frac{1}{T} \int_0^T \rho \left(\frac{d\xi}{dt}\right) \frac{d\xi^*}{dt} \, dt \tag{A8.2.4}$$

$$\frac{d\xi}{dt} = i\omega A \exp\left[-\alpha_e x + i(\omega t - kx)\right] \tag{A8.2.5}$$

$$W = \rho\omega^2 A^2 \exp\left(-2\alpha_e x\right) \tag{A8.2.6}$$

$$|\Delta W| = 2\alpha_e \rho\omega^2 A^2 \exp\left(-2\alpha_e x\right) \Delta x|_{x=0} \tag{A8.2.7}$$

In one cycle $\Delta x = \lambda = c/f$

$$Q^{-1} = \frac{\alpha_e c}{\pi f} \tag{A8.2.8}$$

$$\alpha, \frac{dB}{distance} = 8.686\alpha_e \tag{A8.2.9}$$

If α is proportional to frequency, Q is constant.

A8.3.   REFLECTION AT A LIQUID–SOLID INTERFACE

When a compressional wave is incident on an elastic solid, part of the energy of the incident wave is reflected, part is transmitted as a compressional wave in the solid, and part is converted to a transmitted shear wave in a solid. At the boundary, the vertical displacements in the liquid and in the solid are equal. The vertical stress is continuous. The tangential stress in the solid vanishes at the interface because the (nonviscous) liquid does

not support shear waves. Snell's law applies to both the compressional
waves and the shear waves

$$\frac{\sin \theta_1}{c_1} = \frac{\sin \theta_{2p}}{c_{p2}} = \frac{\sin \theta_{2s}}{c_{s2}} \tag{A8.3.1}$$

where c_1 is the sound speed in the liquid, c_{p2} is the compressional wave
velocity, and c_{s2} is the shear wave velocity.

We omit the details of the derivation, and give the reflection coefficient
derived in Tolstoy and Clay (1966, p. 25). The parameters are as follows;

$$\alpha = \frac{\omega}{c_1} \sin \theta_1$$

$$\gamma_1 = \frac{\omega}{c_1} \cos \theta_1$$

$$\gamma_2 = \frac{\omega}{c_{p2}} \left[1 - \left(\frac{c_{p2}}{c_1} \sin \theta_1 \right)^2 \right]^{1/2} \tag{A8.3.2}$$

$$\delta_2 = \frac{\omega}{c_{s2}} \left[1 - \left(\frac{c_{s2}}{c_1} \sin \theta_1 \right)^2 \right]^{1/2}$$

$$\mathcal{R}_{12} = \frac{4 \gamma_2 \delta_2 \alpha^2 + (\delta_2^2 - \alpha^2)^2 - (\rho_1/\rho_2)(\gamma_2/\gamma_1)(\omega^4/c_{s2}^4)}{4 \gamma_2 \delta_2 \alpha^2 + (\delta_2^2 - \alpha^2)^2 + (\rho_1/\rho_2)(\gamma_2/\gamma_1)(\omega^4/c_{s2}^4)} \tag{A8.3.3}$$

The frequency ω can be factored out of the equation and $\mathcal{R}_{12}$ does not
depend on frequency. Total reflection occurs when c_{s2} and c_{p2} are greater
than c_1. Additional formulas and numerical examples are given in Ewing,
Jardetzky, and Press (1957).

A8.4. REFLECTION AT A LOSSY HALF SPACE

Absorption of energy from the sound wave is an additional cause of
attenuation. When sound reflects from an absorptive half space, the
reflection coefficient is reduced. Even beyond the critical angle, the
reflection coefficient is less than unity.

We introduce absorption loss by letting the sound speed in the second
medium be complex

$$c_2 = c_2' + i c_2'' \tag{A8.4.1}$$

where c_2' and c_2'' are real. We relate the complex sound speed to the usual
expression for attenuation loss by calculating the complex wave number

k_2 for sound waves in the second medium.

$$k_2 = \frac{\omega}{c_2} = \frac{\omega}{c_2' + ic_2''} \tag{A8.4.2}$$

For $|c_2''|$ much less than c_2', we approximate k_2 as follows:

$$k_2 \simeq \frac{\omega}{c_2'} \left(1 - \frac{ic_2''}{c_2'}\right)$$

$$k_2 = k_2' - i\alpha_e \tag{A8.4.3}$$

$$k_2' \equiv \frac{\omega}{c_2'} \tag{A8.4.4}$$

$$\alpha_e \equiv \frac{\omega c_2''}{c_2'^2} \tag{A8.4.5}$$

The substitution of (A8.4.3) into the equation of a plane wave traveling in the $+z$ direction gives

$$p \sim \exp\left[i(\omega t - k_2' z) - \alpha_e z\right] \tag{A8.4.6}$$

Measurements of p as a function of z are used to determine α_e, c_2', and c_2''.

To calculate the reflection coefficient, we assume that Snell's law holds for the complex sound speed and that the satisfaction of the boundary conditions also holds for complex sound speeds. Then the derivation that yields (2.9.5) is valid for the complex c_2. Using (A8.4.1) and Snell's law, the $\cos \theta_2$ becomes complex and is given by

$$\cos \theta_2 = \left[1 - \frac{c_2^2}{c_1^2} \sin^2 \theta_1\right]^{1/2} \tag{A8.4.7}$$

where

$$c_2^2 = c_2'^2 + 2ic_2'c_2'' - c_2''^2 \tag{A8.4.8}$$

The substitution of (A8.4.1) and (A8.4.7) and (2.9.5) gives

$$\mathscr{R}_{12} = \frac{\rho_2 c_2' \cos \theta_1 + i\rho_2 c_2'' \cos \theta_1 - \rho_1 c_1 \cos \theta_2}{\rho_2 c_2' \cos \theta_1 + i\rho_2 c_2'' \cos \theta_1 + \rho_1 c_1 \cos \theta_2} \tag{A8.4.9}$$

Ordinarily c_2'' is very small relative to c_2' and we can approximate

$$c_2' \simeq c_2 \tag{A8.4.10}$$

ARRAYS IN A WAVEGUIDE, MATCHED ARRAYS FOR SIGNAL TO NOISE IMPROVEMENT, AND THE CONTINUOUSLY VARYING MEDIUM

A9.1. ARRAYS OF SOURCES AND RECEIVERS IN A WAVEGUIDE

This extension of the development of the transmission from a single source to a single receiver in a waveguide consists of adding all the transmissions from all sources to all receivers. The arrays of sources and receivers can be horizontal lines of transducers, vertical strings of transducers, combinations of horizontal and vertical arrays, and three-dimensional arrays.

The vertical array and the horizontal line array demonstrate the most important features of the phenomena of transmission and reception by arrays. For our examples we transmit from a vertical array of transducers

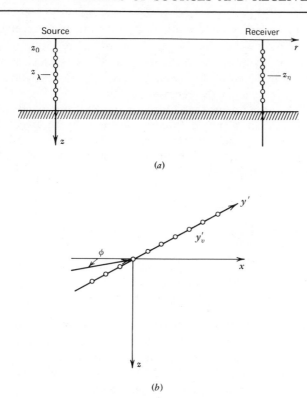

Figure A9.1.1.   Arrays in a waveguide. a) vertical arrays. b) horizontal line array.

to a vertical array of transducers (Fig. A9.1.1). Each of the transducers can be a horizontal array to give the system directional sensitivity.

We start with (9.1.30), sum over all sources and receiver positions, and compress the notation. The output signal of the additive array is for M modes.

$$s(t) = \exp\left[i\left(\omega t + \frac{\pi}{4}\right)\right] \sum_{m=1}^{M} a_m U_m b_m \qquad \text{(A9.1.1)}$$

source:

$$a_m \equiv \sum_{\lambda=0}^{\Lambda-1} \alpha_\lambda Z_m(z_\lambda) \qquad \text{(A9.1.2)}$$

receiver:

$$b_m \equiv \sum_{\eta=0}^{L-1} \beta_\eta Z_m(z_\eta) \qquad \text{(A9.1.3)}$$

transmission: $U_m \equiv \rho(r)^{-1/2} q_m \exp\left(-\delta_m r - i\kappa_m r\right) \qquad \text{(A9.1.4)}$

where Λ is the number of sources, L is the number of receivers, α_λ is the relative pressure amplitude of the λth source at depth z_λ, β_η is the sensitivity of ηth hydrophone at depth z_η, and ρ is assumed to be nearly constant and is included in the range function U_m. Collectively, a_m is the source function and b_m is the receiver function.

The systems can be generalized by letting each source and receiver be horizontal arrays. In a simple example assume the receiver at z_η is a horizontal array and each hydrophone has the sensitivity β'_v and time delay τ_v. For a horizontal line array along y' and the hydrophone positions y'_v, we write

$$\beta_\eta = \sum_{v=0}^{N-1} \beta'_v \exp\left[i(\kappa_m y'_v \sin\phi - \omega\tau_v)\right] \tag{A9.1.5}$$

where ϕ is the steering direction. There are similar expressions for source arrays.

A9.1.1. Vertical Array

The vertical array operates as a "mode filter" and can be used to transmit and receive in selected modes. We demonstrate this by showing how to transmit in the mth mode. We select a set of source depths for which the $|Z_m(z_\lambda)|$ are maximum or large. For M modes we use at least M sources. The α_λ are chosen to satisfy the following equations:

m_0th mode: $\qquad\qquad \displaystyle\sum_{\lambda=1}^{M} \alpha_\lambda Z_{m_0}(z_\lambda) = 1$

all other modes: $\qquad\qquad \displaystyle\sum_{\lambda=1}^{M} \alpha_\lambda Z_{m_0}(z_\lambda) = 0 \tag{A9.1.6}$

In principle the equations can be solved for any set of depths. In practice we suggest spreading the sources from near the surface to the bottom of the water layer.

When many sources are evenly spaced from the bottom to the top of the waveguide, we can use the orthogonality of the eigen functions to choose α_λ. For example, to transmit in the m_0th mode we let

$$\alpha_\lambda = \rho_\lambda Z_{m_0}(z_\lambda) \tag{A9.1.7}$$

and then (where ρ is the average density)

$$a_m = \rho^{-1} \sum_{\lambda=1}^{M} \rho_\lambda Z_{m_0}(z_\lambda) Z_m(z_\lambda) \tag{A9.1.8}$$

For M large and greater than the number of modes, this summation is equivalent to the orthogonality integral (9.1.21) and a_m is nearly zero except for $m_0 = m$.

We have used (A9.1.7) to shade a vertical array of transducers for transmission in the first mode in a laboratory experiment. The waveguide has a pressure release top (air), a 20 cm thick water layer, and a pressure release bottom ("styrofoam"). The transducer array was driven by 220 kHz pings. We used (A9.1.7) for the first mode to determine the driving voltage amplitude at each transducer. The rms sound pressure amplitude $P(z)$ was measured as a function of depth in the water layer at 150 cm range. Fourier analysis of $P(z)$ showed that the first mode was predominant, the amplitudes of the next three modes were less than 10% of the first mode, and higher modes were negligible. Transmission from a small transducer to the shaded vertical array gave predominant reception in the first mode. The transmission was reciprocal and the array was a mode filter for both transmission and reception.

If we think of the modes as being the sum of up- and down-traveling waves, an alternative way to excite selected modes is to use time delays and "steer" the array to transmit at the incident angle of θ_m. Waves could be excited going either up or down.

Numerical Examples. We use the wave guide and source frequency given in the numerical example in Section 9.3. We wish to place the sources and drive them so as to transmit in the second mode, not the first.

The eigenfunctions are given in Fig. 9.3.5. Since the maxima of $|Z_2(z)|$ are at 6 and 21 m, we choose the depths 6 and 21 m. Equations (A9.1.6) become

$$m = 2: \qquad \alpha_1 Z_2(6) + \alpha_2 Z_2(21) = 1$$

$$m = 1: \qquad \alpha_1 Z_1(6) + \alpha_2 Z_1(21) = 0$$

The values of $Z_2(z)$ and $Z_1(z)$ are read from Fig. 9.3.5.

$$\alpha_1 - \alpha_2 = 1$$
$$0.6\alpha_1 + 0.8\alpha_2 = 0$$

The solutions are $\alpha_1 = 0.57$ and $\alpha_2 = -0.43$.

A9.2.   SIGNAL TO NOISE RATIO AND THE MATCHED ARRAY

Acoustic noise in a waveguide such as the ocean may come from many sources. The sources are in all directions and at all ranges. Empirical evidence indicates that most of the sources are near the surface of the ocean. Seismic events (earthquakes, landslides, etc.) are on and beneath the bottom. Other than being at random positions and being random sources, the noise sources excite the modes the same as any other source.

We assume that the mean square noise pressure per hertz in the mth mode is N_m^2. The mean square output of the receiving array is

$$\langle n_b^2 \rangle = \Delta\omega \sum_{m=1}^{M} |b_m|^2 N_m^2 \tag{A9.2.1}$$

where $\Delta\omega$ is a narrow frequency band, $|b_m|^2$ is the mean square response of the array over $\Delta\omega$, and the subscript b is for the output of the array.

The ratio of the mean square signal to noise pressure ratio is found using (A9.1.1) and (A9.2.1):

$$\frac{|s|^2}{\langle n_b^2 \rangle} = \frac{\left| \sum_{m=1}^{M} a_m U_m b_m \right|^2}{\Delta\omega \sum_{m=1}^{M} |b_m|^2 N_m^2} \tag{A9.2.2}$$

We multiply and divide by N_m to write the numerator as follows, then apply Cauchy's inequality [Abramowitz and Stegun, 1964 (3.2.9)]

$$\left| \sum_{m=1}^{M} \left(\frac{a_m U_m}{N_m} \right) (b_m N_m) \right|^2 \le \sum_{m=1}^{M} \left| \frac{a_m U_m}{N_m} \right|^2 \sum_{m=1}^{M} |b_m|^2 N_m^2 \tag{A9.2.3}$$

where the equality holds for

$$\left(\frac{a_m U_m}{N_m} \right)^* \sim b_m N_m \tag{A9.2.4}$$

The substitution of the inequality in (A9.2.2) gives

$$\frac{|s|^2}{\langle n_b^2 \rangle} \le (\Delta\omega)^{-1} \sum_{m=1}^{M} \left| \frac{a_m U_m}{N_m} \right|^2 \tag{A9.2.5}$$

The maximum signal to noise pressure squared ratio occurs for the matched condition (A9.2.4) or

$$b_m \sim \frac{(a_m U_m)^*}{N_m^2} \tag{A9.2.6}$$

where b_m is the response of the *matched array*. This condition matches the receiver to a particular source at a particular range, and gives the maximum signal to noise ratio for the transmission.

If matching to range is impractical, the signal to noise ratio can be improved by letting

$$|b_m| \sim \frac{|a_m U_m|}{N_m^2} \tag{A9.2.7}$$

The *matched array* is the spatial counterpart of the *matched* filter in the time domain (Clay, 1966).

A9.3. CONTINUOUSLY VARYING MEDIUM AND THE WKB APPROXIMATION

We approximate the structure of the ocean by a sound speed that only depends on z. Within the depth range of our analysis, we assume that $c(z)$ is continuous. These are the same conditions used in Chapter 3 for tracing ray paths. Here, instead of calculating ray paths, we calculate approximate eigenfunctions for sound field.

Two approaches for determining wave functions for an arbitrary $c(z)$ are the replacement of $c(z)$ by many layers and the use of the Wentzel–Kramers–Brillouin (WKB) approximation. We choose the latter.

Again we only need to consider the *z dependent wave equation*

$$Z'' + \gamma^2 Z = 0 \tag{A9.3.1}$$

where the primes indicate differentiation with respect to z. Both Z and γ are functions of z. We assume that Z is close to being a harmonic function and express it in the following form

$$Z = A e^{i\Phi} \tag{A9.3.2}$$

where A and Φ are functions of z. The substitution of (A9.3.2) into (A9.3.1) and separation of real and imaginary parts gives

$$A'' - A[(\Phi')^2 - \gamma^2] = 0 \tag{A9.3.3}$$

$$A\Phi'' + 2A'\Phi' = 0 \tag{A9.3.4}$$

By writing (A9.3.4) as follows,

$$\frac{2A'}{A} = -\frac{\Phi''}{\Phi'}$$

the integration gives $\quad -2 \ln A = \ln \Phi' + \text{constant}$

$$A = A_0(\Phi')^{-1/2} \tag{A9.3.5}$$

where A_0 is the constant of integration. Now we consider an approximation to simplify (A9.3.3). By writing Z as $A \exp(i\Phi)$, we separate the dependence of the amplitude A and phase Φ on z. When $c(z)$ changes slowly, we would expect A to change slowly and A'' to be negligible compared to A. The WKB approximation consists of dropping A''. Then

(A9.3.3) becomes

$$\Phi' \simeq \pm \gamma \tag{A9.3.6}$$

$$\Phi \simeq \pm \int \gamma \, dz \tag{A9.3.7}$$

and (A9.3.5) becomes

$$A \simeq A_0(\gamma)^{-1/2} \tag{A9.3.8}$$

Since both signs of (A9.3.7) are solutions, we write the WKB eigenfunction as follows:

$$Z = A_0(\gamma)^{-1/2} \exp\left[i \int \gamma \, dz\right] + B_0(\gamma)^{-1/2} \exp\left[-i \int (\gamma) \, dz\right] \tag{A9.3.9}$$

Here A_0 and B_0 are determined by the boundary conditions at the bottom and top of the layer. The limits of integration are the bottom and top of the layer. We explain the limits of integration in more detail later.

The approximation is not good at depths where γ approaches zero: that is, where the vertical component of the wave number vanishes. Recalling our discussion of ray paths, this corresponds to the ray path becoming horizontal, then turning upward for increasing $c(z)$. Analogously, the depths where γ goes to zero are called "turning depths." Function Z tends to zero below the turning depth. Advanced developments of the WKB approximation give connection formulas to carry the wave function through the turning depths. We do not need them here.

For the mth mode we write

$$\Phi_m = \int \gamma_m \, dz \tag{A9.3.10}$$

and the characteristic equation

$$\Phi_m - \Phi_u - \Phi_l = (m-1)\pi \tag{A9.3.11}$$

When the ray path turns at the lower depth, Tolstoy and Clay give approximate values of the phase shift $\Phi_l = \pi/4$. Similarly, a turn at an upper depth gives $\Phi_u = \pi/4$. The limits of integration for (A9.3.10) are from turning depth to turning depth, or the thickness of the layer, or from reflecting interface to reflecting interface.

A detailed discussion of the WKB approximation and connection formulas appears in Schiff (1949).

Another type of approximation, the *parabolic approximation*, is useful when the sound speed also depends on r. Essentially it involves writing the displacement potential $\varphi \sim u(r, z)r^{-1/2} \exp[i(\omega t - k_0 r)]$, substituting into the cylindrical wave equation and dropping the term $\partial^2 u(r, z)/\partial r^2$. The suggested readings are a place to start a study of this approximation.

SUGGESTED READING

C. S. Clay, "Use of arrays for acoustic transmission in a noisy ocean." *Rev. of Geophysics* **4**, 475-507 (1966). Gives theory for optimum arrays for transmission in a layered wave guide in presence of noise and the coherence of transmission to arrays in a randomly perturbed wave guide.

R. M. Fitzgerald, "Helmholtz equation as an initial value problem with application to acoustic propagation." *J. Acoust. Soc. Am.* **58**, 839–842 (1975).

S. M. Flatte, F. D. Tappert "Calculation of the effect of Internal waves on Oceanic Sound Transmission." *J. Acoust. Soc. Am.* **58**, 1151–1159 (1975).

S. T. McDaniel "Parabolic approximations for underwater sound propagation" *Ibid.* **58**, 1178–1185 (1975).

A. O. Williams, Jr. "Comments on 'Propagation of normal mode in parabolic approximation' [Suzanne T. McDaniel, *J. Acoust. Soc. Am.* **57**, 307–311 (1975)]" *Ibid.* **59**, 1320–1321 (1975).

A10

SCATTERING AND TRANSMISSION OF SOUND AT A ROUGH SURFACE

We derive the Helmholtz–Kirchhoff equation, then use it to calculate the scattered sound for three different problems. The first is the backscattered sound for an impulsive source such as an explosion or very short ping. The scattering surface consists of segments of planes, and we are able to show that each edge of a plane contributes a boundary diffraction wave (or is the source of Huygens wavelets). The second problem is the scattering of CW sound at a time-dependent, randomly rough surface. We demonstrate the dependence of the scattered signal on the statistical description of the surface. The third problem is the transmission of sound through a rough interface.

Our development for the first problem is based on work by Trorey (1970). He used the Laplace transformation to express the transient nature of the boundary diffraction waves. We have changed his development to use the Fourier transformation instead. Our treatment of the other problems starts with Eckart's formulation (1953). His five-page paper is a major contribution to scattering theory and attests to Eckart's insight in the scattering problem. A pair of experimental studies, La Casce and Tamarkin (1956) and Proud, Beyer, and Tamarkin (1960), give laboratory verification of Eckart's theory. Comparisons of Eckart's solution for incident plane waves with marine experiments have revealed the need to extend the theory to include spherically diverging waves and more types of rough surface. We give an extension that is based on our work and that of C. W. Horton, Sr., and his students.

A10.1. HELMHOLTZ–KIRCHHOFF INTEGRAL

The Helmholtz–Kirchhoff (H–K) integral gives a mathematical description of Huygens' principle. The idea is simple. The source insonifies the surface, and each incremental area becomes a source of a Huygens wavelet. The wavelets expand spherically and reach the point Q. The sound pressure at Q is the integral over all the wavelets. It is difficult to specify the source strength for each of the wavelets, and we use an approximation to estimate the source strength. We then evaluate the integral for typical problems that occur in acoustical oceanography.

A10.1.1. Theorems of Gauss and Green

The Helmholtz–Kirchhoff equation is derived from Green's theorem, and Green's theorem uses Gauss' theorem. Therefore we begin by deriving Gauss' theorem. The sources within a volume V cause a field $\mathbf{F}$, where $\mathbf{F}$ is a vector. The net outward flux from the sources is the integral of $\mathbf{F}$ over the surface S that encloses the volume

$$\text{flux} = \int_S \mathbf{F} \cdot d\mathbf{S}$$

where $d\mathbf{S}$ is the incremental area and has a direction normal to the surface. The scalar or dot product $\mathbf{F} \cdot d\mathbf{S}$ is proportional to the cosine of the angle between $\mathbf{F}$ and $d\mathbf{S}$. The divergence operation $\boldsymbol{\nabla} \cdot \mathbf{F}$ is also a measure of the flux due to sources. For the volume enclosed by S, the flux is the volume integral.

$$\text{flux} = \int_V \boldsymbol{\nabla} \cdot \mathbf{F} \, dv$$

Equating the two expressions gives Gauss' theorem

$$\int_V \boldsymbol{\nabla} \cdot \mathbf{F} \, dv = \int_S \mathbf{F} \cdot d\mathbf{S} \qquad (A10.1.1)$$

We used this concept in Chapter 3 in the derivation of the relationship between the source power Π, the acoustic pressure, and the particle velocity.

Green's theorem is derived by placing a pair of rather arbitrary vectors $U_1(\nabla U_2)$ and $U_2(\nabla U_1)$ into Gauss' theorem and manipulating the results. We recall from Section A2.2 that ∇U is a vector, the gradient of U. Applying Gauss' theorem to $U_1(\nabla U_2)$, we have

$$\int_S U_1(\nabla U_2) \cdot dS = \int_V \nabla \cdot (U_1 \nabla U_2)\, dv \qquad \text{(A10.1.2)}$$

and similarly to $U_2(\nabla U_1)$

$$\int_S U_2(\nabla U_1) \cdot dS = \int_V \nabla \cdot (U_2 \nabla U_1)\, dv \qquad \text{(A10.1.3)}$$

The del operation $\nabla \cdot (\ )$, yields

$$\nabla \cdot (U_1 \nabla U_2) = (\nabla U_1) \cdot (\nabla U_2) + U_1 \nabla^2 U_2$$
$$\nabla \cdot (U_2 \nabla U_1) = (\nabla U_2) \cdot (\nabla U_1) + U_2 \nabla^2 U_1 \qquad \text{(A10.1.4)}$$

We obtain Green's theorem by substituting (A10.1.4) into (A10.1.2) and (A10.1.3), and taking the difference. We find

$$\int_S (U_1 \nabla U_2 - U_2 \nabla U_1) \cdot dS = \int_V (U_1 \nabla^2 U_2 - U_2 \nabla^2 U_1)\, dv$$
$$\text{(A10.1.5)}$$

It is convenient to replace the product $\nabla U \cdot dS$ by its equivalent, the normal derivative of U at dS

$$\nabla U \cdot dS = -\frac{\partial U}{\partial n}\, dS$$

where we have chosen the convention of drawing the normal *inward*. The new expression for Green's theorem is

$$-\int_S \left(U_1 \frac{\partial U_2}{\partial n} - U_2 \frac{\partial U_1}{\partial n} \right) dS = \int_V (U_1 \nabla^2 U_2 - U_2 \nabla^2 U_1)\, dv$$
$$\text{(A10.1.6)}$$

The derivatives are evaluated on the surface S. If there are no singularities within the volume enclosed by S, the right-hand side of (A10.1.6) is zero for the time dependence $e^{i\omega t}$ because $U_1 \nabla^2 U_2 = U_2 \nabla^2 U_1 = -(\omega^2/c^2)U_1 U_2$ and the difference is zero.

A10.1.2. Derivation of the Helmholtz–Kirchhoff Integral

We assume the source emits a continuous wave, having frequency ω. The field functions $U_1(x, y, z, t)$ and $U_2(x, y, z, t)$ are expressed as $U_n \exp(i\omega t)$; U_1 and U_2 are solutions of the wave equation. A source or singularity is at point Q within the volume surround by S. For the present we omit the amplitude of the source. In addition, U_1 and U_2 are functions of the geometry and frequency and are proportional to the sound pressure.

As Fig. A10.1.1 indicates, the effect of the singularity is removed by

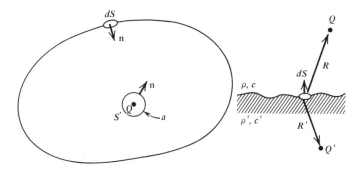

Figure A10.1.1. Region of integration: singularity is at Q; S' has radius a; normals **n** are inward to the volume bounded by S and S'.

enclosing it in the surface S'. Since there are no singularities enclosed in the volume defined by S and S', the right-hand side of (A10.1.6) is zero.

$$\int_S \left(U_1 \frac{\partial U_2}{\partial n} - U_2 \frac{\partial U_1}{\partial n} \right) dS + \int_{S'} \left(U_1 \frac{\partial U_2}{\partial n} - U_2 \frac{\partial U_1}{\partial n} \right) dS' = 0$$

(A10.1.7)

We assume that U_2 is the solution of the spherical wave equation

$$U_2 = \frac{e^{-ikR}}{R}$$

(A10.1.8)

drop the subscripts, let the distance to S be R, and the radius of S' be a. Equation (A10.1.7) becomes

$$\int_S \left[\left(U \frac{\partial}{\partial n} \left(\frac{e^{-ikR}}{R} \right) - \frac{e^{-ikR}}{R} \frac{\partial U}{\partial n} \right) \right] dS$$

$$= -\int_{S'} \left[\left(U \frac{\partial}{\partial n} \left(\frac{e^{-ika}}{a} \right) - \frac{e^{-ika}}{a} \frac{\partial U}{\partial n} \right) \right] dS' \quad \text{(A10.1.9)}$$

The singularity at Q is enclosed by the sphere having radius a. The normal derivative on S' is $\partial/\partial a$ and dS' is $a^2\,d\Omega$. We take the derivative in the right-hand side of (A10.1.9) and evaluate it at the point Q by letting a tend to zero:

$$\lim_{a\to 0} -\int_{4\pi}\left[U\frac{e^{-ika}}{a}\left(-ik-\frac{1}{a}\right)-\frac{e^{-ika}}{a}\frac{\partial U}{\partial a}\right]a^2\,d\Omega = 4\pi U(Q)$$

(A10.1.10)

Thus

$$U(Q)=\frac{1}{4\pi}\int_S\left[U\frac{\partial}{\partial n}\left(\frac{e^{-ikR}}{R}\right)-\frac{e^{-ikR}}{R}\left(\frac{\partial U}{\partial n}\right)\right]_S dS \quad\textbf{(A10.1.11)}$$

$$k=\frac{\omega}{c}$$

This is the integral theorem of Helmholtz and Kirchhoff for harmonic sources. The derivatives and functions are evaluated on S. In our calculations, S is the rough surface on the x–y plane and is joined to a half sphere at infinity. We use the artifice of transmitting a ping that is short enough to separate the direct and scattered arrivals and still long enough to permit the application of CW theory. Thus the field at Q is determined by the sound that is scattered from the rough surface because the contributions from the source and the surface at infinity are not present at Q when we observe the scattered signal.

We also use (A10.1.11) to calculate the transmitted sound. The geometry appears in Fig. A10.1.1. The value on the boundary for the upward-traveling wave is U, and in the lower medium U' is the value on the boundary for the downward-traveling wave.

$$U'(Q')=\frac{1}{4\pi}\int_S\left[U'\frac{\partial}{\partial n}\left(\frac{e^{-ik'R'}}{R'}\right)-\frac{e^{-ik'R'}}{R'}\frac{\partial U'}{\partial n}\right]dS$$

(A10.1.12)

$$k'=\frac{\omega}{c'}$$

where $U(Q)$ and $U'(Q')$ have the units of m^{-1}, They become the sound pressure on multiplication by the source function $P_0 R_0 \exp(i\omega t)$ where P_0 is the sound pressure at distance R_0 from the point source.

A10.2. APPROXIMATIONS

Although the Helmholtz–Kirchhoff integral expresses the wave field $U(Q)$, the problem is still difficult because we have to evaluate U on the surface S. To proceed we make approximations. The first approximation is sometimes called the *Kirchhoff approximation* or *method* and it gives a procedure for estimating U on the surface.

KIRCHHOFF APPROXIMATION

We assume that U is the reflection of the incident wave field U_s and U' is the transmitted wave field. Both U and U' are evaluated on the rough surface S. Designating $\mathcal{R}$ as the pressure reflection coefficient and $\mathcal{T}$ as the pressure transmission coefficient, we write

$$U = \mathcal{R} U_s \tag{A10.2.1}$$

$$\frac{\partial U}{\partial n} = -\mathcal{R} \frac{\partial U_s}{\partial n} \tag{A10.2.2}$$

$$U' = \mathcal{T} U_s \tag{A10.2.3}$$

$$\frac{\partial U'}{\partial n} = \mathcal{T} \frac{\partial U_s}{\partial n} \tag{A10.2.4}$$

where U_s is the wave field from the source that would exist if the scattering surface were not present. The substitution of (A10.2.1)–(A10.2.4) into (A10.1.11) and (A10.1.12) gives

$$U(Q) = \frac{1}{4\pi} \int_S \mathcal{R} \frac{\partial}{\partial n} \left(U_s \frac{e^{-ikR}}{R} \right) dS \tag{A10.2.5}$$

$$U'(Q') = \frac{1}{4\pi} \int_S \mathcal{T} \left[U_s \frac{\partial}{\partial n} \left(\frac{e^{-ik'R'}}{R'} \right) - \frac{e^{-ik'R'}}{R'} \frac{\partial U_s}{\partial n} \right] dS \tag{A10.2.6}$$

The *Kirchhoff approximation* assumes that the coefficients $\mathcal{R}$ and $\mathcal{T}$ which were derived for an infinite plane wave at an infinite plane interface can be used at every point of a rough surface.

REFLECTION COEFFICIENT

Even with the Kirchhoff approximation, the integral is difficult to evaluate because $\mathcal{R}$ is inside the integral. Section 2.9 describes how the reflection of a spherical wave can be approximated by an image source and the plane wave reflection coefficient. We use this approximation when kR is large and the incident angle θ is not too close to critical.

Second, if $\mathcal{R}$ is constant or very slowly varying in the region of major contributions to the integral, we can use a mean value. Also $\mathcal{R}$ is constant when the reflection is due to a density contrast and $c = c'$. We assert that

$\mathcal{R}$ is nearly constant for many geophysical measurements. At the water–air interface, for example, the density contrast is so large that $\mathcal{R} \simeq -1$. At the water-sediment interface, $c \simeq c'$, and $\mathcal{R}$ is nearly constant. This approximation is best near vertical incidence and may be poor at shallow grazing angle.

Thus we remove $\mathcal{R}$ from the integrand and we write

$$U(Q) \simeq \frac{\mathcal{R}}{4\pi} \int_S \frac{\partial}{\partial n} \left(U_s \frac{e^{-ikR}}{R} \right) dS \qquad \textbf{(A10.2.7)}$$

The same approximation applies to sound transmission. We bring $\mathcal{T}$ out of the integral and $U'(Q')$ is

$$U'(Q') \simeq \frac{\mathcal{T}}{4\pi} \int_S \left[U_s \frac{\partial}{\partial n} \left(\frac{e^{-ik'R'}}{R'} \right) - \frac{e^{-ik'R'}}{R'} \frac{\partial U_s}{\partial n} \right] dS \qquad \textbf{(A10.2.8)}$$

SINGLE SCATTERING

We ignore secondary scattering of scattered waves by other features on the rough surface. We also ignore waves that may travel in the second medium and be diffracted back into the first medium by features at the interface.

These three approximations reduce a very difficult problem to one we can handle. The approximations are compatible with many experimental conditions. As we show later, the approximate theory compares very well with experiment.

Although it is beyond our level, we mention that iterative solutions give the Kirchhoff approximation as the first term (Maue, 1949; Meecham, 1956; Noble, 1962). Meecham's (1956) calculations show that the additional terms can be neglected when the minimum radius of curvature of the surface is much greater than the acoustic wavelength. Beckmann and Spizzichino consider this for the scattering of electromagnetic waves (1963). De Santo and Shisha (1974) give a numerical solution that includes the additional terms.

A10.3. BOUNDARY WAVES FROM INFINITE PLANE STRIPS

The geometry of a plane strip is in Fig. A10.3.1. Our purpose is to calculate the diffracted boundary wave, for the edge at x_1. This involves

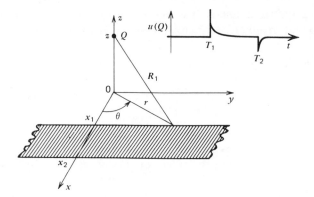

Figure A10.3.1. Geometry for scattering by a plane strip. For an impulsive source $\delta(t)$, the boundary waves from x_1 and x_2 arrive at T_1 and T_2, respectively; $u(Q)$ is the signal.

calculating $d\theta/dt$. From the figure, we define the following times:

$$T_x \equiv \frac{2x_1}{c}$$

$$T_z \equiv \frac{2z}{c}$$

$$T_1^2 \equiv T_x^2 + T_z^2 \qquad \text{(A10.3.1)}$$

$$t \equiv \frac{2R_1}{c}$$

$$t^2 = \frac{T_x^2}{(\cos\theta)^2} + T_z^2$$

Also, we have

$$\frac{d\theta}{dt} = \frac{t\cos^2\theta}{T_x^2 \tan\theta}$$

$$\frac{d\theta}{dt} = \frac{tT_x}{(t^2 + T_x^2 - T_1^2)(t^2 - T_1^2)^{1/2}} \qquad \text{(A10.3.2)}$$

$$\mathscr{D}(t) = \left(\frac{1}{t^2}\frac{d\theta}{dt}\right) \qquad \text{for} \qquad t_1 < t < t_2$$

$$= 0 \qquad \text{otherwise} \qquad \text{(10.2.15)}$$

The boundary wave for the edge at x_1 is generated as R_1 moves from

$\theta = 0$ to $\pi/2$ and simultaneously as R_1 moves from 0 to $-\pi/2$. Hence $\mathfrak{D}_1(t)$ is doubled and, from (10.2.17) for the single edge

$$u_1(Q) = \frac{\mathfrak{R} z}{\pi c^2} \mathfrak{D}_1(t)$$

$$\mathfrak{D}_1(t) = \frac{2 T_x}{t(t^2 + T_x^2 - T_1^2)(t^2 - T_1^2)^{1/2}} \qquad \text{for} \quad t > T_1 \qquad \textbf{(A10.3.3)}$$

$$= 0 \quad \text{for} \quad t < T_1$$

As t tends to T_1, $\mathfrak{D}_1(t)$ becomes infinite and has an impulsive response.

When the source is over the edge, or $T_x = 0$, we use (10.2.9), $R_2 = \infty$, evaluate $U(Q)$, and then $u(Q)$. The boundary wave along the straight edge is not a function of θ. The integral over $d\theta$ is from $-\pi/2$ to $\pi/2$, and the integration gives the factor π. Again the integration over df gives $\delta(t - 2z/c)$. For the source over the edge of a half plane, $u(Q)$ reduces to

$$u(Q) = \frac{\mathfrak{R}}{4z} \delta\left(t - \frac{2z}{c}\right) \qquad \text{(A10.3.4)}$$

This gives an important result. The "image reflection" of a source over the edge of a half plane is half the reflection over a whole plane.

For the origin in the strip, $x_1 < 0$, and $x_2 > 0$

$$u(Q) = \mathfrak{R}\left[\frac{\delta(t - T_z)}{2z} - \frac{z}{\pi c^2} \mathfrak{D}_1(t) - \frac{z}{\pi c^2} \mathfrak{D}_2(t)\right] \qquad \textbf{(A10.3.5)}$$

For the strip in $+x$, $x_1 > 0$ and $x_2 > x_1$

$$u(Q) = \mathfrak{R} \frac{z}{\pi c^2} [\mathfrak{D}_1(t) - \mathfrak{D}_2(t)] \qquad \textbf{(A10.3.6)}$$

As in Fig. A10.3.1, the near edge of the strip gives a $(+)$ boundary diffraction wave, and the far edge gives a $(-)$ boundary diffraction wave. When the origin is in the strip, both edges give $(-)$ boundary diffraction waves. As the strip tends to zero, $u(Q)$ tends to zero.

Numerical evaluations of $u(Q)$ require care because $\mathfrak{D}_1(t)$ and $\mathfrak{D}_2(t)$ are impulsive functions and tend to infinity as t goes to T_1 and T_2. We assume that the evaluations are done at a sequence of time steps that are spaced Δ apart. Using $\mathfrak{D}_1(t)$ as an example, we average the value of $\mathfrak{D}_1(t)$ for the time interval $t = T_1$ to $t = T_1 + \Delta$. We let

$$t = T_1 + \Delta \qquad \text{(A10.3.7)}$$

substitute into (A10.3.3), dropping the second-order terms

$$\mathfrak{D}_1(t) \simeq \frac{2 T_x}{T_1(2 T_1 \Delta + T_x^2)(2 T_1 \Delta)^{1/2}} \qquad \text{(A10.3.8)}$$

where we retain T_x and $2T_1\Delta$ because T_x can go to zero. We average $\mathcal{D}_1(t)$ over Δ to obtain $\mathcal{D}_1(T_1)$.

$$\mathcal{D}_1(T_1) \simeq \frac{1}{\Delta} \int_0^\Delta \mathcal{D}_1(t) \, d\Delta \qquad (A10.3.9)$$

The integration can be done by changing the variable from Δ to x^2 and using an integral table. The result is

$$\mathcal{D}_1(T_1) \simeq \frac{2}{T_1^2 \Delta} \arctan\left[\frac{(2T_1\Delta)^{1/2}}{T_x}\right] \qquad \text{for} \quad T_1 \le t \le T_1 + \Delta$$

$$(A10.3.10)$$

For $t > T_1 + \Delta$, we use (A10.3.3) to calculate $\mathcal{D}_1(t)$.

When the source is over the plane $\mathcal{D}_1(t)$ becomes the delta function and we let

$$\delta(t) = \frac{1}{\Delta} \qquad \text{for} \quad 0 \le t \le \Delta$$

$$= 0 \qquad \text{otherwise} \qquad\qquad (A10.3.11)$$

A10.4. SCATTERING OF AN IMPULSE BY A LONG–CRESTED ROUGH SURFACE

A simple rough surface can be constructed by connecting a number of individual plane strips as in Fig. A10.4.1. Generalization of the diffraction equations to this case is straightforward. For each plane strip, as defined by x_n, z_n and x_{n+1}, z_{n+1}, we rotate and translate the coordinates so that the reference coordinate system contains the strip (Fig. A10.4.2). In the

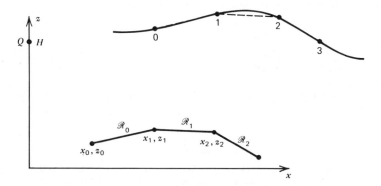

Figure A10.4.1. Profile of a two-dimensional rough surface. Strips are infinite planes in the y direction. Source and receiver are the z axis at $z = H$.

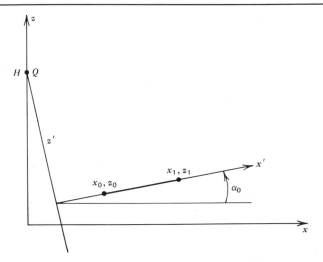

Figure A10.4.2. Rotation and translation.

x', z' coordinates, point Q and the strip are at

$$z' = (H - z_0) \cos \alpha_0 + x_0 \sin \alpha_0 \qquad \text{for} \quad Q$$
$$x_0' = -(H - z_0) \sin \alpha_0 + x_0 \cos \alpha_0 \qquad \text{for} \quad x_0' \qquad (A10.4.1)$$
$$x_1' = -(H - z_1) \sin \alpha_0 + x_1 \cos \alpha_0 \qquad \text{for} \quad x_1'$$

$$\tan \alpha_0 = \frac{z_1 - z_0}{x_1 - x_0}$$

The procedure is to calculate $u(Q)$ for each strip and add them to calculate the complete impulse response. For the zeroth strip, the edge at x_0, z_0 has $\mathcal{D}_0(t)$ and the edge at x_1, z_1 has $-\mathcal{D}_1(t)$. The next strip has an arrival $\mathcal{D}_1(t)$ that coincides with $-\mathcal{D}_1(t)$ from the zeroth strip. The net magnitude of the boundary diffraction wave depends on the *change* of *slope* from strip 0 to strip 1.

For the boundary wave from the nth edge u_n, we define $\mathcal{D}_2(n-1)$ as the signal from the far edge of the $(n-1)$th strip and $\mathcal{D}_1(n)$ as being the signal from the near edge of nth strip having reflection coefficient $\mathcal{R}_n$.

$$u_n = \frac{1}{\pi c^2} [-z_{n-1}' \mathcal{R}_{n-1} \mathcal{D}_2(n-1) + z_n' \mathcal{R}_n \mathcal{D}_1(n) \qquad (A10.4.2)$$

$$u_n = 0 \qquad \text{for} \quad t < \frac{2R_n}{c}$$

$$u(Q) = \sum u_n \qquad (A10.4.3)$$

In our computations, we program the computer to test each strip after the coordinate transformation to determine whether (A10.3.5) or (A10.3.6) applies. We calculate t at time steps of Δ and use (A10.3.10) when t is within Δ of T_n.

A10.5. SCATTERING OF A CW DIRECTIONAL SIGNAL

Let us derive the scattering of a CW signal by a rough surface, including the curvature of the wave front and the directionality of the source.

We assume that a directional transducer insonifies a small patch of the surface. From Section A5.3, a Gaussian shaded transducer has a Gaussian beam pattern. For a rectangular transducer having the shading parameters W_g and L_g, the half beam widths to $D = e^{-1}$ are

$$\Delta\phi \simeq \sin\phi = \frac{2}{kW_g}$$
$$\Delta\chi \simeq \sin\chi = \frac{2}{kL_g} \tag{A10.5.1}$$

We use the beam widths to calculate the surface illumination $\mathscr{I}$. At distance R_1, the amplitude of the illumination on the surface (Figure A10.5.1) is

$$\mathscr{I} = \exp\left(-\frac{x^2}{X^2} - \frac{y^2}{Y^2}\right) \tag{A10.5.2}$$

where

$$X \simeq R_1 \frac{\Delta\chi}{\cos\theta_1} \tag{A10.5.3}$$

$$Y \simeq R_1 \Delta\phi \tag{A10.5.4}$$

This insonification function is convenient because it can be integrated repeatedly.

A10.5.1. Scattering Integral

Our starting place is the Helmholtz-Kirchhoff equation (A10.2.7). The incident wave field on the surface U_s depends on the insonification of the surface and the geometry. We let R_s be the separation of the source and dS (Figure A10.5.1).

$$U_s = \frac{\mathscr{I}e^{-ikR_s}}{R_s}$$

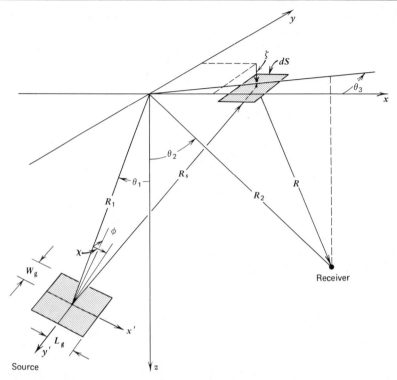

Figure A10.5.1. Geometry: dS is at the position x, y, ζ relative to the origin. The plane of the source is perpendicular to R_1, where R_1 is in the x–z plane. The plane of the transducer x'–y' is normal to R_1.

Since we are using a transducer to insonify the area, the assumptions are as follows.

1.–The source and receiver are far from the insonified area.

2.–The dimensions of the source are small compared to R_1 and R_2.

3.–The source is a Gaussian shaded transducer and is directed along R_1 (Fig. A10.5.1).

4.–No shadows are present.

5.–The value of $\mathfrak{R}$ is constant over the insonified area.

Using (5.2.5) and (5.2.7) for the sound pressure along the axis of the transducer, the incident sound pressure on the surface is written

$$p_s = \mathscr{I}\, BU_s e^{i\omega t}$$

$$= \frac{\mathscr{I}\, B \exp\left[i(\omega t - kR_s)\right]}{R_s} \tag{A10.5.5}$$

where

$$B^2 \equiv \Pi \rho c Q_t (4\pi)^{-1}$$
$$\Pi \equiv \text{source power} \qquad \text{(A10.5.6)}$$
$$Q_t \simeq 2k^2 L_g W_g$$

The substitution of (A10.5.5) into (A10.2.7) yields

$$p = \frac{\mathcal{R}B}{4\pi} \int_S \mathcal{I} \frac{\partial}{\partial n} \left\{ \frac{\exp\left[i\omega t - ik(R + R_s)\right]}{RR_s} \right\} dS \qquad \text{(A10.5.7)}$$

The rough surface ζ is a function of x, y and t. The mean of the surface is the x–y plane. Using Fig. A10.5.1, R^2 and R_s^2 are found to be

$$R_s^2 = (R_1 \sin \theta_1 + x)^2 + y^2 + (R_1 \cos \theta_1 - \zeta)^2$$
$$R^2 = (R_2 \sin \theta_2 \cos \theta_3 - x)^2 + (R_2 \sin \theta_2 \sin \theta_3 - y)^2 + (R_2 \cos \theta_2 - \zeta)^2$$

$$\text{(A10.5.8)}$$

where R and R_s are obtained by using the binomial expansion. We retain ζ and second order terms in x and y.

$$k(R + R_s) \simeq k(R_1 + R_2) + 2(\alpha x + \beta y + \gamma \zeta) + x_f^{-2} x^2 + y_f^{-2} y^2$$
$$\text{(A10.5.9)}$$

$$x_f^{-2} \equiv \frac{k}{2} \left(\frac{\cos^2 \theta_1}{R_1} + \frac{1 - \sin^2 \theta_2 \cos^2 \theta_3}{R_2} \right), \qquad y_f^{-2} \equiv \frac{k}{2} \left(\frac{1}{R_1} + \frac{1 - \sin^2 \theta_2 \sin^2 \theta_3}{R_2} \right)$$

$$\text{(A10.5.10)}$$

$$2\alpha = k(\sin \theta_1 - \sin \theta_2 \cos \theta_3)$$
$$2\beta = -k(\sin \theta_2 \sin \theta_3) \qquad \text{(A10.5.11)}$$
$$2\gamma = -k(\cos \theta_1 + \cos \theta_2)$$

We assume that the illuminated area covers many ups and downs of the rough surface. Correspondingly, the normal to dS points many different directions. Following Eckart (1953), we approximate the average of $\partial/\partial n$ by $\partial/\partial \zeta$, (ζ is along the z direction). This is the small slope approximation. On taking $\partial/\partial \zeta$, the scattering integral becomes

$$p \simeq \frac{iB\mathcal{R}kf(\theta) \exp\left[i\omega t - ik(R_1 + R_2)\right]}{2\pi R_1 R_2} \times$$

$$\times \int\int_{-\infty}^{\infty} \mathcal{I} \exp\left[-i\left(\frac{x^2}{x_f^2} + \frac{y^2}{y_f^2} + 2\alpha x + 2\beta y + 2\gamma \zeta\right)\right] dy \, dx \qquad \text{(A10.5.12)}$$

where

$$f(\theta) = \frac{\cos \theta_1 + \cos \theta_2}{2}, \qquad \text{small slope} \qquad \text{(A10.5.13)}$$

The effect of slopes for incident plane waves is calculated by doing an integration by parts, Beckmann and Spizzichino (1963). We give the form given in Tolstoy and Clay (1966, Eq. 6.25).

$$f(\theta) = \frac{1 + \cos\theta_1 \cos\theta_2 - \sin\theta_1 \sin\theta_2 \cos\theta_3}{\cos\theta_1 + \cos\theta_2} \quad \text{for} \quad \gamma X \gg 1$$

(A10.5.14)

A10.5.2. Mean Square Scattered Signal

We compute the covariance of the signal $\langle p^* p(\tau) \rangle$ using (A10.5.12) to express p. Since the expressions are quite messy, we do the operations symbolically. We let p^* be represented by an integral over dS and $p(\tau)$ be represented by an integral over dS'.

$$p^* = \exp(-i\omega t) \int_S P[x, y]^* \exp[2i\gamma\zeta] \, dS$$

(A10.5.15)

$$p(\tau) = \exp(i\omega t + i\omega\tau) \int_S P[x', y'] \exp[-2i\gamma\zeta'] \, dS'$$

After moving $\langle\ \rangle$ inside the integral and letting it operate on the random part, the covariance is found to be

$$\langle p^* p(\tau) \rangle = \exp(i\omega\tau) \int_S \int_{S'} P[x, y]^* P[x', y'] \langle \exp[2i\gamma(\zeta - \zeta')] \rangle \, dS \, dS'$$

(A10.5.16)

The $\langle\ \rangle$ quantity in the integrand is called the *characteristic function* of the bivariate PDF of the surface. We define it as $\mathcal{W}$.

$$\mathcal{W} \equiv \langle \exp[2i\gamma(\zeta - \zeta')] \rangle = \int\int_{-\infty}^{\infty} w_2 \exp[2i\gamma(\zeta - \zeta')] \, d\zeta' \, d\zeta$$

(A10.5.17)

For a stationary surface, the value of $\mathcal{W}$ depends on the separation of ζ and ζ'; $\mathcal{W}$ is a function of ξ, η, and τ.

We make a change of variables to include our assumed dependence of $\mathcal{W}$ on ξ and η.

$$x = x'' + \frac{\xi}{2} \qquad y = y'' + \frac{\eta}{2}$$

$$x' = x'' - \frac{\xi}{2} \qquad y' = y'' - \frac{\eta}{2}$$

(A10.5.18)

The integrals transform as [use (A1.5.12) and remember the absolute value sign]

$$\int (\ )\, dS \int (\ )\, dS' = \iint (\ )\, dy''\, dx'' \int\limits_{-\infty}^{\infty}\!\!\int\limits_{-\infty}^{\infty} d\xi\, d\eta \quad \text{(A10.5.19)}$$

$$\langle p^2\rangle = \frac{k^2 f^2(\theta) B^2 \mathcal{R}^2 XY \exp(i\omega\tau)}{8\pi R_1^2 R_2^2} \iint \mathcal{G}W\, d\xi\, d\eta \quad \text{(A10.5.20)}$$

where

$$\mathcal{G} \equiv \exp\left(-a_\xi \xi^2 - a_\eta \eta^2 + 2i\alpha\xi + 2i\beta\eta\right) \quad \text{(A10.5.21)}$$

$$a_\xi = \frac{X^2}{2x_f^4} + \frac{1}{2X^2}$$

$$a_\eta = \frac{Y^2}{2y_f^4} + \frac{1}{2Y^2}$$

Next we cast $\langle p^* p(\tau)\rangle$ into the form of a scattering equation

$$\langle pp^*(\tau)\rangle = \frac{B^2 A\mathcal{S}}{R_1^2 R_2^2} \quad \text{(A10.5.22)}$$

where

$$A \equiv \pi XY \quad \text{(A10.5.23)}$$

$$\mathcal{S} \equiv \frac{B_1}{\pi} \int\limits_{-\infty}^{\infty}\!\!\int \mathcal{G}W\, d\xi\, d\eta \quad \text{(A10.5.24)}$$

where $B_1 \equiv k^2 f^2(\theta)\, \mathcal{R}^2 \exp(i\omega\tau)/(8\pi)$.

The scattering function $\mathcal{S}$ has the form of a two-dimensional Fourier integral. In principle, measurements of $\mathcal{S}$ as functions of α, β can be Fourier transformed to give $\exp[-a_\xi \xi^2 - a_\eta \eta^2]W$. Since $\exp[-a_\xi \xi^2 - a_\eta \eta^2]$ tends to zero rapidly for large ξ and η, we would limit the calculation of W to values of ξ and η less than $a_\xi^{-1/2}$ and $a_\eta^{-1/2}$.

A10.6. EVALUATIONS OF THE SURFACE SCATTERING FUNCTION

The integrand of $\mathcal{S}$ is the product of a function of the sonar system and geometry $\mathcal{G}$ and W, $\mathcal{G}$ has the form of an error function and tends to zero rapidly for large ξ and η, whereas W depends on the bivariate PDF of the surface, w_2 in (A10.5.17).

When the PDF is approximately Gaussian, it is reasonable to estimate w_2 by using the bivariate Gaussian PDF (A1.7.15),

$$w_2 = [2\pi\sigma^2(1-C)^{1/2}]^{-1} \exp\left[\frac{-\zeta^2+\zeta'^2-2\zeta\zeta'C}{2(1-C^2)^{1/2}\sigma^2}\right] \quad \text{(A10.6.1)}$$

$$C \equiv \frac{\langle\zeta\zeta'\rangle}{\sigma^2}$$

where C is the spatial and temporal correlation function of the surface. Evaluation of (A10.5.17) using (A1.7.17) gives

$$\mathcal{W} = \exp[-4\gamma^2\sigma^2(1-C)] \quad \text{(A10.6.2)}$$

and $\mathcal{W}$ ranges between 1 and 0: it is 1 at $C=1$ and $\exp(-4\gamma^2\sigma^2)$ at $C=0$.

The temporal dependence of the surface is included in the correlation function, In general, the correlation function is not separable into the product of spatial and temporal functions. Windblown water waves are an example. Several crests of the waves move along the direction of the wind. The envelope of the crests moves at the group velocity of the wave while the crests move at the phase velocity. On the basis of laboratory measurements of the correlation function of windblown water waves (Medwin *et al.*, 1970), we write the following approximation:

$$C \approx \exp\left[\frac{-(\xi/u-\tau)^2}{T^2}\right]\cos\left[\Omega\left(\frac{\xi}{v}-\tau\right)\right]\exp(-b\eta^2) \quad \text{(A10.6.3)}$$

where the waves are moving in the $+x$ direction, u is the group velocity, T is the temporal "length" of the correlation packet, Ω is the angular frequency of the surface wave, v is the phase velocity, b is a parameter for the length of each crest, and C has the form of a traveling wave. The envelope $\exp[-(\xi/u-\tau)^2/T^2]$ moves at the group velocity u.

On expanding the square in (A10.6.3), we obtain a term that involves $\xi\tau$. This shows that correlation function for the windblown waves is not separable into the product of spatial and temporal components. Since τ is a constant in the integration, the lack of separability does not affect the integration. The τ dependence is more complicated than it would be if the τ dependence were separable.

In a confused sea, the waves travel in all directions and may form standing wave patterns for short times. Under these conditions we would expect the spatial correlation function to be isotropic and not to have the form of a traveling wave packet. Then, the temporal and spatial components are expected to be separable. Even so, a preferable analysis is to compute $\mathcal{S}$ for waves that travel in different directions and average $\mathcal{S}$.

When the sound scattered at the bottom is being measured from a moving ship, then relative to the sonar, the bottom is a moving surface. We would expect the effective correlation function of the bottom to have the form of a traveling wave.

Evaluations of $\mathscr{S}$ depend on what is known about the surface. If the surface is very smooth, $\gamma\sigma \ll 1$, or very rough, $\gamma\sigma > 1$, we use simple expansions and approximations. In between, evaluations are more elaborate. All the evaluations are based on devising an integrable form of $\mathscr{W}$. Ordinarily we fit analytical expressions to the measurements of C or $\mathscr{W}$. The fit must be good in the region where $\mathscr{G}\mathscr{W}$ is large, but it can be poor outside.

We give three methods of evaluating the scattering integral. The first uses the exponential expansion of $\mathscr{W}$ and is most useful for small $\gamma\sigma$. The second method, which uses a second-order approximation to express $\mathscr{W}$ in an integral form, is useful for overall values of $\gamma\sigma$. The third method is partly numerical. The numerical values of $\mathscr{W}$ as a function of ξ and η are expanded in a Fourier series, and the resulting expressions can be integrated. This may be the most powerful method.

A10.6.1.   Exponential Expansion for Small Roughness

We use a series expansion of $\mathscr{W}$ for the bivariate Gaussian PDF.

$$\mathscr{W} \simeq \exp\left(-4\gamma^2\sigma^2\right)\left[1 + 4\gamma^2\sigma^2 C + \frac{(4\gamma^2\sigma^2 C)^2}{2} + \ldots\right]$$

$$(A10.6.4)$$

The expansion is valid for all values of $4\gamma^2\sigma^2$; however the convergence is slow for $4\gamma^2\sigma^2 > 1$. After evaluation of the first two terms, substitution of (A10.6.4) in (A10.5.24) gives

$$\mathscr{S} \simeq B_1 \exp\left(-4\gamma^2\sigma^2\right)\left[(a_\xi a_\eta)^{-1/2}\exp\left(-\frac{\alpha^2}{a_\xi}-\frac{\beta^2}{a_\eta}\right)\right.$$
$$\left. +\frac{4}{\pi}\gamma^2\sigma^2 \int\!\!\int_{-\infty}^{\infty} \mathscr{G}C\,d\xi\,d\eta + \ldots\right] \quad (A10.6.5)$$

Equation (A10.6.5) is easy to evaluate for correlation functions such as (A10.6.3). Very careful measurements of $\mathscr{S}$ are required to determine the dependence of $\mathscr{S}$ on C because the scattering effects are small when $4\gamma^2\sigma^2$ is much less than 1.

A10.6.2. Polynomial Approximation for all Roughnesses

This method of polynomial approximation applies to large and small roughness. At small $\gamma\sigma$ we have to be careful about the evaluation of $\mathcal{G}$ at large ξ and η. Again the bivariate Gaussian PDF is used to estimate $\mathcal{W}$.

If the temporal component of the correlation function can be factored, the temporal dependence can be moved to the front of the integral and the spatial dependence remains. When a traveling wave type of correlation function is needed, it is a function of $(\xi - v\tau)$. We can transform to a moving coordinate system $\dot{\xi}' = (\xi - v\tau)$ and the scattering integral retains the same form. For simplicity we ignore the τ dependence, assuming that it is factored or absorbed in the transformation.

Assuming that C is expressed as a quadratic function, we write

$$C = 1 - a_1|\xi| - a_2\xi^2 - b_1|\eta| - b_2\eta^2$$
$$C = 0 \quad \text{for} \quad |\xi| > \xi_1 \quad \text{and} \quad |\eta| > \eta_1 \tag{A10.6.6}$$

where $\mathcal{G}$ has the form of the error function integral. The absolute value signs are in (A10.6.5) because C is symmetric; $\mathcal{W}$ is symmetric and can be factored into the product of ξ and η dependent functions. Expressing $\mathcal{G}$ and $\mathcal{W}$ as products, we find

$$\mathcal{G} = \mathcal{G}_\xi \mathcal{G}_\eta$$

$$\mathcal{W} = \mathcal{W}_\xi \mathcal{W}_\eta \tag{A10.6.7}$$

$$\mathcal{G} = \left(\frac{1}{\pi}\right) B_1 \int_{-\infty}^{\infty} \mathcal{G}_\xi \mathcal{W}_\xi \, d\xi \int_{-\infty}^{\infty} \mathcal{G}_\eta \mathcal{W}_\eta \, d\eta$$

Since both integrals have the same form, we evaluate one and give results for both.

We include the dependence of C on $|\xi|$ by changing the limits of integration.

$$\int_{-\infty}^{\infty} \mathcal{G}_\xi \mathcal{W}_\xi \, d\xi = \int_{0}^{\infty} \mathcal{G}_\xi \mathcal{W}_\xi \, d\xi + \int_{0}^{\infty} \mathcal{G}_\xi^* \mathcal{W}_\xi \, d\xi \tag{A10.6.8}$$

We include $C = 0$ for $\xi > \xi_1$ and write (A10.6.8) as follows

$$\int_{-\infty}^{\infty} \mathcal{G}_\xi \mathcal{W}_\xi d\xi = \int_{0}^{\xi_1} \mathcal{G}_\xi \mathcal{W}_\xi \, d\xi + \exp\left(-4\gamma^2\sigma^2\right) \int_{\xi_1}^{\infty} \mathcal{G}_\xi \, d\xi + \text{complex conjugate}$$

$$\tag{A10.6.9}$$

The same limits apply to the complex conjugates. For $\gamma^2\sigma^2 > 1$, the contribution of the second integral in (A10.6.9) is negligible, and we can let ξ_1 become infinite to simply the algebra.

We change variables to transform (A10.6.9) to the tabulated error function integral, using

$$\mathcal{G}_\xi W_\xi = \exp\left[-(a_\xi + 4\gamma^2\sigma^2 a_2)\xi^2 - 2(2\gamma^2\sigma^2 a_1 - i\alpha)\xi\right]$$

$$\text{(A10.6.10)}$$

The changes of variables are as follows:

$$q = \frac{\xi}{A_x} + r$$

$$r = A_x B_x \qquad \text{(A10.6.11)}$$

$$s = \frac{\eta}{A_y} + t$$

$$t = A_y B_y$$

$$A_x \equiv (a_\xi + 4\gamma^2\sigma^2 a_2)^{-1/2}$$

$$B_x \equiv 2\gamma^2\sigma^2 a_1 - i\alpha$$

$$A_y \equiv (a_\eta + 4\gamma^2\sigma^2 b_2)^{-1/2} \qquad \text{(A10.6.12)}$$

$$B_y \equiv 2\gamma^2\sigma^2 b_1 - i\beta$$

The substitution of these changes of variable gives

$$\int_{-\infty}^{\infty} \mathcal{G}_\xi W_\xi \, d\xi = \sqrt{\pi} A_x \mathcal{U}$$

$$\text{(A10.6.13)}$$

$$\int_{-\infty}^{\infty} \mathcal{G}_\eta W_\eta \, d\eta = \sqrt{\pi} A_y \mathcal{V}$$

where

$$\mathcal{U} \equiv \pi^{-1/2}\left(e^{r^2}\int_r^\infty e^{-q^2}\, dq + e^{r^{*2}}\int_{r^*}^\infty e^{-q^2}\, dq\right) \qquad \text{(A10.6.14)}$$

$$\mathcal{U} \equiv \frac{e^{r^2}\operatorname{erfc} r + e^{r^{*2}}\operatorname{erfc} r^*}{2}$$

$$\text{(A10.6.15)}$$

$$\mathcal{V} = \frac{e^{t^2}\operatorname{erfc} t + e^{t^{*2}}\operatorname{erfc} t^*}{2}$$

These are standard error integrals for real r. For complex r and t, we use the complex error integral $w(z)$, Abramowitz and Stegun (1964), and (A1.5.1) to (A1.5.19).

$$w(z) \equiv e^{-z^2}\operatorname{erfc}(-iz) \qquad \text{(A10.6.16)}$$

Letting r and t be $-iz$, $\mathcal{U}$ and $\mathcal{V}$ are found to be

$$\mathcal{U} = \frac{w(ir) + w(ir^*)}{2}$$

$$\mathcal{V} = \frac{w(it) + w(it^*)}{2} \tag{A10.6.17}$$

Abramowitz and Stegun give tables of $w(z)$. When $r = x + iy$ is pure imaginary or pure real

$$\mathcal{U} = \mathcal{U}(iy) = e^{-y^2} \tag{A10.6.18}$$

$$\frac{2\pi^{-1/2}}{x + (x^2 + 2)^{1/2}} < \mathcal{U}(x) \le \frac{2\pi^{-1/2}}{x + (x^2 + 4/\pi)^{1/2}} \tag{A10.6.19}$$

We use the asymptotic expansion of $w(z)$, (A1.5.19) and (A10.6.15) to write an approximate expression

$$\mathcal{U} \simeq (2\sqrt{\pi})^{-1}\left[\frac{2x}{x^2 + y^2} + \frac{x(3y^2 - x^2)}{(x^2 + y^2)^3}\right] \tag{A10.6.20}$$

The approximations for $\mathcal{V}$ are similar.

The substitution of (A10.6.13) into (A10.6.7) gives

$$\mathcal{S} = \frac{A_x A_y k^2 f^2(\theta)\mathcal{U}\mathcal{V}}{8\pi}\mathcal{R}^2 e^{i\omega\tau} \tag{A10.6.21}$$

In the specular direction ($\alpha = \beta = 0$), the parameters r and t are real. Then we use (A10.6.19) for numerical evaluations of $\mathcal{S}$. For other geometries, such as the backscattering direction, B_x in (A10.6.12) is complex and correspondingly r is complex. In the backscattering direction $\beta = 0$ and t is real. We use (A10.6.19) to calculate $\mathcal{V}$.

A10.6.3. Numerical Evaluations of the Scattering Integral

Our purpose is to make numerical evaluations of (A10.5.24). We assume that $\mathcal{W}$ is known at a rectangular grid of values of ξ and η. When C is known and the surface has a Gaussian PDF, the bivariate PDF can be used to calculate numerical values of $\mathcal{W}$ at the grid of points. Presumably the grid of points obey the sampling theorem and are more closely spaced than half the shortest wavelengths of features on the surface. The grid of points must include the region for which $\mathcal{G}$ in (A10.5.24) is significantly different from zero. We use a_ξ and a_η to choose the dimensions of the region in the $\xi\eta$ plane

$$L > 2a_\xi^{-1/2}$$

$$W > 2a_\eta^{-1/2}$$

The cost of numerical evaluation is roughly proportional to the number of grid points. If one coordinate can be evaluated by using a polynomial expression as in Section A10.6.2, the cost may be reduced by an order of magnitude. For example, the crests of windblown waves are much wider than the crest-to-crest distances. If the crests are along the y direction, the polynomial approximation may be adequate for the η dependence. Alternatively, the spacings of the grid points along the y direction can be wider than along the x direction. This also reduces the total number of grid points.

The integral can be transformed by replacing $\mathcal{W}$ by its expansion as a Fourier series. We have chosen to use the Fourier series because the numerical computations can be done by using the FFT algorithm. The FFT expansion introduces periodicity in $\mathcal{W}$, and we choose L and W for the spatial periods.

Since $\mathcal{G}$ is negligible for ξ and η larger than L and W, we can ignore the periodicity of the Fourier series expansion of $\mathcal{W}$. The sampling intervals ξ_0 and η_0 are chosen to be less than half the shortest "wavelength" in $\mathcal{W}$ or C. The numbers of points M and N, are

$$M = \frac{L}{\xi_0}$$

$$N = \frac{W}{\eta_0}$$

(A10.6.23)

The two-dimensional Fourier expansion of $\mathcal{W}$ is

$$\mathcal{W} = \sum_{m=-M/2+1}^{M/2} \sum_{n=-N/2+1}^{N/2} H_{mn} \exp\left(\frac{2\pi im\xi}{L} + \frac{2\pi in\eta}{W}\right) \quad \text{(A10.6.24)}$$

$$H_{mn} = (LW)^{-1} \int_{-W/2}^{W/2} d\eta \int_{-L/2}^{L/2} \mathcal{W} \exp\left(\frac{-2\pi im\xi}{L} - \frac{2\pi in\eta}{W}\right) d\xi$$

(A10.6.25)

One would probably use FFT methods to calculate H_{mn}.

The series expansion of $\mathcal{W}$, (A10.6.24), is substituted in (A10.5.24), the equation of $\mathcal{G}$. We retain the infinite limits in (A10.5.24) because the contributions are negligible for ξ and η greater than L and W. The scattering coefficients $\mathcal{G}_{mn}$ that correspond to the Fourier coefficients H_{mn}

are

$$\mathcal{S}_{mn} \equiv \frac{H_{mn}B_1}{\pi} \int\!\!\int_{-\infty}^{\infty} \mathcal{G} \exp\left(\frac{2\pi i m \xi}{L} + \frac{2\pi i n \eta}{W}\right) d\xi\, d\eta \quad \text{(A10.6.26)}$$

The integrations follow directly by using (A1.5.6), and the result is

$$\mathcal{S}_{mn} = B_1(a_\xi a_\eta)^{-1/2} H_{mn} \exp\left[-\frac{(\alpha + m\pi/L)^2}{a_\xi} - \frac{(\beta + n\pi/W)^2}{a_\eta}\right]$$
$$\text{(A10.6.27)}$$

The scattering function is the sum over all components

$$\mathcal{S} = \sum_{m=-M/2+1}^{M/2} \sum_{n=-N/2+1}^{N/2} \mathcal{S}_{mn} \quad \text{(A10.6.28)}$$

This is a powerful method of evaluating the scattering integral because it does not depend on special properties of C or limited ranges of $\gamma\sigma$. We like it because it encourages the direct use of measured data points for $\mathcal{W}$ or C and does not interpose the fitting of data points to an arbitrary function.

A10.7. TRANSMISSION

The transmission equations follow directly from (A10.2.8) and the expressions for R_s and R' (A10.5.8). We place the source above the x–y plane in Fig. A10.5.1 and measure θ_1 from the minus z axis; c is the sound speed above and c' is the sound speed below the interface. With these changes, the primed quantities are the corresponding transmission parameters and functions, using $k' = ck/c'$ and letting R_s be the source separation and R' be the receiver separation from dS.

$$kR_s + k'R' = k\left(R_s + \frac{c}{c'}R'\right) \quad \text{(A10.7.1)}$$

$$kR_s + k'R' = k\left(R_1 + \frac{c}{c'}R_2\right) + 2(\alpha'x + \beta'y + \gamma'\zeta) + x_f^{-2}x^2 + y_f^{-2}y^2 \quad \text{(A10.7.2)}$$

$$2\alpha' \equiv k\left(\sin\theta_1 - \frac{c}{c'}\sin\theta_2\cos\theta_3\right)$$

$$2\beta' \equiv -k\frac{c}{c'}\sin\theta_2\sin\theta_3 \quad \text{(A10.7.3)}$$

$$2\gamma' \equiv k\left(\cos\theta_1 - \frac{c}{c'}\cos\theta_2\right)$$

$$x_f'^{-2} \equiv \frac{k}{2}\left[\frac{\cos^2\theta_1}{R_1}+\frac{c(1-\sin^2\theta_2\cos^2\theta_3)}{c'R_2}\right]$$

$$\hspace{6cm}(A10.7.4)$$

$$y_f'^{-2} = \frac{k}{2}\left[\frac{1}{R_1}+\frac{c(1-\sin^2\theta_2\sin^2\theta_3)}{c'R_2}\right]$$

$$f'(\theta) \equiv \cos\theta_1 + \frac{c}{c'}\cos\theta_2 \hspace{2cm}(A10.7.5)$$

$$B_1' = k^2 f'(\theta)^2 \mathcal{F}^2\frac{e^{i\omega\tau}}{8\pi} \hspace{2cm}(A10.7.6)$$

Substitution of these quantities in the expression for p' (A10.5.12) gives

$$p' = iB\mathcal{F}kf'(\theta).\exp\frac{i\omega t - ik[R_1+(c/c')R_2]}{2\pi R_1 R_2}$$

$$\times \int\!\!\!\int_{-\infty}^{\infty} \mathscr{I}\exp\left[-i\left(\frac{x^2}{x_f'^2}+\frac{y^2}{y_f'^2}+2\alpha'x+2\beta'y+2\gamma'\zeta\right)\right]dy\,dx \quad (A10.7.7)$$

The roughness of the interface scatters sound waves into the second medium. The derivation of the mean square scattered sound pressure is the same for scattered transmitted pressures as it is for the reflected components. Using (A10.5.20) to (A10.5.24) as models, the transmission scattering function is found to be

$$\mathscr{S}' = \frac{B_1'}{\pi}\int\!\!\!\int_{-\infty}^{\infty}\mathscr{G}'W\,d\xi\,d\eta \hspace{2cm}(A10.7.8)$$

$$\mathscr{G}' \equiv \exp\left[-a_\xi'\xi^2 - a_\eta'\eta^2 + 2i\alpha'\xi + 2i\beta'\eta\right] \hspace{1cm}(A10.7.9)$$

$$a_\xi' \equiv \frac{X^2}{2x_f'^4}+\frac{1}{2X^2}$$

$$\hspace{6cm}(A10.7.10)$$

$$a_\eta' \equiv \frac{Y^2}{2y_f'^4}+\frac{1}{2Y^2}$$

The procedures for evaluating the effects of the rough interface on the transmitted sound pressure are the same as for the reflected and scattered sound pressures.

SUGGESTED READING

P. Beckmann and A. Spizzichino, *The Scattering of Electromagnetic Waves from Rough Surfaces*, Macmillan, New York, 1963. This is an advanced monograph on scattering theory. The first part is theoretical and gives an extensive review of the plane wave theory. The second part is primarily experimental.

M. Born and E. Wolf, *Principles of Optics, 3rd Edition*, Pergamon Press, New York, 1965. The historical introduction gives an outline of the development of diffraction and optical theory. Chapter 8 begins with the contributions of Huygens, Fresnel, Kirchhoff, and others; it ends with Gabor's use of holograms to reconstruct images. Rigorous diffraction theory and the diffraction of light by ultrasonic waves are in chapters 11 and 12.

J. F. MacDonald, F. B. Teuter and J. G. Zornig, "Spatial interfrequency correlation effects in a surface-scatter channel," *J. Acoust. Soc. Am.* **59**, 1284–1293 (1976). The correlation of sound scattered to a pair of spaced receivers.

David Middleton, "A statistical theory of reverberation and similar first-order scattered fields—Part I: Waveforms and the general process," *Trans. IEEE Inf. Theory* **IT-13**, 372–392 (1967); "Part II: Moments, spectra, and special distributions," **IT-13**, 393–414 (1967). Middleton uses a random distribution of "point scatterers" to represent the irregularities of the surface. It is quasi-phenomenological in that each point has its own impulse response and directivity pattern. Physically, points scatter zero amounts of energy; thus his points represent scattering features of some finite size. He suggests that the "point scatterers" be quantified by experiment. The papers are very elegant.

T. D. Plemons, J. A. Shooter, and David Middleton. "Underwater acoustic scattering from lake surfaces. I. Theory, experiment, and validation of the data. II. Covariance functions and related statistics," *J. Acoust. Soc. Am.* **52**, 1487–1515 (1972). An extensive statistical study of reverberation processes. The experimental measurements were made at the Lake Travis Test Station of the Applied Research Laboratories of the University of Texas at Austin.

G. A. Sandness, "A numerical evaluation of the Helmholtz integral in acoustic scattering," Ph.D. thesis, Department of Geology and Geophysics, University of Wisconsin, Madison, 1973. Available through University Microfilms, Ann Arbor, Mich. Tests the Fresnel approximation and the zero slope with numerical evaluation of the integral and a series of laboratory experiments show that the numerical results fit the data very well. The Fresnel approximation, Gaussian illumination function, and a slope correction factor also fit the data.

R. G. Williams, "Estimating the ocean wind wave spectra by means of underwater sound," *J. Acoust. Am.* **53**, 910–920 (1973). Comparisons of acoustic measurements, theory, and sea surface measurements.

REFERENCES

Abramowitz, M. and I. A. Stegun, "Handbook of Mathematical Functions," *Natl. Bur. Stand.*, Applied Mathematical Series, 55. Government Printing Office, Washington, D.C. 1964.

Anderson, A. L. and L. L. Hampton, "A Method for Measuring *in-Situ* Acoustic Properties During Sediment Coring," in L. L. Hampton, Ed., *Physics of Sound in Marine Sediments*, Plenum Press, New York, 1974.

Anderson, V. C., "Sound scattering from a fluid sphere," *J. Acoust. Soc. Am.* **22**, 426–431 (1950).

Andreeva, I. B., "Scattering of sound by air bladders of fish in deep sound-scattering ocean layers," *Sov. Phys.—Acoust.* **10**, 17–20 (1964).

Backus, M. M., "Water reverberations—Their nature and elimination," *Geophysics*, **24**, 233–261 (1959).

Barham, E. G., "Siphonophores and the deep scattering layer," *Science*, **140**: 3568, 826–828 (1963).

Barraclough, W. E., R. J. LeBrasseur, and O. D. Kennedy, "Shallow scattering layer in the subarctic Pacific Ocean: Detection by high-frequency echo sounder," *Science*, **166**, 611–613 (1969).

Batzler, W. E. and G. V. Pickwell, "Resonant Acoustic Scattering from Gas-Bladder Fish," pp. 168–179 in G. Brooke Farquhar, Ed., *Proceedings of an International Symposium on Biological Sound Scattering in the Ocean*, Maury Center for Ocean Science, Department of the Navy, Washington, D.C., 1970. Government Printing Office stock no.: 0851-0053; price, $6.50.

Beckmann, P. and A. Spizzichino, *The Scattering of Electromagnetic Waves from Rough Surfaces*, Macmillan, New York, 1963.

Berkson, J. M. and C. S. Clay, "Microphysiography and possible iceberg grooves on the floor of Western Lake Superior," *Geol. Soc. Am. Bull.* **84**, 1315–1328 (1973).

Beyer, R. T., *Nonlinear Acoustics*, Naval Sea Systems Command, 1974. Government Printing Office, Washington, D.C., stock no.: 0-596-215, 1975.

Beyer, R. T. and S. V. Letcher, *Physical Ultrasonics*, Academic Press, New York (1969).

Bezdek, H. F., "Pressure dependence of sound attenuation in the Pacific Ocean," *J. Acoust. Soc. Am.* **53**, 782–788 (1973).

Biot, M. A., "General theory of three-dimensional consolidation," *J. Appl. Phys.* **12**, 155–164 (1941).

Biot, M. A., "Theory of elastic waves in a fluid-saturated porous solid. I. Low-frequency range," *J. Acoust. Soc. Am.* **28**, 168–178 (1956); "II. Higher-frequency range," **28**, 179–191 (1956).

Biot, M. A., "Mechanics of deformation and acoustic propagation in porous media," *J. Appl. Phys.* **33**, 1482–1498 (1962a).

Biot, M. A., "Generalized theory of acoustic propagation in porous dissipative media," *J. Acoust. Soc. Am.* **34**, 1254–1264 (1962b).

Black, L. J. "Physical analysis of distortion produced by the nonlinearity of the medium," *J. Acoust. Soc. Am.* **12**, 266–267 (1940).

Blackstock, D. T., "Connection Between the Fay and Fubini Solutions for Plane Sound Waves of Finite Amplitude," *J. Acoust. Soc. Am.* **39**, 1019 (1966).

Born, M. and E. Wolf, *Principles of Optics*, Pergamon Press, Oxford (1965).

Bradbury, M. G. et al., "Studies of the Fauna Associated with the Deep Scattering Layers in the Equatorial Indian Ocean Conducted on *R/V Te Vega* During October and November 1964," pp. 409–452 in G. Brooke Farquhar, Ed., *Proceedings of an International Symposium on Biological Sound Scattering in the Ocean*, Maury Center for Ocean Science, Department of the Navy, Washington, D.C., 1970. Government Printing Office stock no.: 0851-0053; price, $6.50.

Brandt, S. B., "Acoustic determination of fish abundance and distribution in Lake Michigan with special reference to temperature," MS thesis, University of Wisconsin, 1975.

Brekhovskikh, L. M., *Waves in Layered Media*, Academic Press, New York (1960).

Bryan, G. M., "Sonobuoy Measurements in Thin Layers," in L. L. Hampton, Ed., *Physics of Sound in Marine Sediments*, Plenum Press, New York, 1974, pp. 119–130.

Bucker, H. P., "Sound propagation calculations using bottom reflection functions," in L. L. Hampton, Ed., *Physics of Sound in Marine Sediments*, Plenum Press, New York, 1974, pp. 223–240.

Cagniard, L., *Reflection et refraction des ondes progressives seismiques*, Gauthier-Villars, Paris, 1939; transl E. A. Flinn and C. H. Dix, McGraw-Hill, New York, 1962.

Chapman, R. P., "Sound scattering in the ocean," in V. M. Albers, Ed., *Underwater Acoustics*, Vol. 2, Plenum Press, New York, 1967.

Chapman, R. P., et al., "Geographic Variations in the Acoustic Characteristics of Deep Scattering Layers," pp. 306–317 in G. Brooke Farquhar, Ed., *Proceedings of an International Symposium on Biological Sound Scattering in the Ocean*, Maury Center for Ocean Science, Department of the Navy, Washington, D.C., 1970. Government Printing Office stock no.: 0851-0053; price, $6.50.

Clay, C. S., "Effect of a slightly irregular boundary on the coherence of waveguide propagation," *J. Acoust. Soc. Am.* **36**, 833–837 (1964).

Clay, C. S., "Use of arrays for acoustic transmission in a noisy ocean," *Rev. Geophys.* **4**, 475–507 (1966).

Clay, C. S. and P. A. Rona, "Studies of seismic reflections from thin layers on the ocean bottom in the Western Atlantic," *J. Geophys. Res.* **70**, 855–869 (1965).

Clay, C. S. and G. A. Sandness, "Effect of beam width on acoustic signals scattered at a rough surface," Advisory Group for Aerospace Research and Development, *North Atlantic Treaty Organization Conference Proceedings*, **21**: 90, 1–8 (1971).

Clay, C. S., H. Medwin, and W. M. Wright, "Specularly scattered sound and the probability density function of a rough surface," *J. Acoust. Soc. Am.* **53**, 1677–1682 (1973).

Colladon, J. D. and J. K. F. Sturm, "The Compression of Liquids" (in French), *Ann. Chim. Phys.* Series 2, **36,** Part IV, "Speed of Sound in Liquids," 236–257 (1827).

DeSanto, J. A. and O. Shisha, "Numerical solution of a singular integral equation in random rough surface scattering theory," *J. Comput. Phys.* **15,** 286–292 (1974).

Devin, Charles, Jr., "Survey of thermal, radiation and viscous damping of pulsating air bubbles in water," *J. Acoust. Soc. Am.* **31,** 1654–1667 (1959).

Dix, C. H., "Seismic velocities from surface measurements," *Geophysics,* **20,** 67–86 (1955).

Dolph, C. L., "A current distribution of broadside arrays which optimizes the relationship between beam width and side lobe level," *Proc. Inst. Radio Eng.* **34,** 335 (1946).

Dunlap, C. R., "A Reconnaissance of the Deep Scattering Layers in the Eastern Tropical Pacific and the Gulf of California," pp. 395–408 in G. Brooke Farquhar, Ed., *Proceedings of an International Symposium on Biological Sound Scattering in the Ocean,* Maury Center for Ocean Science, Department of the Navy, Washington, D.C., 1970. Government Printing Office stock no.: 0851-0053; price, $6.50.

Dunn, D. J., "Turbulence and its effect upon the transmission of sound in water," *J. Sound Vib.* **2:** 3, 307–327 (1965).

Durbaum, H., "Zur Bestimmung von Wellengeschwindkeiten aus reflexionsseismischen Messungen," *Geophys. Prospect.* **2,** 151–167 (1954).

Eckart, C. "Vortices and streams caused by sound waves" *Phys. Rev.* **73,** 68–76 (1948).

Eckart, C., "The scattering of sound from the sea surface," *J. Acoust. Soc. Am.* **25,** 566–570 (1953).

Eckart, C., Ed., *Principles and Applications of Underwater Sound,* Department of the Navy, NAVMAT P-9674, 1968. Government Printing Office, Washington, D.C., price, $4.75.

Eigen, M. and K. Tamm, "Sound absorption in electrolyte solutions due to chemical relaxation," *Z. Electrochem.* **66,** 93–121 (1962).

Elder, S. A., "Cavitation microstreaming," *J. Acoust. Soc. Am.* **31,** 54–64 (1959).

Eller, A. I., "Damping constants of pulsating bubbles," *J. Acoust. Soc. Am.* **47,** 1469–1470 (1970).

Epstein, P. S. and R. R. Carhart, "The absorption of sound in suspensions and emulsion. I. Water, fog in air," *J. Acoust. Soc. Am.* **25,** 553–565 (1953).

Ewing, J. I., "Elementary Theory of Seismic Refraction and Reflection Measurements," in M. N. Hill, Ed., *The Sea,* Vol. 3, Wiley-Interscience, New York, 1963, pp. 3–19.

Ewing, J. I., "Seismic Model of the Atlantic," in P. J. Hart, Ed., *The Earth's Crust and Upper Mantle,* Geophysical Monograph 13, American Geophysical Union, Washington, D.C., 1969.

Ewing, W. M., W. S. Jardetzky and F. Press, *Elastic Waves in Layered Media*, McGraw-Hill, New York (1962).

Ewing, M. and J. L. Worzel, "Long-Range Sound Transmission," in *Propagation of Sound in the Ocean*, Memoir 27, Geological Society of America, New York, 1948, Fig. 5, p. 19.

Farquhar, G. Brooke, Ed., *Proceedings of an International Symposium on Biological Sound Scattering in the Ocean*, Maury Center for Ocean Science, Department of the Navy, Washington, D.C., 1970. Government Printing Office, Washington, D.C., stock no.: 0851-0053; price, $6.50.

Fisher, F. H. and S. A. Levison, "Dependence of the low frequency (1 kHz) relaxation in seawater on boron concentration," *J. Acoust. Soc. Am.* **54,** 291 (1973).

Flynn, H. G., "Physics of Acoustic Cavitation in Liquids," pp. 57–172 in W. P. Mason, Ed., *Physical Acoustics*, Vol. I, Part B, Academic Press, New York, 1964.

Folds, D. L., "Speed of sound and transmission loss in silicone rubbers at ultrasonic frequencies," *J. Acoust. Soc. Am.* **56,** 1295(L) (1974).

Forbes, S. T. and O. Nakken, Eds., *Manual of Methods for Fisheries Resource Survey and Appraisal*, Part 2. *The Use of Acoustic Instruments for Fish Detection and Abundance Estimation*, Food and Agricultural Organization of the United Nations, FAO Manuals in Fisheries Science No. 5, Rome, 1972.

Fortuin, L., "Survey of literature on reflection and scattering of sound waves at the sea surface," *J. Acoust. Soc. Am.* **47,** 1209–1228 (1970).

Fox, F. E. and K. F. Herzfeld, "On the forces producing the ultrasonic wind," *Phys. Rev.* **78,** 156–157 (1950).

Fox, F. E. and K. F. Herzfeld, "Gas bubbles with organic skin as cavitation nuclei," *J. Acoust. Soc. Am.* **26,** 984–989 (1954).

Gassmann, F., "Über die Elastizität poröser Medien," *Vierteljahrsschr. Naturforsch. Ges. Zürich*, **96,** 1–23 (1951).

Gavrilov, L. R., "On the size distribution of gas bubbles in water," *Sov. Phys.—Acoust.* **15,** 22–24 (1969).

Glotov, V. P., et al., "Investigation of the scattering of sound by bubbles generated by an artificial wind in seawater and the statistical distribution of bubble sizes," *Sov. Phys.—Acoust.* **7,** 341–345 (1962).

Grant, F. S. and G. F. West, *Interpretation Theory in Applied Geophysics*, McGraw-Hill, New York, 1965.

Gruber, J. and R. Meister. "Ultrasonic attenuation in water containing brine shrimp in suspension," *J. Acoust. Soc. Am.* **33,** 733–740 (1961).

Hamilton, E. L., "Sediment sound velocity measurements made *in situ* from the bathyscaphe *Trieste*," *J. Geophys. Res.* **68,** 5991–5998 (1963).

Hamilton, E. L., "Sound Velocity, Elasticity and Related Properties of Marine Sediments, North Pacific," Naval Undersea Research and Development Center," Technical Publication 143 (1969).

Hamilton, E. L., "Sound Velocity, Elasticity, and Related Properties of Marine Sediments, North Pacific," Naval Undersea Research and Development Center, NUC TP 144, San Diego, Calif., October 1969. Also *J. Geophys. Res.* **75**, 4423–4446 (1970).

Hamilton, E. L., "The elastic properties of marine sediments," *J. Geophys. Res.* **76**, 579–604 (1971a).

Hamilton, E. L., "Prediction of *in-situ* acoustic and elastic properties of marine sediments," *Geophysics*, **36**, 266–284 (1971b).

Hamilton, E. L., "Compressional-Wave Attenuation in Marine Sediments," *Geophysics*, **37**, 620–646 (1972).

Hampton, L. L., Ed., *Physics of Sound in Marine Sediments*, Plenum Press, New York, 1974.

Harvey, E. N., et al., "Bubble formation in animals, 1. Physical factors," *J. Cell. Comp. Physiol.* **24**, 1–22 (1944).

Heelan, P. A., "On the theory of head waves," *Geophysics*, **18**, 871–893 (1953).

Heezen, B. C., and D. C. Hollister, *The Face of the Deep*, Oxford University Press, New York, 1971.

Hersey, J. B. and R. H. Backus, "Sound Scattering by Marine Organisms," in M. N. Hill, Ed., *The Sea*, Vol. 1, John Wiley & Sons, New York, 1962.

Hersey, J. B., "Continuous Reflection Profiling," in M. N. Hill, Ed., *The Sea*, Vol. 3, Wiley-Interscience, New York, 1963, pp. 47–72.

Hickling, R., "Analysis of echoes from a hollow metallic sphere in water," *J. Acoust. Soc. Am.* **36**, 1124–1137 (1964).

Hill, M. N., Ed., *The Sea*, Vol. 3, Wiley-Interscience, New York, 1963.

Hill, M. N., "Single-Ship Seismic Refraction Shooting," in M. N. Hill, Ed., *The Sea*, Vol. 3, Wiley-Interscience, New York, 1963, pp. 39–46.

Hilterman, F. J., "Three-dimensional seismic modeling," *Geophysics*, **35**, 1020–1037 (1970).

Hole, W. L., *Final Summary Report, Hudson Laboratories of Columbia University*, Vol. 1, 1951–1969 (1969).

Horton, C. W., Sr., "A review of reverberation, scattering, and echo structure," *J. Acoust. Soc. Am.* **51**, 1049–1061 (1972).

Horton, C. W., Sr., "Dispersion relations in sediments and sea water," *J. Acoust. Soc. Am.* **55**, 547–549 (1974).

Horton, C. W., Sr., and D. R. Melton, "Importance of the Fresnel correction in scattering from a rough surface, II. Scattering coefficient," *J. Acoust. Soc. Am.* **47**, 299–303 (1970). (See Melton and Horton for Part I.)

Houtz, R. E., "Preliminary study of global sediment sound velocities from sonobuoy data," Card files at Lamont-Doherty Geological Observatory. Also in L. L. Hampton, Ed., *Physics of Sound in Marine Sediments*, Plenum Press, New York, 1974, pp. 519–536.

LaCase, E. O. and P. Tamarkin, "Underwater sound reflection from a corrugated surface," *J. Appl. Phys.* **27**, 138–148 (1956).

Laughton, A. S., "Sound propagation in compacted ocean sediments," *Geophysics,* **22,** 233–260 (1957).

Lebedeva, L. P., "Measurement of the dynamic complex shear modulus of animal tissues," *Sov. Phys.—Acoust.* **11,** 163–165 (1965).

Leonard, R. W., P. C. Combs, and L. R. Skidmore, "Attenuation of sound in sea water," *J. Acoust. Soc. Am.* **21,** 63 (1949).

Le Pichon, X., J. I. Ewing, and R. E. Houtz, "Deep-sea sediment velocity determinations made while reflection profiling," *J. Geophys. Res.* **73,** 2597–2614 (1968).

Levin, F. K., "The seismic properties of Lake Maracaibo," *Geophysics,* **27,** 35–47 (1962).

Liebermann, L. N., "Second viscosity of fluids," *Phys. Rev.* **75,** 1415–1422 (1949).

Liebermann, L. N., "Analysis of rough surfaces by scattering," *J. Acoust. Soc. Am.* **35,** 932 (1963).

Lineback, J. A., D. L. Gross, and R. P. Meyer, "Glacial tills under Lake Michigan," Environmental Geology Notes, No. 69, Illinois State Geological Survey, Urbana, July 1974.

Litovitz, T. A. and E. H. Carnevale, "Effect of pressure on sound propagation in water," *J. Appl. Phys.* **26,** 816–820 (1955).

Lockwood, J. C. and J. G. Willette, "High speed method for computing the exact solution for the pressure variations in the near field of a baffled piston," *J. Acoust. Soc. Am.* **53,** 735–741 (1973).

Lowrie, A. and E. Escowitz, *Kane 9. Global Ocean Floor Analysis and Research Data Series.* U.S. Naval Oceanographic Office, 1969.

Marsh, H. W., "Exact solution of wave scattering by irregular surfaces," *J. Acoust. Soc. Am.* **33,** 330–333 (1961).

Marshall, N. B., "Swimbladder Development and the Life of Deep-Sea Fish, pp. 69–78 in G. Brooke Farquhar, Ed., *Proceedings of an International Symposium on Biological Sound Scattering in the Ocean,*" Maury Center for Ocean Science, Washington, D.C., 1970. Government Printing Office stock no.: 0851-0053, price, $6.50.

Maue, A. W., "Zur Formulierung eines allgemeinen Beugungsproblems durch eine Integralgleichung," *Z. Phys.* **126,** 601–618 (1949).

Maynard, G. L., G. H. Sutton, D. M. Hussong, and L. W. Kroenke, "The Seismic Wide Angle Reflection Method in the Study of Ocean Sediment Velocity Structure," in L. L. Hampton, Ed., *Physics of Sound in Marine Sediments,* Plenum Press, New York, 1974, pp. 89–118.

Mayne, W. H. and R. G. Quay, "Seismic signatures of large air guns," *Geophysics,* **36,** 1162–1173 (1971).

McCartney, B. S. and A. R. Stubbs, "Measurement of the Target Strength of Fish in the Dorsal Aspect, Including Swimbladder Resonance," pp. 180–211 in G. Brooke Farquhar, Ed., *Proceedings of an International Symposium on Biological Sound Scattering in the Ocean,* Maury Center for Ocean Science, Department of the Navy, Washington, D.C., MC Report 005, 1970. Government Printing Office stock no.: 0851-0053; price, $6.50.

Medwin, H., "Acoustic streaming experiment in gases," *J. Acoust. Soc. Am.* **26**, 332–340 (1954).

Medwin, H., "*In-situ* acoustic measurements of bubble populations in coastal waters," *J. Geophys. Res.* **75**, 599–611 (1970).

Medwin, H., "Acoustic fluctuations due to microbubbles in the near-surface ocean," *J. Acoust. Soc. Am.* **56**, 1100–1104 (1974).

Medwin, H., "Speed of sound in water: A simple equation for realistic parameters," *J. Acoust. Soc. Am.* **58**, 1318–1319 (1975).

Medwin, H., C. S. Clay, J. M. Berkson, and D. L. Jaggard, "Traveling correlation function of the heights of windblown water waves, *J. Geophys. Res.* **75**, 4519–4524 (1970).

Medwin, H., J. Fitzgerald, and G. Rautmann, "Acoustic miniprobing for ocean microstructure and bubbles," *J. Geophys. Res.* **80**, 405–413 (1975).

Medwin, H., "Counting bubbles acoustically; a review," *Ultrasonics* **15**, 7–13 (1977).

Meecham, W., "On the use of the Kirchhoff approximation for the solution of reflection problems," *J. Rational Mech. Anal.* **5**, 323–333 (1956).

Meister, R. and R. St. Laurent, "Ultrasonic absorption and velocity in water containing algae in suspension," *J. Acoust. Soc. Am.* **32**, 556–559 (1960).

Mellberg, L. E. and O. M. Johannessen, "Layered oceanic microstructure—Its effect on sound propagation," *J. Acoust. Soc. Am.* **53**, 571–580 (1973).

Mellen, R. H. and D. G. Browning, "Low frequency attenuation in the Pacific Ocean," *J. Acoust. Soc. Am.* **57**, Suppl. 1, S65 (1975).

Melton, D. R. and C. W. Horton, Sr., "Importance of Fresnel correction in scattering from a rough surface. I. Phase and amplitude fluctuations," *J. Acoust. Soc. Am.* **47**, 290–298 (1970). (See Horton and Melton for Part II.)

Messino, D., D. Sette, and F. Wanderlingh, "Statistical approach to ultrasonic cavitation," *J. Acoust. Soc. Am.* **35**, 1575–1583 (1963).

Millikan, R. A., D. Roller, and E. C. Watson, *Mechanics, Molecular Physics, Heat, and Sound*, M.I.T. Press, Cambridge, Mass., 1965.

Muir, T. G., "Non-linear acoustics and its role in the sedimentary geophysics of the sea," pp. 241–287, in L. L. Hampton, Ed., *Physics of Sound in Marine Sediments*, Plenum Press, New York, 1974.

Neubauer, W. G. and L. R. Dragonette, "Observation of waves radiated from circular cylinders caused by an incident pulse," *J. Acoust. Soc. Am.* **48**, 1135–1149 (1970).

Noble, B., in R. E. Langer, Ed., *Electromagnetic Waves*, University of Wisconsin Press, Madison, 1962, pp. 332–360.

Northrop, J. and J. G. Colborn, "Sofar Channel axial sound speed and depth in the Atlantic Ocean," *J. Geophys. Res.* **79**, 5633–5641 (1974).

Nyborg, W., "Acoustic streaming," pp. 265–331 in W. P. Mason, Ed., *Physical Acoustics*, Vol. II, Part B, Academic Press, New York, 1965.

Parsons, T. R. and M. Takahashi, *Biological Oceanic Processes*, Pergamon Press, New York, 1973.

Pekeris, C. L., "Theory of Propagation of Explosive Sound in Shallow Water," in *Propagation of Sound in the Oceans*, Geological Society of America Memoir 27, New York, 1948.

Peterson, M. L., C. S. Clay, and S. B. Brandt, "Acoustic estimates of fish density and scattering function," *J. Acoust. Soc. Am.* **60,** 618–622 (1976).

Phillips, O. M., *The Dynamics of the Upper Ocean*, Cambridge University Press, New York, 1966, p. 158.

Pierson, W. J., Jr., G. Neumann, and R. W. James, *Practical Methods for Observing and Forecasting Ocean Waves by Means of Wave Spectra and Statistics*, U.S. Navy Hydrographic Office, Pub. 603, 1955.

Pinkerton, J. M. M., "The absorption of ultrasonic waves in liquids and its relation to molecular constitution," *Proc. Phys. Soc. London*, **B62,** 129 (1949).

Proud, J. M., R. T. Beyer, and P. Tamarkin, "Reflection of sound from randomly rough surfaces," *J. Appl. Phys.* **31,** 543–552 (1960).

Rayleigh, Lord [J. W. Strutt], *The Theory of Sound*, Vols. 1 and 2 (2nd Eds., 1894 and 1896), published in one volume by Dover, New York, 1945.

Ricker, N., "The form and laws of propagation of seismic wavelets," *Geophysics*, **18,** 10 (1953).

Roderick, W. I. and B. F. Cron, "Frequency spectra of forward-scattered sound from the ocean surface," *J. Acoust. Soc. Am.* **48,** 759–766 (1970).

Rogers, P. H. and W. J. Trott, "Acoustic slow waveguide antenna," *J. Acoust. Soc. Am.* **56,** 1111–1117 (1974).

Rusby, J. S. M., "The onset of sound wave distortion and cavitation in seawater," *J. Sound Vib.* **13,** 257–267 (1970).

Sandness, G. A., "A Numerical Evaluation of the Helmholtz Integral in Acoustic Scattering," Ph.D. thesis, Department of Geology and Geophysics, University of Wisconsin, Madison, 1973. Available through University Microfilms, Ann Arbor, Mich.

Sanford, T. B., "Observations of strong current shears in the deep ocean and some implications on sound rays," *J. Acoust. Soc. Am.* **56,** 1118–1121 (1974).

Schiff, L. I., *Quantum Mechanics*, McGraw-Hill, New York, 1949, pp. 178–187.

Schulkin, M. and H. W. Marsh, "Sound absorption in seawater," *J. Acoust. Soc. Am.* **35,** 864–865 (1962).

Shah, P. M. and F. K. Levin, "Gross properties of time-distance curves," *Geophysics*, **38,** 643–656 (1973).

Shooter, J. A., et al., "Acoustic saturation of spherical waves in water," *J. Acoust. Soc. Am.* **55,** 54–62 (1974).

Shor, G. G., "Refraction and Reflection Techniques and Procedures," in M. N. Hill, Ed., *The Sea*, Vol. 3, Wiley-Interscience, New York, 1963, pp. 20–38.

Shumway, G., "Sound velocity versus temperature in water-saturated sediments," *Geophysics*, **23,** 494–505 (1958).

Shumway, G., "Sound speed and absorption studies of marine sediments by a resonance method, Part I; Part II," *Geophysics* **25,** 451–467, 659–682 (1960).

Simmons, V. P., "Investigation of the 1 kHz Sound Absorption in Sea Water," Ph.D. thesis, University of California, San Diego, 1975.

Skretting, A. and C. C. Leroy, "Sound attenuation between 200 Hz and 10 kHz," *J. Acoust. Soc. Am.* **49,** 276–282 (1971).

Slotnik, M. M., *Lessons in Seismic Computing*, Society of Exploration Geophysicists, Box 3098, Tulsa, Oklahoma, 1959.

Smith, D. T., "Acoustic and Mechanical Loading of Marine Sediments," in L. L. Hampton, Ed., *Physics of Sound in Marine Sediments*, Plenum Press, New York, 1974, pp. 41–62.

Spindel, R. C. and P. M. Schultheiss, "Acoustic surface-reflection channel characterization through impulse response measurements," *J. Acoust. Soc. Am.* **51,** 1812 (1972).

Spindel, R. C. and P. M. Schultheiss, "Two-dimensional probability structure of wind-driven waves," *J. Acoust. Soc. Am.* **52,** 1065–1068 (1972).

Stenzel, H., "On the disturbance of a sound field brought about by a rigid sphere," in German, *Elektr. Nachr. Tech.* **15,** 71–78 (1938). Transl. G. R. Barnard and C. W. Horton, Sr., and republished as Technical Report No. 159, Defense Research Laboratory, University of Texas, Austin, 1959.

Stoll, R. D., "Acoustic Waves in Saturated Sediments," in L. L. Hampton, Ed., *Physics of Sound in Marine Sediments*, Plenum Press, New York, 1974, pp. 19–40.

Strutt, J. W. [Lord Rayleigh], *The Theory of Sound*, Vol. II (2nd ed., 1896), Dover, New York, 1945, p. 89, 282.

Tannaka, Y. and T. Koshikawa, "Solid-liquid compound hydroacoustic lens of low aberration," *J. Acoust. Soc. Am.* **53,** 590–595 (1973).

Thorp, W. H., "Deep-ocean sound attenuation in the sub- and low-kilocycle-per-sec region," *J. Acoust. Soc. Am.* **38,** 648–654 (1965).

Tolstoy, I., *Wave Propagation*, McGraw-Hill, New York, 1973.

Tolstoy, I. and C. S. Clay, *Ocean Acoustics*, McGraw-Hill, New York, 1966.

Trorey, A. W., "A simple theory for seismic diffractions," *Geophysics*, **35,** 762–784 (1970).

Turner, W. R., "Microbubble persistence in fresh water," *J. Acoust. Soc. Am.* **33,** 1223–1232 (1961).

Wagner, R. J., "Shadowing of randomly rough surfaces," *J. Acoust. Soc. Am.* **41,** 138–147 (1967).

Warren, B. A., "General circulation of the South Pacific," in W. S. Wooster, Ed., *Scientific Exploration of the South Pacific*, National Academy of Sciences, Washington, D.C., 1970.

Watson, J. D., and R. Meister, "Ultrasonic absorption in water containing plankton in suspension," *J. Acoust. Soc. Am.* **35,** 1584–1589 (1963).

Westervelt, P. J., "Parametric acoustic array," *J. Acoust. Soc. Am.* **35,** 535–537 (1963).

Weston, D., "Acoustic interaction effects in arrays of small spheres," *J. Acoust. Soc. Am.* **39,** 316–322 (1966).

Weston, D., "Sound Propagation in the Presence of Bladder Fish," in V. M. Albers, Ed., *Underwater Acoustics*, Vol. 2, Plenum Press, New York, 1967, pp. 55–88.

Wildt, R., Ed., "Acoustic Properties of Wakes," Part IV of *Physics of Sound in the Sea*, Vol. 8, Summary Technical Report of Div. 6, National Defense Research Committee, Department of the Navy, Washington, D.C., 1946. Reissued by Naval Material Command, 1969.

Woodward, P. M., *Probability and Information Theory with Applications to Radar*, 2nd ed., Pergamon Press, New York, 1964.

Worzel, J. L. and M. Ewing,"Explosion Sound in Shallow Water in *Propagation of Sound in the Ocean*," Geological Society of America Memoir 27, New York, 1948.

Yeager, E., F. H. Fisher, J. Miceli, and R. Bressel, "Origin of the low-frequency sound absorption in seawater," *J. Acoust. Soc. Am.* **53,** 1705–1707 (1973).

INDEX